Lecture Notes in Mobility

Series Editor

Gereon Meyer, VDI/VDE Innovation + Technik GmbH, Berlin, Germany

More information about this series at https://link.springer.com/bookseries/11573

Marin Marinov · Janene Piip
Editors

Sustainable Rail Transport 4

Innovate Rail Research and Education

Springer

Editors
Marin Marinov
Engineering Systems and Supply Chain Management
Aston University
Birmingham, UK

Janene Piip
JP Research and Consulting
Port Lincoln, SA, Australia

ISSN 2196-5544 ISSN 2196-5552 (electronic)
Lecture Notes in Mobility
ISBN 978-3-030-82094-7 ISBN 978-3-030-82095-4 (eBook)
https://doi.org/10.1007/978-3-030-82095-4

This Springer imprint is published by the registered company Springer Nature Switzerland AG
The registered company address is: Gewerbestrasse 11, 6330 Cham, Switzerland

Foreword by Prof. Mark Robinson

I am honoured to provide this Foreword for *Sustainable Rail Transport 4*. The fact that we are now on volume 4 and the quality has been maintained and even improved is a testament to Dr. Marin Marinov's stamina and dedication. He is incredibly hard working and thorough in everything he does, and it is a pleasure working with him. It is also true that Marin's drive and personality help the authors honour their commitments to produce these fine articles.

The breadth of the papers is astonishing, and it is fair to say that the importance of rail as the backbone of an intermodal "Mobility as a Service" within cities and beyond, for both passengers and goods, meeting the needs of customers and society is absolutely supported by the articles. Inclusivity and accessibility are considered for all, from persons with reduced mobility to storing luggage.

The challenges of climate and environmental change are recognised with reference to energy saving, emission reduction and alternative zero carbon propulsion technologies. Due to climate change, more sustainable passenger and freight transport are in need. Rail is considered a more sustainable mode of transport compared to others such as road transport. Usage of rail lines that are currently under-utilised could help increase the sustainability of transport through enhanced utilisation of them. Aspects such as logistics are considered in terms of economic growth in relation to the world market, and comparisons are made that can support individual countries growth. Rail is demonstrated as the environmental situation in Europe if it was to be at the centre of sustainability-driven policies. In particular, it provides a relative picture of the CO_2 emissions generated by short-distance air passenger transportation in Europe which could have been transferred to high-speed rail and produce less CO_2. Russia is used as a case study sustainability of Russia's railway infrastructure, whose development faces innumerable challenges including its geography, demographic density and specific political considerations.

Rail is the safest form of travel, and these articles support this theme and develop it further by addressing issues such as level crossings where there is potential for cross-mode accidents. This volume has not shirked these issues and also addresses the aspect such as noise that can be generated by heavy haul, the different types of noise that originate from railroads were identified, the squeal noise generated at curves, the impact noise generated at rail joints and the rolling noise stand out.

There are a number of factors addressed by these articles including opening up the rail passenger market to compete the ambition to achieve European technological leadership, increasing urbanisation and strategic autonomy and ensuring a sustainable and inclusive recovery from the COVID crisis requiring the convergence of dispersed research and innovation efforts within a shared vision of system transformation.

Addressing these issues needs long-term action plans, based on the Sustainable Development Goals of the United Nations. Rail and public transport are part of the solution, offering innovated services that provide mobility for passengers and delivery of goods and ensure sustainable socio-economic evolution. The volume identifies the importance of rail traffic flows from China and Europe and explores the potential of the railway network of the Republic of Kazakhstan as an effective and competitive transit transport route. There is also consideration of the TEN-T corridors.

I commend Dr. Marin Marinov in the production of *Sustainable Rail Transport 4*, as it highlights and builds on rail's existing credentials as the most environmentally friendly form of mass land transport and provides the latest discussions for consideration from the challenge of Hyperloop to the transportation of dangerous goods.

Newcastle upon Tyne, UK

Prof. Mark Robinson
Director of NewRail and Professor of
Rail Systems Engineering, Group Lead
Future Mobility
Newcastle University

Foreword by Prof. Ed Sweeney

I am delighted to see that Volume 4 of *Sustainable Rail Transport* has been published. The fact that another volume has been developed is itself a testament to the quality of the earlier volumes and the contribution that they made to the ongoing discourse in this hugely important field. I am also honoured to again be invited to provide the Foreword to what is another volume of extremely high quality. This high quality suggests that it will become an indispensable work for experts in the fields, as well as a valuable source of fascinating insights for non-expert rail enthusiasts like myself. For rail specialists involved in research and scholarship, as well as for those in professional practice, this volume provides a wealth of information and knowledge about key developments and trends impacting the sector.

The railway industry is a dynamic and challenging one. The COVID-19 pandemic of 2020 has created a range of new challenges for the industry to grapple with. This is in addition to the challenges of operating in a market place has come more sophisticated, comprising customers with increasingly demanding value expectations. The wider business and regulatory environment is a complex one and subject to a high degree of change and unpredictability. Technology has continued to develop at a rapid rate, presenting rail professionals with an array of opportunities and threats. The anthropogenic impact of transport and logistics processes is now widely understood with climate change representing an existential threat. The rail industry is well positioned to support wider policy initiatives—including but not limited to modal shift—aimed at creating more environmentally sustainable passenger and freight transport systems. In this context, it is becoming clear that more integrated approaches are required, particularly in the context of transport system design with rail, and play a more critical role. The publication of this volume is very timely in this context.

It is important that rail policy-making is evidence-based and that innovation in the sector is based on the best available knowledge about operations, systems, technology and management. The high-quality research described in this volume is critical in this regard. As with previous volumes in this series, the research that has been undertaken is characterised by high levels of rigour in the research methodologies and approaches adopted by scholars. This is vital as the development of deeper and richer insights into the complex phenomena under investigation needs to be based on research designs that are logical and systematic. However, this research excellence and academic rigour alone is not sufficient; it needs to be combined with a deep understanding of the evolving needs of the industry for the outputs to be truly valuable and impactful. I am delighted that this volume clearly demonstrates both academic excellence and practical relevance, with much of the work representing the fruits of effective academic/industry collaboration. The volume also demonstrates clearly that high-quality rail research requires truly interdisciplinary approaches. There is extensive use of multi-phase research approaches that adopt multiple methods of data collection and analysis. This presents challenges, but the material in this volume provides many excellent examples of good research practice in this regard. In tandem with this, there are illustrations of innovative practices from a range of different geographical settings. This is very welcome in the context of the increasingly international complexion of transportation and logistics systems, and the attendant needs to develop world-class levels of performance.

As I noted in my Foreword to volume 3, rich tomes of this kind require immense dedication and commitment on the part of a range of individuals—authors, editors, publishers and others. In this context, I would again like to commend my Aston University colleague, Dr. Marin Marinov, for his energy and enthusiasm in bringing this work to fruition. Marin is a passionate advocate of rail research and its critical role in developing a sustainable future for this strategically critical industry. Thanks to his work, as well as that of his collaborators, *Sustainable Rail Transport 4* will play an important role in our ongoing discourse and debate in this field.

December 2020

Prof. Ed Sweeney
Professor of Logistics and Systems
Engineering Systems and Supply
Chain Management
Aston University
Birmingham, UK

Preface

Welcome to this important book, *Sustainable Rail Transport 4*, a collection of papers written by academics and students with a passion for rail transport. Since 2014, a core group of enthusiastic writers has presented works that have inspired others to explore some of the latest thinking, research and developments in rail transport. Led by Dr. Marin Marinov, the latest and fourth volume of Sustainable Rail Transport continues to inspire all who will read it. The call for global papers met with great interest and readers will not be disappointed by the contributions, so the opportunity to again be involved with the development of this vital piece of work has been an absolute pleasure.

The book comes at a time in history like no other. COVID-19 has upended our lives, causing rail companies to rethink how and when they offer services. New approaches to costing and service delivery have been vital in the ever-changing situation of the COVID-19 pandemic. Ensuring services remain viable while incorporating strict health and safety measures, unheard of even a year before 2020, has been a battle for many rail companies.

Amidst the need for social distancing, personal protective equipment and personal hygiene, the demand for rail passenger transport has declined for the time being. The industry has realised new ways of tracking customers who use rail services will be essential. As a result, innovations such as QR code scanning to track passengers have become a standard practice in some countries. Additionally, there has been a rise in opportunities for new types of jobs including the importance of once considered menial roles such as cleaning and disinfecting to ensure passengers remain safe—a positive outcome for millions of people working in unskilled roles in rail companies.

Freight transport has risen to the fore as an essential service delivering essential goods.

Other realisations lean towards new train designs to accommodate social distancing using materials that prevent viruses from living for long periods. While the pandemic is global, the rail industry will need to consider culturally and ethically appropriate practices for each country where they operate. All these realities highlight that the world in which we live cannot be controlled, and therefore, an open mind is needed to explore new ideas and break away from past mindsets. Recent developments indicate the need for creativity and innovation for the rail sector to survive and "pivot" quickly.

The ideas offered by authors who have contributed to *Sustainable Rail Transport 4* embrace these new ways of looking at problems in a novel way.

Topics in this volume include alternative energy sources in urban guided transport systems and the effects of CO_2 production in high-speed rail. Regarding transport corridors, there are topics on the Scandria corridor, the hyperloop concept, Kazakh rail and the Bedford to Bletchley simulation.

Safety topics include investigating the need for multi-resource allocation to address safety issues at level crossings rather than looking at singular issues in isolation. Noise issues and dangerous goods feature in two chapters, while station boarding systems for people with reduced mobility and luggage on passenger trains are topics that look to the future of rail transport.

Economics of countries and their impacts on the rail industry are featured in chapters from Russia, the Slovak Republic and Brazil.

Thank you to all the authors and contributors who have made this volume possible. I commend this volume as a worthy contribution to new rail industry knowledge and trust you will benefit from the papers within *Sustainable Rail Transport 4*.

Port Lincoln, Australia Dr. Janene Piip

Contents

Energy and Emissions Saving in Urban Guided Transport Systems: Effectiveness and Assessment of Hydrogen-Based Solutions

Matt Davoudizavareh and Stefano Ricci

Abstract Global warming and climate change are indisputable facts. Since the Industrial Revolution, the mean temperature of the planet has increased by 1 °C. Now, temperatures are approaching a higher stage of +1.5 °C, and the attention is on both CO_2 emissions and energy consumption. Transportation is a significant component of the environmental impact, accounting for approximately 30% of air pollution and energy consumption. Due to the rapid urbanization in the EU, with an estimated 74.3% of the population living in cities, which is forecasted to rise to 80% by 2050, urban mobility is dramatically increasing its relevance. Therefore, a reduction in energy consumption and pollutant emissions is a crucial factor to consider in the development of urban transportation and particularly rail-based systems to provide energy-saving transport services by improving the urban environment. Several methods and techniques are under development to improve the energy performance of Light Rail Transport (LRT), which spread from different typologies of power supply to improve energy efficiency. This paper aims to start with the latest developments and innovative energy sources for LRT systems. The focus is on two parts: (a) trams running on hydrogen in parallel with onboard batteries with energy-saving control techniques, (b) potential renewable energy sources to meet the demand power. The comparison is with traditional power sources and equipment (e.g., catenary-based). The development of the methods, based on selected indicators, is ongoing, and the paper describes results of calculations and simulations for a relevant case study: the new tramline in the city of Brescia (Italy).

Keywords Railways · Tram · Urban transit · Renewable energy · Fuel cell · Hydrogen

M. Davoudizavareh (✉) · S. Ricci
Sapienza Università di Roma—DICEA, Via Eudossiana, 18, 00184 Roma, Italy
e-mail: mdavoudi@kth.se

S. Ricci
e-mail: stefano.ricci@uniroma1.it

M. Marinov and J. Piip (eds.), *Sustainable Rail Transport 4*, Lecture Notes in Mobility,
https://doi.org/10.1007/978-3-030-82095-4_1

1 Problem Positioning

Energy demand globally increased by 2.1% in 2017, according to *Global Energy and CO_2Status Report* issued by Organization for Economic Co-operation and Development (OECD) and the International Energy Agency (IEA) (OECD-IEA 2018), more than twice the average growth rate over the previous five years, which was 0.9%. Moreover, energy-related CO_2 emissions grew by 1.4% in 2017, a historical record, after three years of emissions remaining flat worldwide.

On the other hand, renewable energies had the highest growth rate of any energy source in 2017, meeting a quarter of global energy demand growth last year. Together, China and the United States accounted for half of the increase in renewables-based electricity generation, followed by the European Union (8%), Japan, and India (6% each).

The EU set an ambitious target of 40% greenhouse emission reduction by 2030 and 80% by 2050 (EC 2016a). Based on the Paris agreement, adopted on 12 December 2015, at COP21 and signed by 195 states in 2016, the EU is promoting the following target (European Environmental Agency 2016): *Holding the increase in the global average temperature to well below 2 °Cabove pre-industrial levels and limiting the temperature increase to 1.5* °C. To foster low carbon transition, a framework strategy for a resilient Energy Union links the transport and energy systems (EC 2016b,2016c). Its key features are:

- Reduction of the dependency on particular fuels, energy suppliers, and routes;
- Full integration of the internal energy market and more efficient energy consumption;
- Decarbonization of the economy.

Mobility within cities and between suburban areas and towns is significantly important since transport represents more than 30% of the final energy consumption in Europe (EC 2016b), and the majority of the EU population is urbanized. According to UITP (2016), in 2014, urban rail accounted for 44.3% of all local public transport journeys in Europe (13.6% suburban rail, 16.2% metro, 14.5% tram/LRT).

Besides the environmental issues, health impacts of transport-related air pollution have become one of the most critical concerns. The cause is directly related to emissions such as sulfur dioxide (SO2), carbon monoxide (CO) and hydrocarbons (HC), particulate matter (PM) and nitrogen oxides (NOx). Particulate matter (PM), for instance, is a significant risk factor for mortality and morbidity associated with asthma, lung cancer, heart disease, etc. PM is a crucial concern in underground stations, also for running electric trains due to its generation by train brake and wheel wear, catenary electrical arc and so on. Reportedly, underground PM exposure could be more dangerous than in open air. Exposure to ambient airborne particulate matter is a major risk factor for mortality and morbidity, associated with lung cancer, heart disease, myocardial infarction, and stroke, and more recently type 2 diabetes, dementia and loss of cognitive function and asthma. Reportedly, some asthma inhalers release greenhouse gases linked to global warming.

According to the House of Commons Environmental Audit Committee, metered-dose inhalers account for nearly 4% of greenhouse gas emissions. The researchers estimate replacing 1/10 of these inhalers with an environmentally friendly type (dry powder inhalers) would reduce carbon dioxide equivalent emissions by 58 kilotons (National Health Service 2018). That is similar to the carbon footprint of 180,000 return car journeys from London to Edinburgh.

As the European economy and transport demand are continuing to grow, the mentioned aims are only achievable if attentions of policymakers, local companies' and authorities are considered. In this framework, the article promotes:

- Firstly, a comparative assessment of the traditional power supply (catenary based) in LRT systems with modern renewable energy sources.
- Secondly, new methods for a better climate adaptable enhanced passenger comfort and improved urban environment by removing catenary-based infrastructure must be considered.

The innovative concept includes the possibility to either transfer energy supply to street ground surface with such systems like third rail electrification and magnetic fields or moving the energy source onboard to practically remove the catenary infrastructure.

The most common onboard source nowadays are batteries, but they are expensive, heavy, require the extensive use of rare-earth metals, and lithium-ion batteries which are energy-intensive to be produced. Furthermore, charging them take a long time. Another way to carry a clean energy source is by using the most abundant element of the universe, hydrogen, already used in automobile industry but rarely in railway sector, especially in an urban context. Hydrogen has specific energy up to 40,000 Wh/kg, comparing to only 278 Wh/kg for batteries, which is 236 times more and makes using fuel cell vehicles a feasible choice.

Toyota Mirai (2014), a fuel cell-powered 113 kW car, with only 5 kg of hydrogen for a total range of almost 500 km, was one of the first mass-production Fuel Cell Vehicle (FCV) capable of running on both Hydrogen and Battery with variable energy consumption thanks to a combination of power trains (Yoshida and Kojima 2015). The development of FCVs is followed by other automotive companies (Honda, Hyundai) and Locomotive companies.

In September 2018, Alstom commercialized the world's first hydrogen-powered train, the Coradia iLint that entered passenger service in Lower Saxony, Germany (Alstom 2018). The two pre-series trains, homologated by the German Federal Railway Association in July, are now running in Cuxhaven, Bremerhaven, Bremervörde, and Buxtehude. The train can operate over a daily range of 1000 km. Alstom and the local transport authority of Lower Saxony (LNVG) signed a contract for the delivery of 14 hydrogen fuel cell trains by 2021. San Bernardino County Transportation Authority (SBCTA) has awarded a contract to Stadler for a hydrogen fuel cell train, FLIRT (Fast Light Intercity and Regional Train), which will be the first type to operate in the USA. It will enter into service in 2024 on the Redlands Passenger Rail Project (the *Arrow*), in California, a nine-mile line linking Redlands and the *Metrolink* regional/commuter rail station in San Bernardino. Moreover, the most relevant FCV in mass urban transit ran in October 2017 in China: CRRC Tangshan Railway Company unveiled a prototype low-floor LRV powered by Canadian supplier Ballard Power Systems with hydrogen fuel cell technology, FCveloCity, on trial on the new 14 km light rail line in Tangshan (International Railway Journal 2018).

Ballard's fuel cell technology works in combination with batteries and supercapacitors over a range of 40 km, a top speed of 70 km/h, and a capacity of 336 passengers to offer entirely catenary-free operation on the line. The LRVs have a range of 40 km on a single 12 kg hydrogen fill-up, which takes 15 min to complete. The four-station line includes a 100 kg capacity hydrogen refilling station.

Finally, for better understanding of differences, potentials and drawbacks of the traditional catenary based system and fuel cells; the comparison refers to the new tramline project in Brescia (Italy).

The new tramway network includes three main sections (Fig. 1) of double-track lines: T1, T2, and T3 for a total extension of about 23 km (46 km of single equivalent track), 65% shared with the current urban road including approximately 41 signalized intersections. The project cost would be 450,000,000 €, forecasted to be operated by 2026 with a total number of max 14,040 passengers/h in rush hours with estimated yearly demanded power of 9,000,000 kWh.

The paper will forecast possible environmentally friendly, clean local power sources to cover at least 20% of demanded power. Moreover, it promotes low-cost technology to make renewable hydrogen using sunlight and any water source (Hypersolar.com 2018) directly at or near the depot area, to make a self-sustained renewable zero-carbon hydrogen-powered urban transit combined with intelligent energy management (Fig. 2).

Comparisons are followed by discussing the most critical issues concerning using hydrogen, such as safety, infrastructure, and cost. Simulations are carried out firstly by Opentrack© on section T1 (Fig. 3), and later the simulation output has been used as an input for the OPEUS simulator, which is a result of the Shift2Rail funded research project *Modelling and strategies for the assessment and OPtimisation of Energy Usage aspects of rail innovation* (European Union's Horizon 2020 research and

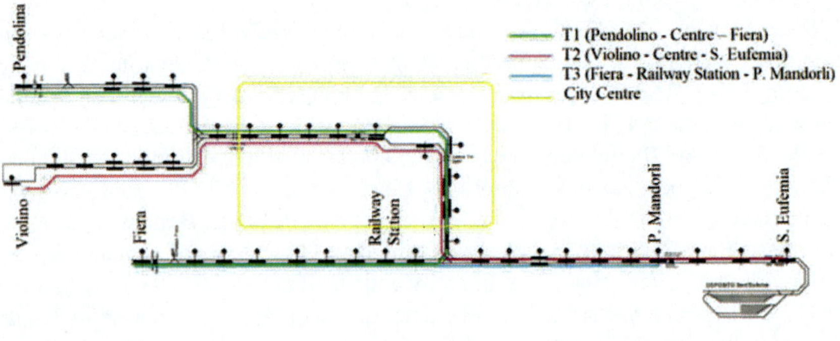

Fig. 1 Track layout

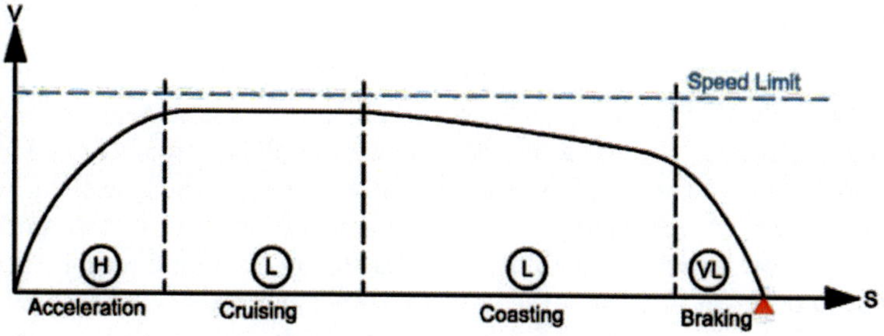

Fig. 2 Energy-saving driving strategies (Energy demand level H: High, L: Low, VL: Very Low)

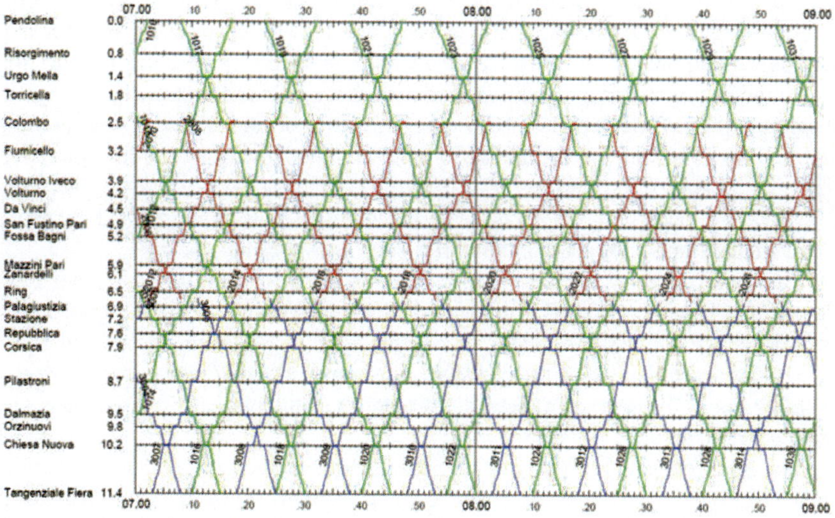

Fig. 3 Opentrack © simulated timetable of T1 section in the morning peak hour

innovation program 2019). OPEUS is aiming to develop a simulation methodology and accompanying modelling tools to evaluate, improve, and optimize the energy consumption of rail systems with a particular focus on in-vehicle innovation. The OPEUS concept is basing on the need to understand and measure the energy used by each of the relevant components of the rail system and, in particular, the vehicle. This includes the energy losses in the traction chain, the use of technologies to reduce these, and optimize energy consumption. Specifically, the OPEUS approach has three core components: the energy simulation model, the energy use requirements (e.g. duty cycles), and the energy use outlook and optimisation strategies recommendation.

2 Hydrogen Potential Role

Electricity production is carbon-intensive and releases massive heat and noise at local power plants, giving rise to negative impacts on the environment and human health throughout its lifecycle, from resource extraction to electricity use. Impacts on climate change, air, water quality, direct and indirect impacts on land resources, etc. must also be considered. Furthermore, impacts stemming from electricity production depend on fossil fuel employed, how to extract and process it, along with the actual technology to produce electricity, and its efficiency. The use of abatement technologies must be considered, as well. Almost full decarbonization of the electricity sector will be necessary to meet the EU's objective of reducing greenhouse gas emissions by 90% by 2050. Increasing electricity generation and use throughout Europe without reforming the current energy system will lead to higher overall health and environmental impacts. Nevertheless, an increase in the transport sector's electricity consumption might bring a positive modal shift towards rail transport or a higher penetration of electric vehicles. The carbon intensity of total electricity generation in the EU in 2016 was 296 g CO2/kWh. In 2013, the total emission from electricity generation was 1207 million tons of CO2, led by Germany (332 Mt), United Kingdom (163 Mt), and Italy (111 Mt). An increase in electricity consumption in the transport sector (mainly railways) arose in countries such as France and Italy.

Instead, Hydrogen (Appleby and Foulkes 1989) (Stolten 2010) and Fuel Cell technology (Hoogers 2003) (Larminie and Dicks 2003) can contribute significantly towards reducing emissions and facilitating the necessary green energy transition in EU regions and cities. The word *hydrogen* is a combination of two words in Greek, *to make water*. The characteristics of hydrogen as an alternative to classic fuels are listed below (Table 1).

Fuel cell technology usage can improve air quality and create positive health impacts for local population, hence enhancing life quality. The fuel cell is an electric cell that, unlike battery cells, can be continuously fed with fuel (in this case hydrogen) so that the electrical output can be maintained indefinitely. Therefore, a fuel cell converts hydrogen through an electrochemical reaction with oxygen directly into electricity and heat, which is the inverse of electrolysis (hydrogen generation):

$$2H_2 + O_2 \rightarrow 2H_2O + \text{energy}$$

Moreover, the fuel cell classification is based on fuel type (gas, solid, liquid), the electrolyte used (liquid or solid) and fuel consumption (direct or indirect), operational temperature, which is the most common criteria:

- Low temperature, operating from 20 to 100 °C (cold);
- Medium temperature, operating from 200 to 300 °C (hot);
- High temperature, operating from 600 to 1500 °C.

Table 1 Hydrogen characteristic

Characteristics	Unit	Value
Density	kg/m^3	0.0838
Higher Heating Value (HHV)/liquid hydrogen (LH2)	MJ/kg	141.90–119.90
HHV/cryogenic hydrogen gas (CGH2)	MJ/m^3	11.89–10.05
Boiling point	K	20.41
Freezing point	K	13.97
Density (liquid)	kg/m^3	70.8
Air diffusion coefficient	cm^2/s	0.61
Specific heat	kJ/kg K	14.89
Ignition limits in air	% (volume)	4–75
Ignition energy in the air	Millijoule	0.02
Ignition temperature	K	585.00
Flame temperature in air	K	2318.00
Energy in explosion	kJ/g TNT	58.823
Flame emissivity	%	17–25
Stoichiometric mixture in air	%	29.53
Air/fuel stoichiometry	kg/kg	34.30/1
Burning speed	cm/s	2.75
Power reserve factor	–	1.00

Regarding the recent study, about 90 cities in the EU planned to invest about 1.8 billion euros in next five years to deploy different H2 transport modes and electrolyzers for H2 production and power generation. This conversion will produce both environmental and local economic effects.

For example, according to Unione Petrolifera, in 2017, Italy imported 15.9 million tons of refined petroleum products (Unione Petrolifera 2017). However, the petroleum industry employs relatively few people; historical data shows that 1 million Euro of the value added in Italy's petroleum sector created only 3.5 jobs in 2017, while hydrogen sectors are almost five times more labor-intensive. Overall, the transition towards a low zero-carbon economy has a net positive impact on employment (19,225 additional jobs in 2030). Moreover, it will create opportunities for the adaptation and transformation of workers. As another example, for the German state of Baden-Württemberg FCV and H2 for green energy, the estimation is of an added value of around 680 million Euro by the year 2030 (Roland Berger GmbH 2018). The Hydrogen Council's vision is to create generally around 30 million additional jobs by 2050, making hydrogen a global energy carrier that could serve up to 18% of global energy demand. Nonetheless, there are negative aspects and barriers to overcome, some of which are:

- Hydrogen burns quickly in the presence of oxygen, that can cause operational safety problems;

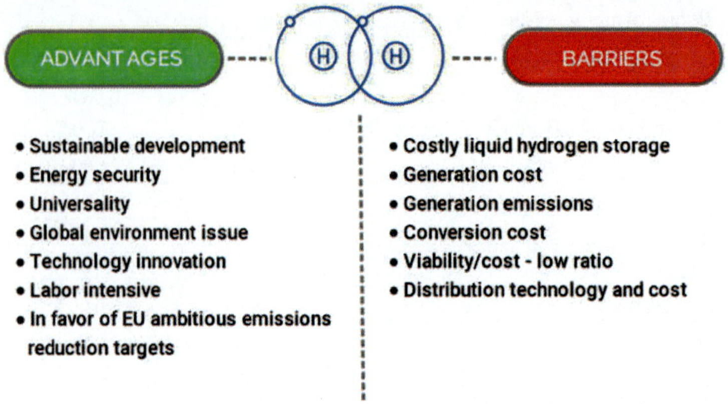

Fig. 4 Advantages and barriers of hydrogen as an energy storage

- Storing hydrogen is not easy and also expensive because very low temperatures are required to convert the hydrogen into liquid;
- The cost of hydrogen technologies and processes is high, and the energy conversion technologies through fuel cells. Hence, the viability/cost ratio of hydrogen and fuel cell technologies is relatively low;
- Finally, apart from the generation and storage, hydrogen distribution is also not cheap, considering the current lack of logistics and transport infrastructure.

Figure 4 summarizes the positive and negative aspects of hydrogen (Momirlan and Veziroglu 2005) (Veziroglu and Sahin 2008) (Iordache and Stefanescu 2010) (Felseghi et al. 2019) schematically.

3 Fuel Cell Tram

- Fuel cells have achieved enough maturity to support the railway sector as they are already on tracks in Germany and considered for operation in France and the United Kingdom for inter-city trains. In urban areas, a fuel cell approach can have the following characteristics and benefits, compared to overhead catenary systems:
- Less expensive, due to much less infrastructure;
- Less impacting on urban area surroundings and operations;
- Visually more attractive, especially in old cities centers and tourist attractions;
- Extendable to additional locations without additional infrastructure;
- Operable independently, in power outage;
- Enabling other independently powered alternatives in future;
- Operating with zero local emissions and less pollution at electricity production plants;

- Best alternative for underground stations.

This approach, compared to the traditional Overhead Catenary System (OCS), eliminates the need for the overhead wires, support poles, notching the existing tunnels and electrical substations. Costs related to removing and trimming trees and relocating existing electrical and telecommunication infrastructure would no longer be existing or drastically lower. In the case study, it means eliminating 13 electrical substations, with a pitch of about 2 km each, providing power in two electrical zones through 750 V power supply with one dedicated transformer in each substation for auxiliaries. The estimated saving would be 30 million Euros (cost of OCS infrastructure). Additionally, the catenary-free approach could make use of abandoned existing tram tracks with reduced or no additional costs for technological upgrades.

A Fuel-cell tram operates as long as stored hydrogen fuel is available during a power outage. Integrations could be available from on-site hydrogen's production during periods of low electricity demand or emerging methods like solar hydrogen's generation with pumped in wastewater and other solar-powered equipment to address potential renewable energy sources to meet the power demand. Also, considering Brescia's geographical coordinates, the solar power production potential is 50.17 kwh/m^2/year. With today's commercial solar technology, an available depot area of about 30,000 m^2 can generate approximately 1.5 GWh of electricity, which could run green hydrogen production with zero CO2 emissions (Fig. 5).

In addition to the main one, two more planned train depots are located in Fiera and Pendolina areas (Fig. 6). Each building contains a three-track tramway and some rooms of service. Each garage is approximately 1900 m^2, another likely future source of renewable electricity generation, thanks to high-efficiency solar panels located on the roofs.

Moreover, a Fuel-cell approach would be an excellent alternative for underground metro stations. As such, the mass concentration of airborne Particulate Matter (PM) underground usually is significantly greater than the outside environment, mainly influenced by specific source materials and generation methods: wear of wheels and brakes, as well as arcing of electrical current between catenary and third rail and the corresponding collecting equipment (Loxham and Nieuwenhuijsen 2019). As a result, PM generated in underground railway systems tends to be rich in metals, especially Fe, but also Cr, Ni, Co, Mn, and Cd.

4 Powertrain Technology

By early 2018, the Alstom Coradia iLint fuel cell entered service on a 100 km route in Germany. This train has a maximum velocity of 140 km/h, a range between 800 and 1000 km, and a capacity of 150 sitting or 300 standing passengers.

In Brescia, an urban light fuel-cell tram would meet these requirements. On line T1 (Pendolina-Tangenziale Fiera) with a length of 11.4 km, 23 stops, 143 trips per

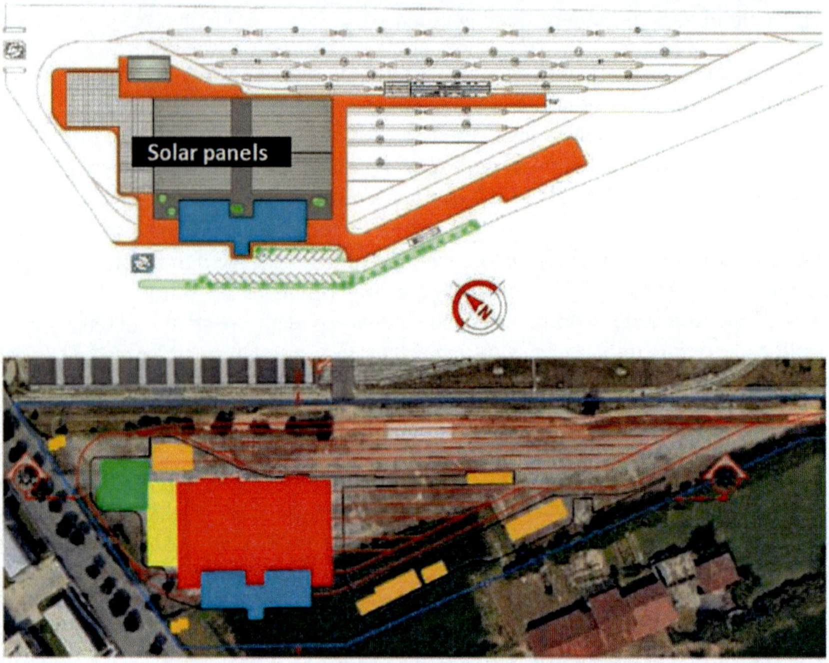

	(A)	Tram maintenance area	mq. 3.440
	(B)	Workshops for tram maintenance	mq. 520
	(C)	General warehouse and deposits	mq. 460
	(D)	Storage and minor buildings	mq. 830
	(E)	Office and services	mq. 770

Fig. 5 Functional layout of the main deposit featuring renewable electricity generation—color legend

Fig. 6 Second and third depot area—Pendolina and Fiera

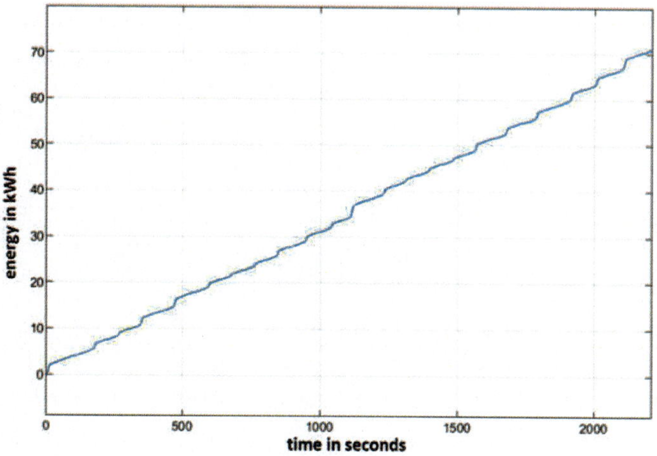

Fig. 7 Cumulative energy consumption of a single run on studied line

day on both directions in the most critical scenario, 5 units are running with expected range of 350 km a day each and max speed of 50 km/h (Fig. 7).

The consumed energy for a single run is 71 kWh (Fig. 8), considering total trips and the number of operating units. Each tram should be capable of storing at least 2350 kWh per day. The detailed simulated output is in Table 2.

The energy consumption of rail transport mainly depends on rolling stock features, stop spacing, and track profile. The local trains are heavier, but they stop less frequently than the trams. The higher stop density includes more accelerations and braking than longer stop spacing (Fig. 9). HyPM HD30 (33 kW) with efficiency map and technical data (Table 1) is the considered fuel cell component on the catenary-free tram (Table 3).

Battery and super-capacitors together are providing enough energy and power for traction and auxiliaries. On the other side, fuel cells and recovered braking energy (about 6.8 kWh) (Fig. 10) charge the battery with a steady trend, and without a sudden variation in output. By providing a balanced state of charge on our single branch battery, we need 11 fuel cells onboard. Hydrogen consumption of simulated tram is 0.3 kg/km and approximately 3.3 kg for one run.

Carrying hydrogen enough for a day would require massive hydrogen tanks, which would make the tram too heavy and will affect the autonomy. Hence a refuel approach of fewer than 15 min at terminus after each cycle is a promising solution. Therefore, 10 kg of hydrogen compressed to 350 bar can be stored, preferably in dual tanks for longer charging lifecycle and reliability with a total weight of 80 kg and a volume of 120 l each. Tanks are on the top and provide mechanical safety valves, which let the tram to release H2 into atmosphere in case of high temperatures. Furthermore, they can indicate any leaking in the system.

In Fig. 11, the suggested hybrid propulsion system, with a combination of super-capacitors, batteries (B), and fuel cell (FC). A real-life application is in Fig. 12. The

Table 2 OPEUS energy simulation output

Station	Travel time (s)	Stop time (s)	Distance (m)	Traction energy at the wheel (kWh)	Total braking energy at the wheel (kWh)	ED-braking energy at the wheel (kWh)	Traction energy at the catenary (kWh)	Recuperated energy at the catenary (kWh)
TAN	0	0	0	0.000	0.000	0.000	0.000	0.000
CHI	108	65	1170.2	1.569	0.824	0.824	5.500	0.522
ORZ	57	35	436.6	0.870	0.507	0.507	2.958	0.284
DAL	46	34	301.8	0.501	0.427	0.427	2.280	0.229
PIL	81	41	780.8	1.215	0.719	0.719	3.992	0.440
COR	83	42	823.9	1.392	0.769	0.769	4.255	0.480
REP	49	34	330.5	0.703	0.271	0.271	2.606	0.121
STA	52	33	345.9	0.328	0.738	0.738	2.156	0.458
PAL	46	35	304.4	0.491	0.478	0.478	2.287	0.267
RIN	52	53	391.6	0.860	0.585	0.585	3.229	0.342
ZAN	56	35	421.5	0.863	0.475	0.475	2.938	0.265
MAZ	37	33	212.4	0.727	0.162	0.162	2.369	0.056
FOS	67	54	654.2	2.137	0.427	0.427	5.043	0.226
SAN	46	32	296.8	0.902	0.234	0.234	2.737	0.098
DAV	53	34	376.6	0.489	0.639	0.639	2.395	0.383
VOL	53	34	365.6	0.483	0.573	0.573	2.395	0.334
VOL 1	47	33	308.9	0.341	0.504	0.504	2.087	0.286
FIU	77	35	717.3	1.217	0.653	0.653	3.784	0.392
COL	76	36	702.1	1.024	0.773	0.773	3.549	0.483
TOR	71	55	638.2	1.035	0.623	0.623	3.894	0.372
URA	57	34	460.7	1.248	0.393	0.393	3.384	0.203
RIS	64	34	549.9	1.244	0.499	0.499	3.517	0.277
PEN	81	30	783.9	1.810	0.450	0.450	4.455	0.244

Table 3 Parameters of HyPM HD30 (33 kW) Fuel Cell (cold)

Performances parameters	Unit	Value
Rated (Max continuous) power	kW	30 (33)
Dimensions (L × W × H)	mm	950 × 1630 × 265
Mass	kg	≤ 70
Gravimetric power density Operating current operating voltage Peak efficiency Stack operating pressure	kW/kg A_{dc} V_{dc} %LHV kPa	0.5 0–500 60–120 55 < 120

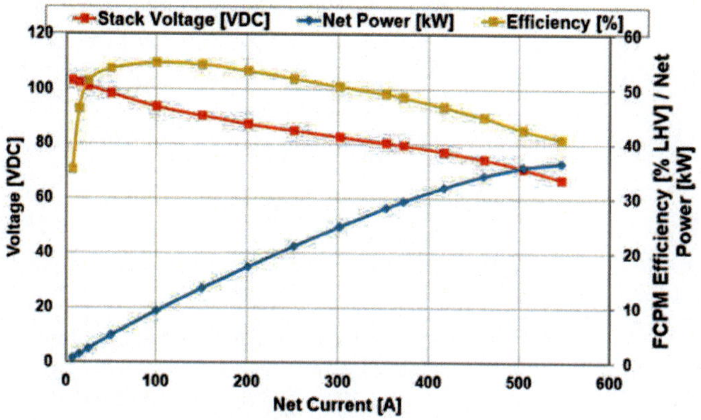

Fig. 8 Fuel-cell component efficiency map

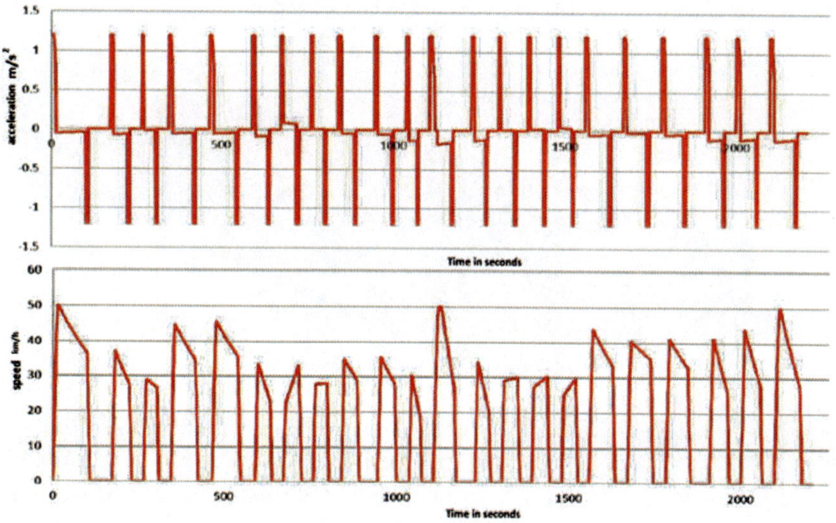

Fig. 9 Speed-time and acceleration-time diagrams of a single run on studied line

energy management system in a hybrid powertrain enables the amount of needed power from each energy source to achieve high efficiencies, high performance, and low consumption and take advantage of the components' features. Batteries have high specific energy and super-capacitors high specific power (B). Moreover, Sc provides energy for more charge/discharge cycles. In high demand for energy, Sc and B provide enough power and energy to supply traction motors (Fig. 2). In low and very low energy demand phases recovered energy and surplus provided by the Fc with steady trend and no sudden output variation, charge B and Sc in cycles. The

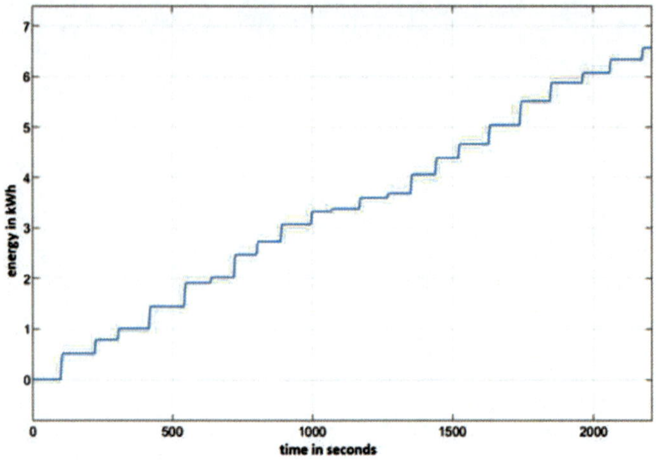

Fig. 10 Energy recuperation in braking of a single run on studied line

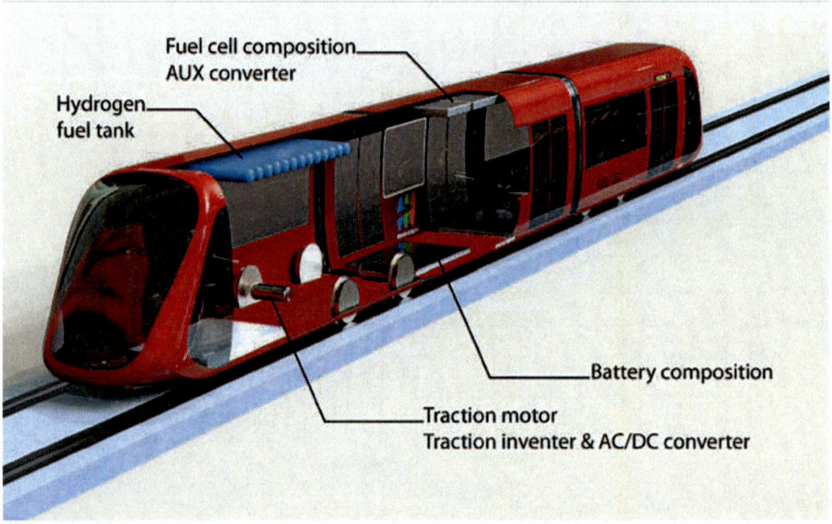

Fig. 11 Schematic demonstration of conceptual simulated hybrid fuel-cell tram

Fig. 12 Real-world application of hydrogen tanks on top (a bus in London)

state of charge of the battery is kept constant by the energy from the fuel cell. Only hydrogen is consumed. This mode is ideal for steady cruising over long distances.

5 Conclusions

The paper discusses new ideas for an integrated hydrogen approach in guided urban transport systems by considering the Brescia tram project as a case study to compare fuel cells with the traditional catenary-based system. Generally, a transition from pollutant electricity generation towards hydrogen and fuel cell technologies would have direct and indirect benefits.

Direct benefits are zero local pollutants, reduced noise level, zero CO2 emissions, and increased use of renewable energies. Moreover, this approach would be an excellent alternative for underground, enclosed stations.

Indirect benefits would be boosting research and innovation, attracting new businesses, creating new jobs, attracting a skilled workforce, boosting local tourism, improving image as a green city, and increasing life quality.

Moreover, the Hydrogen approach is entirely in line with the EU energy road map for 2050, reducing emissions by at least 80% from their 1990s level, improving both energy supply security, and the flexibility of energy systems and infrastructure.

References

Alstom (2018) Coradia iLint—the world's 1st hydrogen powered train. https://www.alstom.com/coradia-ilint-worlds-1st-hydrogen-powered-train

Appleby AJ, Foulkes FR (1989) Fuel Cell Handbook. Van Nostrand, New York, USA

EC (2016a) The implementation of the 2011 White Paper on Transport "Roadmap to a Single European Transport Area—towards a competitive and resource-efficient transport system" five years after its publication: achievements and challenges (Working document). SWD (2016) 226 final. https://ec.europa.eu/transport/sites/transport/files/themes/strategies/doc/2011_white_paper/swd%282016%29226.pdf

EC (2016b) Clean energy for all Europeans (Communication).COM (2016) 860 final. http://eurlex.europa.eu/resource.html?uri=cellar:fa6ea15b-b7b0-11e6-9e3c-01aa75ed71a1.0001.02/DOC_1&format=PDF

EC (2016c) A European strategy for low-emission mobility (Communication).COM (2016) 501 final. http://eurlex.europa.eu/resource.html?uri=cellar:e44d3c21-531e-11e6-89bd-01aa75ed71a1.0002.02/DOC_1&format=PDF

European Environment Agency (2016) Overview of electricity production and use in Europe. Council Decision (EU) 2016/1841 of 5 October 2016 on the conclusion of the Paris Agreement adopted under the United Nations Framework

European Union's Horizon 2020 research and innovation programme (2019) - *Agreement n. 730827 (2019) OPEUS project*. http://opeus-project.eu/

Felseghi R, Carcadea E, Raboaca M, Trufin C, Filote C (2019) Hydrogen fuel cell technology for the sustainable future of stationary applications. Energies 12(23):4593

Hoogers G (2003) Fuel cell technology handbook CRC Press: Boca Raton. FL, USA

Hypersolar.com (2018) HyperSolar solutions. http://www.hypersolar.com/solutions.php

International Railway Journal (2018) CRRC Tangshan trials new hydrogen-fueled tram. https://www.railjournal.com/passenger/light-rail/crrc-tangshan-trials-new-hydrogen-fueled-tram/

Iordache I, ¸Stefanescu I (2010) Tehnici Avansate De Tratare a Apei ¸ Si Separare a Hidrogenului (Advanced Techniques for Water Treatment and Hydrogen Separation); Editura AGIR: Bucuresti, Romania

Larminie J, Dicks A (2003) Fuel cell systems explained, 2nd edn. Wiley, New York, USA

Loxham M, Nieuwenhuijsen MJ (2019) Health effects of particulate matter air pollution in underground railway systems—a critical review of the evidence. Part Fiber Toxicol 16(1)

Momirlan M, Veziroglu TN (2005) The properties of hydrogen as fuel tomorrow in sustainable energy system for a cleaner planet. Int J Hydrogen Energy 30:795–802

National Health Service (2018) UK Progress on reducing F-gas Emissions Fifth Report of Session 2017–19 Report, together with formal minutes relating to the report by authority of the House of Commons. https://publications.parliament.uk/pa/cm201719/cmselect/cmenvaud/469/469.pdf

OCDE-IEA (2018) Global energy and CO_2 status report

Roland Berger GmbH (2018) Fuel cells and hydrogen for green energy in European Cities and Regions. Munich Germany

Stolten D (2010) Hydrogen and fuel cells, fundamentals, technologies and applications (hydrogen energy). Wiley-VCH GmbH & Co. KgaA: Weinheim, Germany

UITP (2016) Statistics Brief: Local Public Transport in the European Union http://www.mobiuitp.org/PermaLinkRecord.htm?archive=164530298271&file=UITP_Statistics_PT_in_EU_DEF_print+.pdf

Unione Petrolifera (2017) Annual Report, Italy
Veziroglu TN, Sahin S (2008) 21st Century's energy: Hydrogen energy system. J. Energy Conversion Management 49:1820–1831
Yoshida T, Kojima K (2015) Toyota MIRAI fuel cell vehicle and progress toward a future hydrogen society. https://www.electrochem.org/dl/interface/sum/sum15/sum15_p45_49.pdf

Rail Station Boarding Systems for Persons with Reduced Mobility (PRMs)

Emmanuel Matsika and Ning Guo

Abstract In modern times, trains have become one of the most popular modes of transportation. However, there are still many problems to be solved, one of which is the gap issues between the platform and vehicle. Since the construction of the railway, the gap between the platform and the vehicle has been regarded as the main cause of fatality and weighted injury risk in the railway station. This project firstly investigates different solutions used all over the world that enable persons with reduced mobility (PRMs) to board independently. It applies mechanical engineering design principles, with strong input from primary and secondary data collected about the needs and preferences of PRMs. Finally, it then develops solutions of gap issues to reduce train dwell time, improve crowd flow and overall safety. An evaluation is conducted to select the most appropriate solution.

Keywords Rail stations · Passengers · Passenger trains · Reduced mobility · Platform design · Boarding systems

1 Introduction

Major railways now carry 16 billion passengers almost every year. On each trip, passengers must cross at least twice between the platform and the train—once on board and once off. Although the number of accidents that occur at the PTI is low, 21% of passenger weighted injuries risk and 48% of passenger fatality risk occurs at the PTI (RSSB 2015). In addition to safety, platform design should also consider cost, capacity, performance, perception, accessibility, and so on.

A station platform is an important part of railway system. It is located next to the rail tracks and is used for passengers to wait and get on and off the train. The platform should be designed to make passengers comfortable, safe and be convenient for waiting, such as providing seating space. However, in the actual use process, passengers often feel unsafe, which is mainly due to the design restrictions of trains

E. Matsika (✉) · N. Guo
Future Mobility Group, Newcastle University, Newcastle upon Tyne, UK
e-mail: emmanuel.matsika@newcastle.ac.uk

M. Marinov and J. Piip (eds.), *Sustainable Rail Transport 4*, Lecture Notes in Mobility,
https://doi.org/10.1007/978-3-030-82095-4_2

and platforms at the passenger train interface (PTI). In most cases this is due to the large horizontal and vertical gaps between the vehicle and the platform. Although this affects the general travailing public, the most adversely affected are persons with reduced mobility (PRMs). Table 1 demonstrates the horizontal and vertical gaps between the platform and vehicle (Fig. 1).

EU regulations require that public transport systems should be accessible for all people, especially those with impaired mobility. PRMs are not only persons with disabilities, but also many other people with reduced mobility, such as the elderly, pregnant women, travelers carrying large luggage and people pushing strollers. By this definition, a study found that about 50% of passengers in a typical EU city can be classified as PRMs (Lemmerer et al. 2018). Therefore, the gap between the platform and the train represents a big problem for the railway system.

The horizontal gap represents the distance from the vertical edge of the platform to the side surface of the train entrance. On the other hand, the vertical gap is the distance from the horizontal edge of the platform to the upper surface of the train entrance. The Railway Vehicle Accessibility Regulations 2010, Part 1.1.2 states that, for wheelchair autonomous access, the vertical gap should not exceed 50 mm and the horizontal gap should not exceed 75 mm. Table 1 shows the most important specialised and operational requirements that must be taken into account when designing train on-board and off-board gap fillers.

Table 1 Specialized and operational requirements (Rüger et al. 2010)

Requirements	Limits
Total Duration for use	<2 min
Platform width	>130 cm
Accessible door width	>=80 cm
Vertical gap	<110 cm
Capacity	Max. 350 kg

Fig. 1 Horizontal and vertical gaps. *Source* Emmanuel Matsika, author

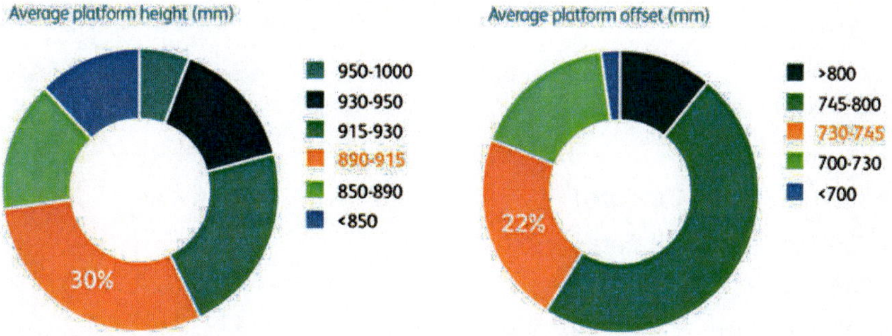

Fig. 2 Average platform height and offset (Rüger et al. 2010)

The challenge is that, during the years of railway expansion, the main railway lines were built at different times by different companies using different design standards. Therefore, most railway infrastructure today, including platforms, are not in line with current railway standards. For example, Fig. 2 shows the average platform height and average platform offset. Only about 30% of existing stations meet the height standard, and 22% meet the offset criteria. Even worse, only 7% of stations meet both the height and offset criteria (RSSB 2015).

2 Background

2.1 Current PTI Conditions

This section explains some of the challenges that passengers face at the PTI.

2.1.1 Standard Platform Dimensions

Figure 3 shows platform dimensions which exist in Europe. The common guidelines inside the Europe Union states is 550 mm and 760 mm. Numerous trains, particularly regional trains, have a floor height of around 550 mm. This allows for level boarding. Moreover, double deck vehicles offer level boarding for those platforms. They can likewise be utilised for long InterCity-movement. Typically, the platform height for highspeed trains is 760 mm. Some of high speed vehicles have lower floor, much the same as the TGV or Spanish Talgo-trains. Indeed, there are many difference platform designs and dimensions in Europe, which gives rise to majority of the challenges for trains and passenger boarding.

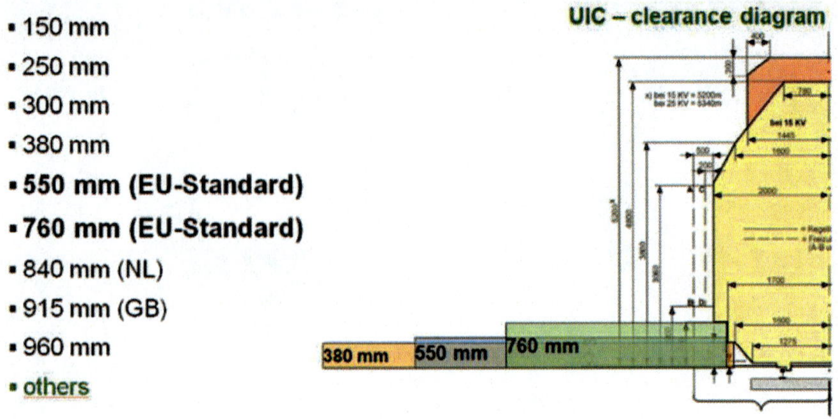

Fig. 3 Different platform heights in Europe (Helmut-Franz et al. 2010)

2.1.2 Challenges for Passengers When Boarding and Alighting

The PTI presents a major safety challenge for both the general public and PRMs. Potentially, they may fall in the gap between the platform and the train. This is particularly prominent for wheelchair users and the blind.

2.2 *The Gap Between Platform and Train*

In most cases, there is a vertical (step) and horizontal (gap) between the vehicle and station platform. This can be inevitable, especially on curved platforms, and those served by trains with a wide range of floor/step heights. The gap is required to permit for horizontal movement of the train as it passes through the station.

In Europe, over 30,000 rail stations have a gap problem (EC FAIR Stations 2020). Only a small proportion have step free access, as outlined in legislation (not exceeding 75 mm horizontal and 50 mm vertical gaps). Among the countries with the biggest problem is the UK, where bringing its 2500 stations with 6000 platforms up to current standards in one go would be prohibitively expensive (Shirres 2019). This applies to the rest of Europe. Where necessary boarding aids are used to improve the PTI safety (Matsika et al. 2013).

2.3 *Stepping Distance*

Many elderly and disabled passengers are able to manage a normal step. However, at several platforms the step and/or the gap between the platform and therefore the train is incredibly large. This issue affects all passengers and additionally has a bearing on train dwell times and safety.

2.3.1 Platform Design and Layout

Many platforms serve different types and lengths of trains. However, there is no standard way of informing passengers the location of the doors when the train stops at the platform. Studies have shown that the greatest problem faced by rail passengers is lack of or ineffective information and signage (Lemmerer et al. 2018). The search for the correct doors sometimes leads to unsafe crowd flows around and across the PTI. Some trains are longer than the platforms. This creates several problems (e.g. wherever wheelchair-accessible doorways are located at the top of the platform, or wherever passengers need to walk down the train to access safe doorways).

3 Review of Rail Passenger Boarding Systems

To improve the safety on the PTI, various solutions have been developed. The overall aim of each solution is to compensate for the height and/or horizontal distance differential between the rail vehicle and station platform. Solutions to close the gap between the vehicle and the platform are categorised as 'platform' or 'vehicle' based. In some cases, the optimum resolution is also a mix of strategies like raising the platform associated with an extendible train footstep.

A further differentiation is the use of 'fixed' and 'active' systems. Fixed gap fillers do not move once installed (for instance physical alterations to the platform edge). Active solutions can move, either purely mechanically or automated. Examples include deployable footsteps or platform edge solutions that may actively extend once a train is in position. The following sections review platform-based mostly solutions, incorporating platform edge devices and track alterations, and followed by train mounted solutions.

3.1 Platform-Based Systems

3.1.1 Fixed Gap Filler

Fixed gap fillers join directly to the platform edge and have been deployed in many places globally. These gadgets run from essential hard-edged fixings that decrease the inclination of the platform edge without needing major infrastructure works, to novel rubberised gadgets that are designed to resist contact with rolling stock. Some of these products are described below.

1. ***Delkor Rail***

Delkor Rail makes equipment that fills the PTI gap using plastic edges as shown in Fig. 4, used on the Heathrow Express Terminal 5. This was viewed as a success (HEX 2016), and there are plans to utilize it at other platforms at Heathrow airport. It has been applied internationally too (Delkor Rail 2015). This design and those comparable in nature can be fitted to any platform edge, on both straight and curved track sections. The adequacy of gap closure reduces as radius of curvature reduces.

At some sections, the design has been combined with a platform hump to aide level boarding as shown in Fig. 5.

Heathrow Express found that fitting the gap filler reduced PTI safety incidents at Terminal 5 from 15 to zero per year (HEX 2016).

Fig. 4 Delkor Rail gap filler (Delkor Rail 2015)

Fig. 5 Combined hump and gap filler arrangement (HEX 2016)

2. ***Pipex px gap filler***

Pipex px have created a filler that provides level access with negligible gap between the platform edge and the rail vehicle floor (Pipex 2015). The device is shown in Fig. 6. The material utilised is 'Treadmaster' TM7 vulcanised floor covering which offers a high degree of horizontal adaptability while remaining firm for stack bearing by passengers crossing the PTI gap. Pipex px have utilized 3D laser technology to examine platform conditions which helps with the positioning and installation of the gap filler.

3. ***Platform edge device***

(1) Safety gap filler for railway platforms

This development provides a gap filler that creates a variable PTI gap (see Figs. 7 and 8) (Lee 2006).

This innovation overcomes the drawback of commercially accessible PTI gap fillers. The inclined space filler diminishes the frictional constrain between the vehicle

Fig. 6 Pipex px gap filler (Pipex 2015)

Fig. 7 Gap filler (Lee 2006)

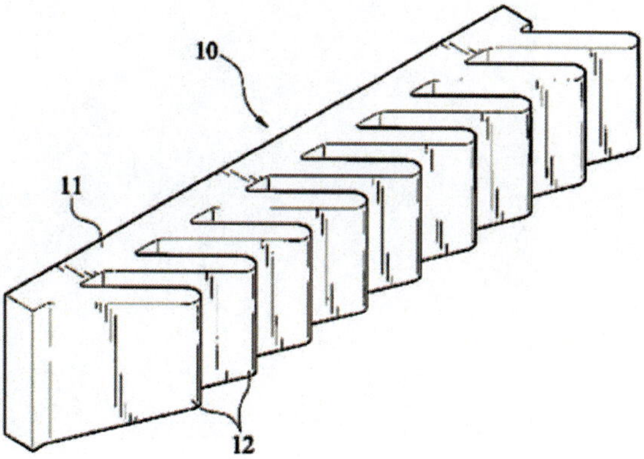

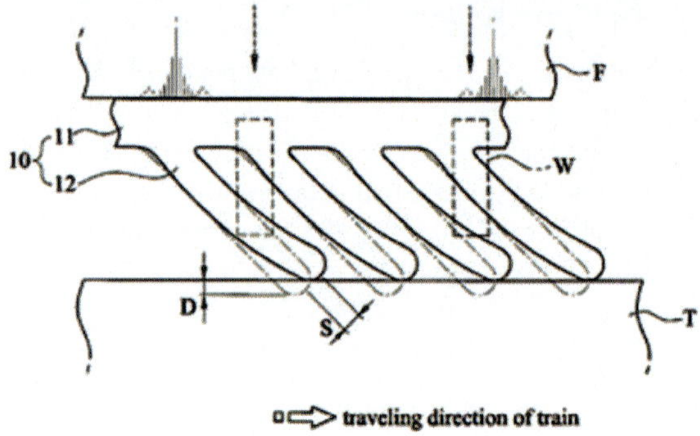

Fig. 8 Friction reduction (Lee 2006)

and device. It guarantees more weight-transfer limit per unit length than standard gap fillers of similar materials. It is simple to install (Lee 2006).

(2) *CDM Flexibord*

The 'CDM Flexibord' can be fitted directly to the platform edge. Compared with Delkor Rail and Pipex px devices, it has high vertical rigidity, but lower horizontal rigidity to be able to take impacts from rolling stock. It can be fitted along the complete length of the rail stage (Fig. 9).

(3) *STRAIL edge*

STRAIL has designed an elastic gap filler 'STRAIL edge' (Fig. 10). It can be fitted to the edge of existing platforms and has already been installed at tramways in various areas in France (STRAIL 2016). The advantages of this solution are:

Fig. 9 CDM Flexibord (CDM 2015)

Fig. 10 STRAIL edge (STRAIL 2016)

individually adjusted to passengers' requirements; Gap-free boarding and alighting; anti-slip and durable cover; quick and easy installation, and the special fastening provides stability and durability.

(4) *Corrugated platform edge*

The BRT in Cali, Colombia, utilises an innovative layered platform edge with wavy material. The photographs in Fig. 11 give a closer perspective of the material being utilised by MetroCali, the operators of the network. There is still need to assess the extent of wear and tear on the folded edge, as this may become a problem over time (Rickert 2010).

4. Raised platform

Raising a platform to improve level access boarding can have the extra impact of decreasing PTI gap. Humps of up to 1100 mm length are utilised at a few platforms

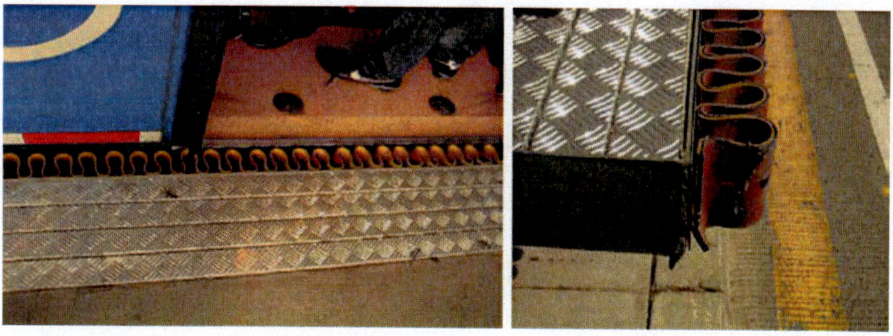

Fig. 11 Corrugated platform edge (Rickert 2010)

Fig. 12 Platform level access hump (Pipx px 2019)

in the UK. However, this may be risky if rolling stock with difference floor heights use the platform.

Platform Level Access hump

Pipex px have developed a pre-fabricated platform framework which aims to diminish the vertical gap between the platform surface and train step or floor (Pipex px 2019; Pipex px 2015; Network Rail 2015). The design, known as the 'Platform Level Access HUMP' (PLA-HUMP), can be built off-site to a bespoke plan. Platform humps are mounted to the surface of the platform (Fig. 12). The platform itself does not present any limitations to installation of the hump.

5. Railway platform gap filler

This is a device that dynamically fills the gap between the edge of the platform and the train. It includes several flexible sheet members that are mounted on the vertical edges of the platform and oriented toward the track (Fig. 13). Their top edges are aligned with the top surface of the platform. When the train is in the station, the flexible gap-filling members in the device dynamically flex as the train contacts. These curved components may be a single uniform unit or may be made up of two parts joined by an elastic material.

Depending on the choice of material and length, the device can be sufficiently curved to carry the weight of the passengers and increase the comfort of the passenger as they step over (Muller 1998).

6. CACOLAC

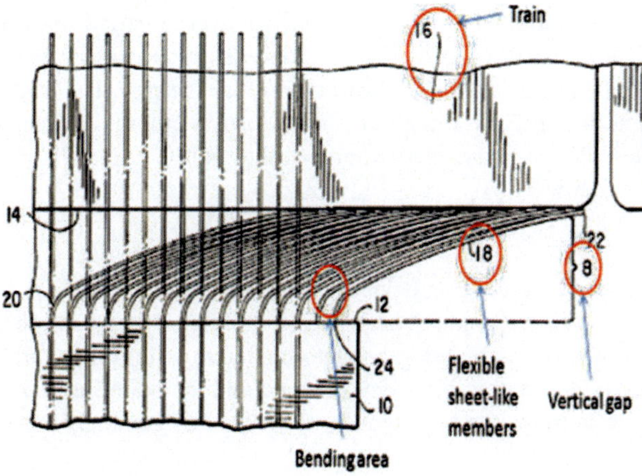

Fig. 13 Platform gap filler (Muller 1998)

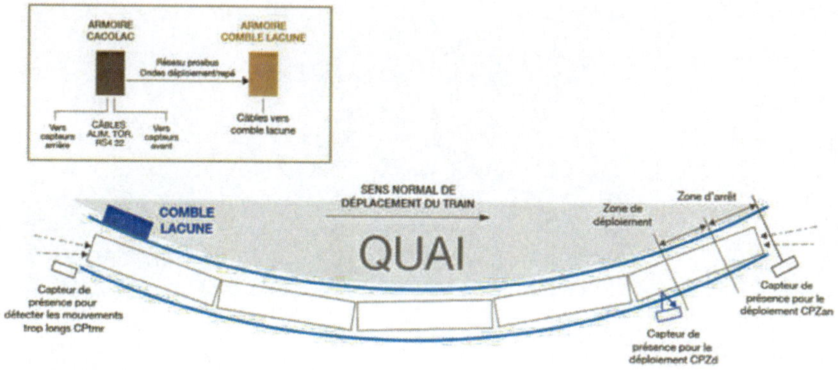

Fig. 14 CACOLAC (ClearSy 2007)

This is a device designed for PRMs. Keeping in mind the end goal to achieve this, it applies an intelligent control system. The platform is controlled by seven laser, infra-red and radar sensors, which measure specifically the distance between the vehicle and the stage as shown in Fig. 14 (ClearSy 2007).

3.1.2 Active Gap Fillers

Active platform gap fillers automatically extend from the platform edge upon the arrival of the train are being applied at several accessible stations. The system is active for the most part of the period of boarding and landing.

1. *Bigorre Ingenierie device*

Bigorre Ingenierie designed a dynamic gap filler for the Paris Public Transport Authority, which fills the PTI gap, utilising mobile steps (Fig. 15). The platform gap filler slides outwards on a straight bearing to close the gap. The RATP BI gap filler was trialed at first in 2007 at the Put d'Italie station in Paris. It has since been introduced at the Noisy le Grand station, and furthermore five installations are set up on the Orleans tramway.

2. *Gap-Bridging Device for Train Platforms*

This system is based on bridging devices. It is composed of a guide rail, a belt, a pulley, a motor and a bridging plate. The guide rail is designed as a ball linear guide or a roller linear guide so that the bridging plate can move on the guide rail (Fig. 16). It is then mounted on the station platform. The belt and the pulley are driven by the power of the motor to move the bridge plate outwards and gradually extend on the guide rail.

Fig. 15 Bigorre Ingenierie device (Bigorre Ingenierie 2016)

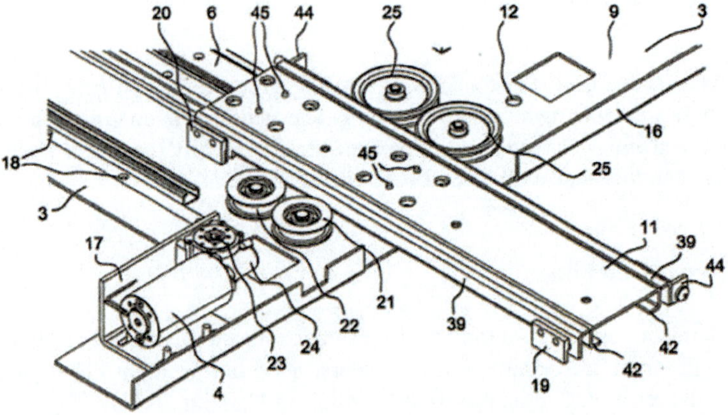

Fig.16 Structure of the gap-bridging system (Winkelmann and Hug 2010)

The problem with this system is that it can only solve the problem of horizontal gaps and cannot effectively reduce vertical gaps and takes much time to deploy (Winkelmann and Hug 2010).

3. *Platform edge extender*

In some curved stations, a movable platform is generally used to reduce the gap. This technology has been implemented at several railway stations. When the train stops at the platform, the movable platform extends from the edge of the platform to the train compartment as shown in Fig. 17.

4. *Platform edge warning ramp*

This is basically a horizontal station platform, as shown in Fig. 18. It is a fixed tilting platform mounted on the edge of the station platform and includes a platform surface, platform edges, and a means for alerting personnel to keep a safe distance. The sloping surface extends a portion of the edge of the platform to a specified

Fig. 17 HongKong Station. *Source* http://en.wikipedia.org/wiki/Platform_gap_filler

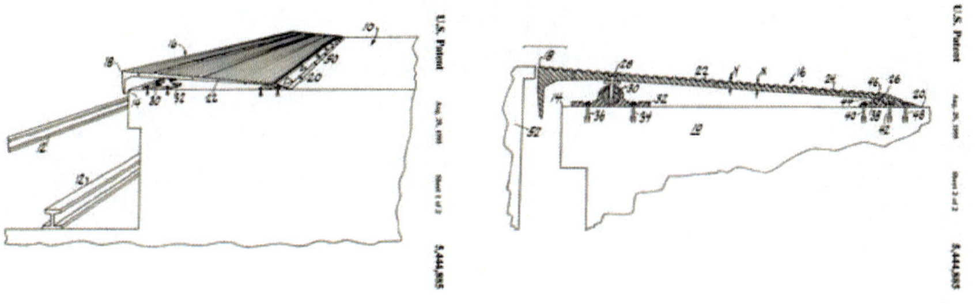

(*a*) (*b*)

Fig. 18 **a** perspective view of the warning platform **b** cross-sectional view of the warning platform (Hanrahan et al. 1995) distance. It can be raised or lowered by bolts to facilitate adjustment of different vertical and horizontal gaps.

The benefit of this warning device is that it can improve the safety of passengers. Especially for passengers with visual impairments, the surface of the platform has different thicknesses, they can check their position according to different footsteps or different sounds of the feet hitting the surface of the platform. However, there are inherent delays in the operation of the system (Hanrahan et al. 1995).

5. *Step apparatus for platform*

This platform-based device includes steps, rotating mechanisms and sensors (Fig. 19). The steps are placed on the platform and rotated horizontally to fill the gap between the platform and vehicle. One end of the arm is connected to the step and the other end is connected to the driving portion of the rotating mechanism, so that the step can be moved to a predetermined extended position. The tolerances on dimensions and form of the step can be changed to suit different trains. A sensor is used to detect the position of the train to respond accordingly (Yamaguchi et al. 2001).

6. *Device for reducing gaps*

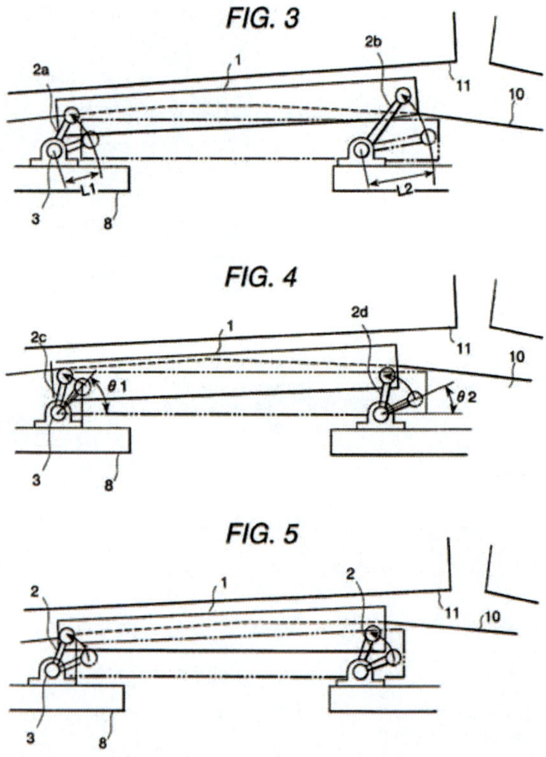

Fig. 19 Step apparatus for platform (Yamaguchi et al. 2001)

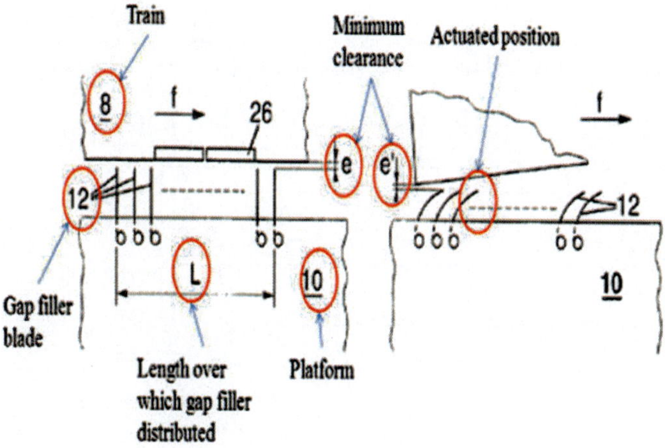

Fig. 20 Automatic gap filler (Vincent-Genod 2003)

This is an automatic device that reduces the PTI gap. It consists of horizontal rows of parallel elastic blades. The ends of these blades are rigidly fixed to the platform and thus have an elastic bending stiffness in the vertical direction. Each blade is in turn made of an elastomeric material, including one or more inserts that are resistant to bending (Fig. 20). The bending resistance in the vertical direction is made larger than the horizontal direction. As the train approaches the station, the induction and control framework determine the amount of bending required to fill the gap and controls the blade to bend to the desired position. A control system can detect vehicle position and control the door operation (Vincent-Genod 2003).

7. Retractable station platform extension

This is a device that can rotate the edge of the telescopic platform (Fig. 21). It has a fixed platform with an upper surface. An extension of this platform can be connected to the station. There is also at least one rotating member that connects the rotating part to the pivot pin of the fixed platform. The pivot pin, pivot member and platform are interconnected, and the extension of the platform is rotatable about the pivot pin due to the action of the pivot pin and the pivot member such that the extension can be moved. When the train arrives at the platform, there is a sensor that can detect the train and signal the control system and then move the extension of the platform down by manual or electric operation to fill the PTI gap (Klohn 1996).

8. *Removable platform edge*

Collapsible edges likewise are being used t0 give safe boarding of passengers. The mobile platform edge is associated with a pivot bolster (See Fig. 22).

9. *Manual ramp*

Manual ramps are for the most part the best and least costly boarding systems. They are operated by station or train staff. When there are passengers in need, the

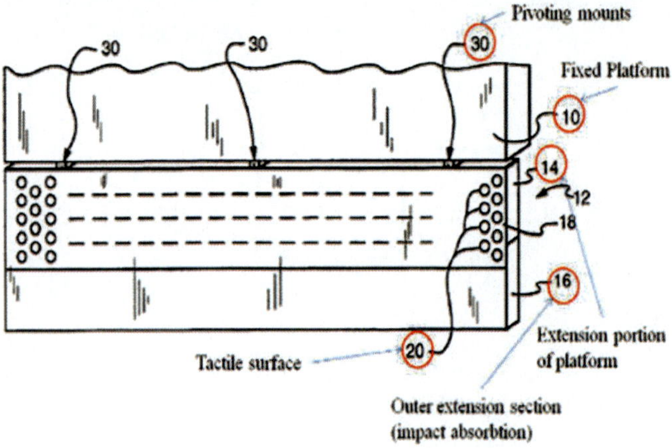

Fig. 21 Top view of retractable platform (Klohn 1996)

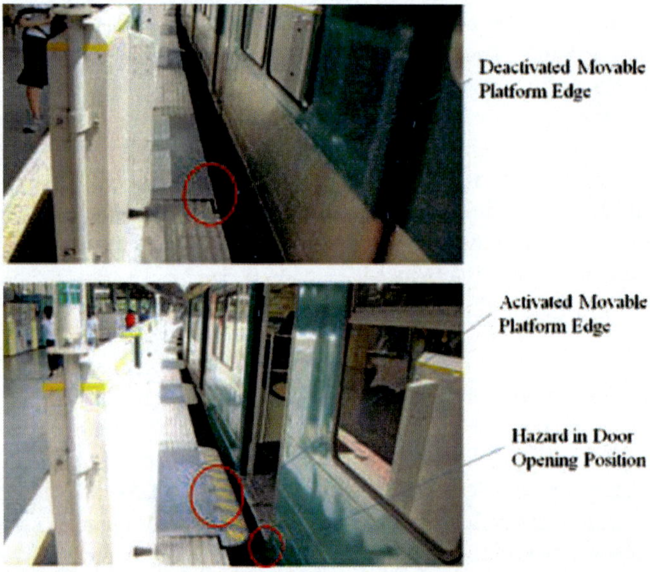

Fig. 22 Removable platform edge

ramp will be deployed to form a long incline to connect the train and platform for passengers to get on and off. Figure 23 shows the use of manual ramp.

Fig. 23 Platform-based manual ramp. *Source* Bernhard Rüger, Vienna University of Technology

3.1.3 Gauntlet (Multiple) Track

This is when a track uses a switch system in a multiple track section to move the train closer to the platform, thereby closing the PTI gap (Atkins 2004), as illustrated in Fig. 24. The benefits of such an arrangement are more prominent on straight track, rather than on curves.

Fig. 24 Gauntlet Track at Station in Portland, OR (Kirse 2010)

Fig. 25 Platform-based elevators (Access lifts 2020)

3.1.4 The Platform-Based Lift

These lifts are operated by station staff and are manually pushed on the platform to the vehicle doorway. Comparable to manual ramps, these elevators need ergonomic arrangement, not only for use by wheelchair users, but also the staff operators. Figure 25 shows an illustration of platform-based lifts.

3.2 Vehicle-Based Systems

3.2.1 Footsteps

The footstep system has been fitted on mainline vehicles worldwide for metro vehicles in cities such as London, Lyon, Vienna and Nuremberg.

1. *Pendolino footstep*

The main application in the UK are the Class 390 'Pendolino' (Fig. 26) and Class 373 'Eurostar'.

2. *Stadler deployable footstep*

Stadler is one of the train manufacturers that fit deployable steps to their vehicles (Fig. 27). The system employs ultrasonic sensors.

3. *Interchangeable step*

This arrangement, proposed by Morlok (2001), carters for the filling of vertical and horizontal gaps on both low and high-level platforms. The step can be pivoted about the hub A–A to give access for various platform heights (Fig. 28).

Fig. 26 Pendolino door step. *Source* https://www.youtube.com/watch?v=-lDvUcD33h8

Fig. 27 Stadler deployable footstep (Rail Insider 2020)

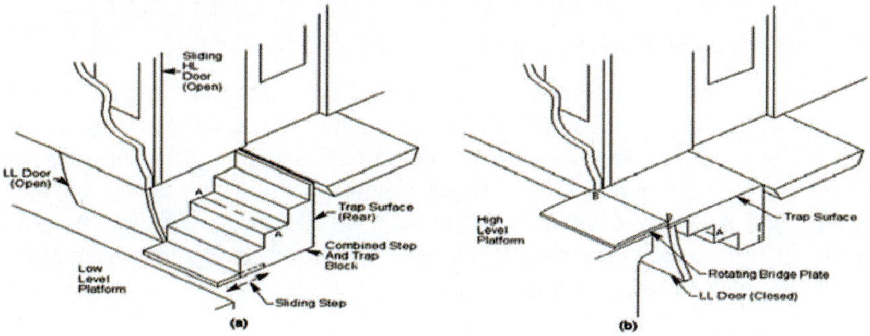

Fig. 28 Interchangeable step, **a** low level platform **b** high level platform (Morlok 2001)

Fig. 29 Bright line train door ramp (Mass Transit 2016)

3.2.2 Automated Ramps

Installation of ramps at doorways of rail vehicles usually completely closes gap. This is particularly advantageous for wheelchair users as there is no ridge for the wheels to overcome.

1. *Brightline train door ramp*

A retractable gap filler is utilised on Florida's 'Brightline' trains (Fig. 29). This was designed in conjunction with Siemens and Bode. The slope rises to 300 mm rotating to meet the platform edge (Mass Transit 2016).

2. *Glidelok ramp*

Kevin Fullerton is a designer who submitted the Glidelok ramp for New Trendsetters in 2005. The multi-plate ramp in Fig. 30 deploys and locates the platform level. Safety features are incorporated in the actuation system using sensors.

3. *Train-to-platform gap mitigator*

The system includes diverse gap filling mechanisms between vehicle door and rail station platform. It consists of several plates and controllers (Fig. 31). When the train arrives at the platform, based on the platform information and data provided by the sensors, the PC program will decide the required gap filling technique, and then the plate will extend from below the rail vehicle vestibule floor to fill the gap. It can handle different station conditions and various changes in level and perpendicular clearances on curved tracks (Chisena 2007).

3.2.3 Manual Ramps

Vehicle-based manual ramps have storage on board. When the train arrives at a platform, staff fit the ends to the entrance floor using a pin and hole securement

Fig. 30 Glidelok ramp (Fullerton 2005)

Fig. 31 Train-to-platform gap mitigator (Chisena 2007)

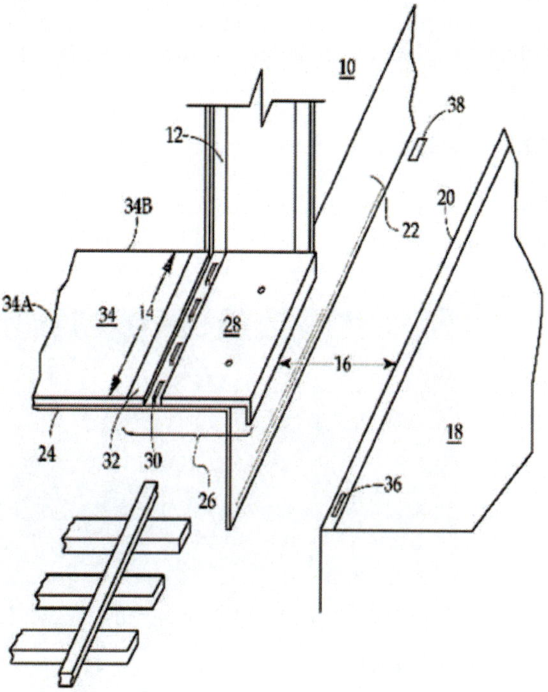

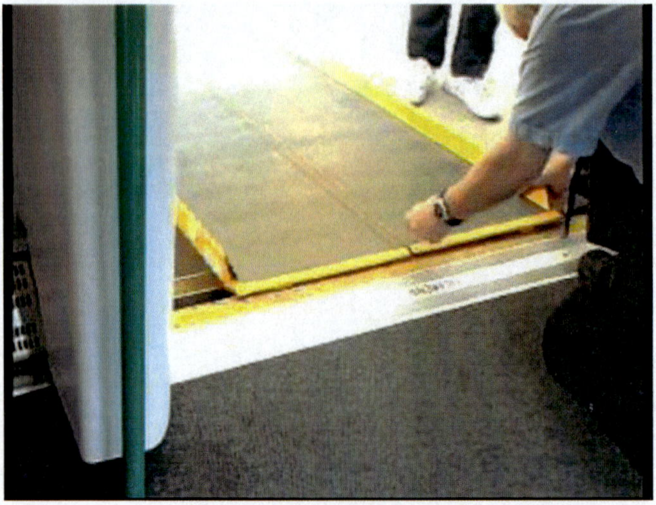

Fig. 32 Vehicle-based manual ramp. *Source* Emmanuel Matsika, Author

(Fig. 32). The greatest advantage of manual ramps (vehicle-based or platform-based) is that passengers, other than PRMs, prefer using them. They automatically offer different angles of inclination (Devadoss et al. 2012).

3.2.4 Vehicle-Based Lift

These comprise lift mechanisms that are fitted to the vehicle (see example in Fig. 33). When the train arrives at a station, this is deployed, operated by the train staff. These

Fig. 33 Vehicle-based lift in some countries (Palfinger 2020)

tend to have a vertical operation of up to 1100 mm, which is much higher than platform-based ones.

3.3 *Summary of Rail Passenger Boarding Solutions*

The horizontal and vertical gap between the platform and vehicle is influenced by many factors. Sections 3.1 and 3.2 discussed various designs and innovations that have been developed to resolve safety concerns at the PTI. Most of the solutions aim to cover the horizontal gap, when really, the vertical gap poses injury potential too. As shown in Figs. 34 and 35 the designs can either be platform-based or vehicle based, respectively.

3.3.1 Platform-Based Systems

Platform-based designs vary from significant platform adjustment to simply as a device on the platform. They are divided into fixed or movable devices. Fixed designs are for the most part flexibly used to partially or fully fill the PTI gap. This sort of arrangement can be utilised both for straight and curved platforms. Their advantages are:

- Minimal vehicle service delay during installation.

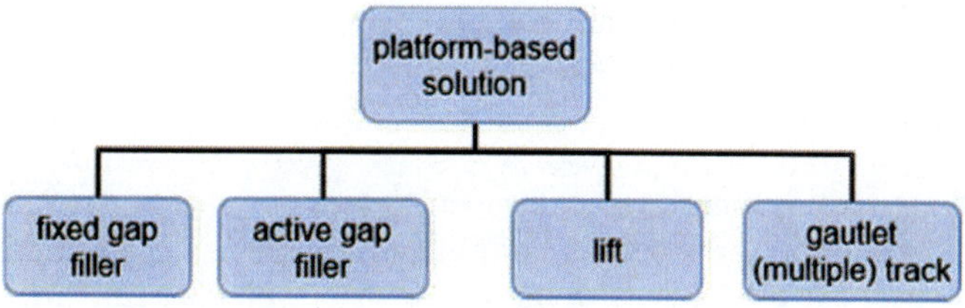

Fig. 34 Platform-based solutions

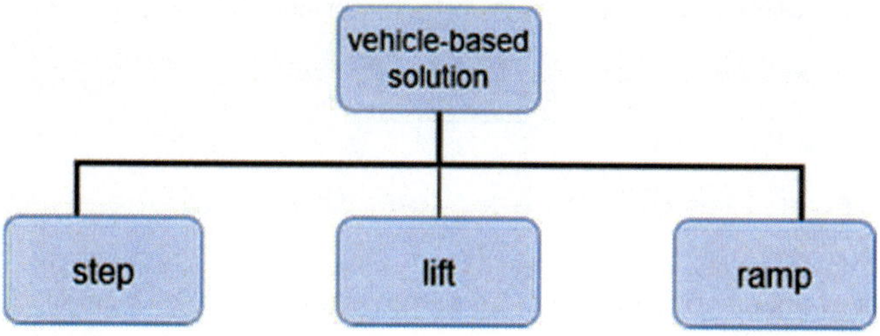

Fig. 35 Vehicle-based solutions

- Less installation and maintenance costs.
- No significant platform alterations are required

The disadvantages of fixed solutions are:

- More detailed need to match user needs.
- Difficulties in addressing the vertical gap.
- Abrasion because of device-vehicle contact.

Active or moveable systems utilise sensors to detect the location of vehicle doorway in reference to the platform and adjust their device to the predefined position. Like fixed designs, moveable systems are not suitable for vertical gaps. Their main advantage is application of sensors to improve the accuracy of the device operation, and reliability. Disadvantages include delays in vehicle service because the detecting parts principally actuate once the vehicle stops; costly installation and maintenance; and requires significant platform alterations.

3.3.2 Vehicle-Based System

These systems can be fixed, or moveable (operated manually or semi-automated). Some of them are suitable for compensating both horizontal and vertical gaps. Most of the vehicle-based designs are intended for horizontal (flat) applications, while a few deploy step type gap fillers.

Fixed designs have the following advantages:

- Minimal vehicle service delay during installation.
- Less installation and maintenance costs.
- No significant platform alterations are required.

Their disadvantages are:

- Limitations in dimensions to suit the vehicle space
- Difficulties in meeting the dual aim of filling the horizontal and vertical gaps.
- Prone to wear and tear because of physical contact with the train.

Moveable designs utilise sensors to detect moment and location to deploy the gap filler. The advantages of these designs are their suitability for horizontal/vertical gaps. Disadvantages include delays in vehicle service, and costly installation and maintenance.

4 Evaluation of Rail Passenger Boarding Solutions

Considering so many options to pick from, this section provides a guide to selection of the gap filling solution that meets some pre-determined criteria. An essential expert subjective investigation was performed within this scientific project. The expert group

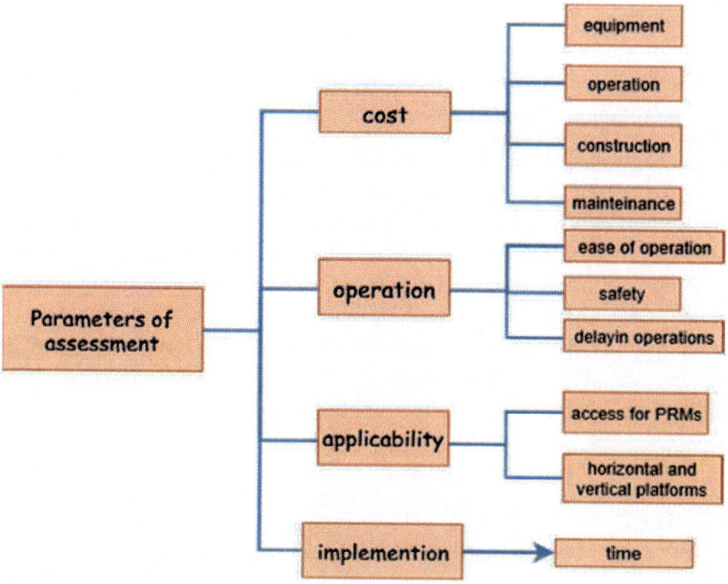

Fig. 36 Factors used for the evaluation criterion

consisted of 10 mobility researchers and transport operators. Subjective evaluation is outlined as an emotional judgment supported non-quantifiable data. For the evaluation criterion, four factors were considered namely cost, applicability, operation, and implementation. This is further sub-divided as shown in Fig. 36.

4.1 Factors for Evaluation

4.1.1 Cost

Cost is a critical component in choosing an ideal solution. In this case, the following cost components are considered:

- *Equipment:* this refers to the actual initial cost of the device and associated components.
- *Operation:* these are costs associated with the exploitation of the system. Utilising electric and control systems is thought to increase costs compared with fixed and mechanical installations.
- *Construction:* From a development perspective, the solution with the least alteration to the current setup would be cheapest one.
- *Maintenance:* Preventive or planned checks and repairs at regular intervals accepted to be less costly than breakdowns and unplanned works.

4.1.2 Operations

Boarding device solutions were additionally evaluated for factors such as, simplicity of actual operation and safety of passengers and delays in vehicle activities:

- *Ease of actual operation:* it is usually desirable to have a system that is simple, and easy to operate. This saves time and reduces train dwell time.
- *Safety:* this is the primary purpose of developing gap fillers and boarding systems at the PTI. The solution should adequately compensate for the horizontal and vertical gaps sufficiently enough to avoid any PRMs falling or being injured when boarding or alighting.
- *Delay in prepare operations:* The actual actuation of the system requires a time period, particularly for moveable systems. There are devices which do not require actuation time, such as fixed platform-based designs.

4.1.3 Applicability

Applicability is a factor which alludes to several aspects such as,

- *Access for PRMs:* The gap filler should be positioned primarily to provide easy and safe boarding and alighting of persons with reduced mobility. Ideally, it should provide a step-free or level boarding.
- *Horizontal and vertical gap filling:* The goal of installing a gap filler or boarding system is to fully compensate for the horizontal and vertical gaps. However, this is not always easy or possible as has been demonstrated by the many different designs in Sect. 3.

4.1.4 Implementation

The time needed to install the solution is important. This will determine how long the platform would be out of service while construction is taking place. It may also indirectly infer the labour cost of installation.

4.2 Weighting and Rankings

Design solutions were evaluated using the factors discussed earlier. The criteria applied a Likert scale of 1 (low performance or less desirable) to 5 (high performance or most desirable), using evaluation of 10 experts. Applicability, deemed to be most important for PRMs, accounted for 40% while cost, operations and implementation each accounted for 20% of the score. Presented in Table 2 are the overall results of the evaluation.

Presented below is the summary of the outcomes of the evaluation.

Ranking based on cost

Table 2 Evaluation results of the design solutions

	cost(20%)						applicability(40%)				operation(20%)					implemention(20%)			overall
	equipment	construction	maintanence	operation	total		access for PRMs	horizontal and vertical platforms	total		ease of operation	safety	delay in operations	total		time	total		100
	5	5	5	5	actual	converted	5	5	actual	converted	5	5	5	actual	converted	5	actual	converted	
platform-based solution																			
Delkor Rail	5	5	4	5	19	19	1	1	2	8	5	2	5	12	16	3	3	12	55
Pipex px gap filler	2	3	2	3	10	10	2	1	3	12	4	3	4	11	14.7	2	2	8	44.667
Safety gap filler	5	5	2	5	17	17	2	1	3	12	5	3	5	13	17.3	3	3	12	58.333
CDM Flexibord	5	4	3	5	17	17	1	2	3	12	5	3	4	12	16	2	2	8	53
STRAIL edge	4	3	3	5	15	15	3	1	4	16	5	4	5	14	18.7	3	3	12	61.667
Corrugated platform edge	4	4	3	5	16	16	3	1	4	16	5	2	3	10	13.3	3	3	12	57.333
Platform level access hump	5	4	3	5	17	17	4	1	5	20	4	2	4	10	13.3	3	3	12	62.333
Railway platform gap filler	4	3	4	4	15	15	1	1	2	8	4	1	4	9	12	2	2	8	43
CACOLAC	1	2	2	5	10	10	3	2	5	20	5	2	3	10	13.3	2	2	8	51.333
Bigorre Ingenierie device	3	3	2	4	12	12	2	1	3	12	4	2	4	10	13.3	2	2	8	45.333
Gap-Bridging Device	3	3	2	1	9	9	3	2	5	20	3	4	2	9	12	2	2	8	49
Platform edge extender	4	3	3	5	15	15	3	1	4	16	5	3	4	12	16	3	3	12	59
Platform edge warning ramp	3	2	2	2	9	9	4	3	7	28	3	3	2	8	10.7	2	2	8	55.667
tep apparatus for platform	2	3	2	3	10	10	4	3	7	28	2	3	2	7	9.33	1	1	4	51.333
Retractable station platform extension	2	2	3	3	10	10	3	1	4	16	4	4	2	10	13.3	1	1	4	43.333
Movable platform edge	3	3	3	2	11	11	4	3	7	28	4	3	1	8	10.7	2	2	8	57.667
Gauntlet (multiple) track	4	4	3	4	15	15	2	2	4	16	4	2	1	7	9.33	2	2	8	48.333
vehicle-based solution																			
Pendolino footstep	4	4	4	5	17	17	3	1	4	16	5	1	5	11	14.7	4	4	16	63.667
Stadler deployable footstep	5	4	4	4	17	17	3	1	4	16	4	4	5	13	17.3	5	5	20	70.333
Interchangeable step	3	2	2	3	10	10	4	3	7	28	2	3	2	7	9.33	2	2	8	55.333
Brightline train door ramp	3	3	4	3	13	13	4	3	7	28	3	4	4	11	14.7	3	3	12	67.667
Glidelok ramp	2	4	3	2	11	11	4	5	9	36	4	4	2	10	13.3	3	3	12	72.333
Train-to-platform gap mitigator	2	2	2	2	8	8	3	3	6	24	2	4	2	8	10.7	2	2	8	50.667
Manual Ramps	5	5	5	5	20	20	5	5	10	40	4	4	2	10	13.3	4	4	16	89.333
Lifts	2	4	3	2	11	11	5	5	10	40	3	5	1	9	12	1	1	4	67

1. Manual ramps
2. Delkor Rail

Ranking based on applicability

1. Manual ramps/Lifts
2. Glidelok ramp

Ranking based on operations

1. STRAIL edge
2. Safety gap filler/Stadler deployable footstep

Ranking based on implementation

1. Stadler deployable footstep
2. Pendolino footstep/Manual ramps

Overall ranking

1. Manual ramps (Devadoss et al. 2012)
- Appropriate for both straight and curved platforms.
- Accessibility of PRMs.
- Horizontal and vertical gap filling.
- Minimal costs of installation, operations and maintenance.
2. Glidelok ramp (Fullerton 2005)
   - No vehicle alteration required.
   - Appropriate for both straight and curved platforms.
   - Accessibility of PRMS.
   - Horizontal and vertical gap filling.
3. Stadler deployable footstep (Stadler 2015)
- Appropriate for both straight and curved platforms.
- Accessibility of PRMs.
- Horizontal and vertical gap filling.

5 New Proposed Solution

The review of the gap filler solutions and boarding systems revealed that they have advantages and disadvantages. In order to provide guidance on the most suitable solution for future development, an evaluation criterion was applied. For accessibility and safety, results show that the manual ramp has more advantages of meeting the needs of PRMs than the other designs. Although they have many advantages, the greatest disadvantage is the need for staff to operate the ramp, which requires additional labour time. Moreover, manual operation is time consuming and laborious, which is very inconvenient. It also causes delays in the train dwell time. Therefore, based on the manual ramp solution, Fig. 37, shows a proposed conceptual design of an automated ramp.

This automatic ramp solution solves the problem of extra manpower hours. However, it increases cost and complexity. Automated ramps use equipment such as motors and sensors, which will undoubtedly increase the cost of installation and maintenance. In addition, without staff, safety precautions need to be considered. Nevertheless, the additional cost could be compensated by a reduction in train dwell time, improved crowd flow, and safety.

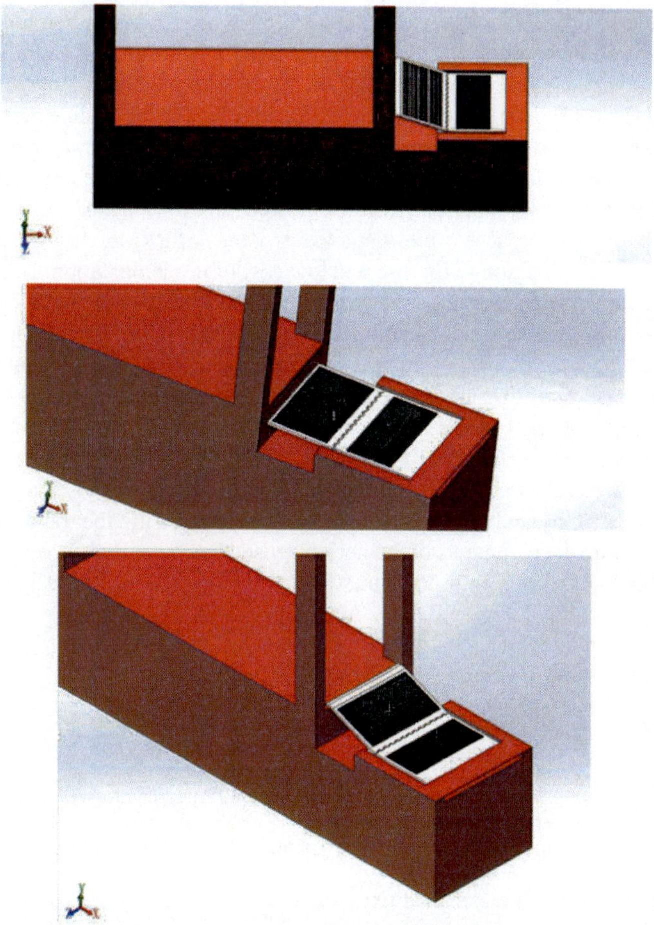

Fig. 37 Automated ramp

6 Conclusions and Future Prospects

The gap at the platform train interface (PTI) has always been a safety concern in general, and particularly for persons with reduced mobility (PRMs). Regulations require that the horizontal and vertical gaps should not exceed 75 mm and 50 mm, respectively. At present many countries in Europe do not meet these requirements since the stations are old, and therefore need other means of filling the gap. New builds, however, incorporate step free designs as far as possible.

From the vehicle dynamics point of view, it makes sense to make the gap large enough for the vehicle body not to contact the platform as the vehicle oscillates horizontally. On the other hand, the gaps create a safety hazard for the general public, but particularly for PRMs. Especially at platforms where trains pass at high speeds, enough gap should be allowed. Therefore, the need for installing a gap filler device becomes inevitable since the presence of the gap is also inevitable.

This paper has shown that current gap filler solutions and boarding systems have shortcomings when it comes to vertical gaps and operating on curved platforms. In order to provide guidance on the most suitable solution for future development, an evaluation criterion was applied. Results showed that the manual ramp has more advantages for meeting accessibility and safety needs of PRMs, than most existing designs. This is in addition to its ease and low cost of implementation. Despite their many advantages, manual ramps the greatest disadvantage is the need for staff to operate the ramp, which requires additional labour. Moreover, manual operation is time consuming and laborious, which is very inconvenient. Its deployment also adds to the train dwell time. To harness the advantages of the manual ramp, yet reduce the train dwell time, an automated design based on the manual ramp is recommended for future work.

The review has shown that operationally, a PTI solution should aim at improving both the accessibility and the PRM safety, while also reducing the train dwell time. Therefore, future developments of the new proposed design or derivatives of existing designs should take this into account, with a high weighting in design decision making.

References

Access Lifts Ltd (2020) https://access-lifts.co.uk/applications/transport-lifts/web-train/

Atkins (2004) Department for transport technical report, significant steps

Bigorre Ingenierie (2016) Passenger accessibility gap filler, received in private correspondence with Bigorre Ingenierie 18th January 2016

CDM (2015): CDM Flexibord, retrieved 10th October 2015. http://vibratec.se/Datasheet/FLEXIBOARD.pdf

Chisena M (2007) Train-to-platform gap mitigator. US Grant Patent no. US7784406B2. Available at: https://patents.google.com/patent/US7784406. Accessed 25 Apr 2018

ClearSy SE (2007) CACOLAC system: Automatic gap filler control device. Retrieved from http://www.Fersil-railway.Com/dl/systeme_cacolac_en.Pdf

Delkor Rail (2015) Private correspondence with Business Development Manager on 5th October 2015

Devadoss R, Ahmad S and Dhamodharan DR (2012) Station platform-Train Interface for Rail passengers-A Technology review". CRC for Rail Innovation

EU FAIR stations (2020): http://www.fairstations.eu/

Hanrahan PJ, Keegan EK, Goffe AT (1995) Platform edge warning ramp, US Patent 5444885 A

Fullerton K (2005) Glidelok ramp, Fresh Innovators 2005, Australia, retrieved 28 Aug 2010

Helmut-Franz W et al (2010) Engineering Metros & Coaches, Conference Public Transport—Accessibility for All, Vienna, May 2010

HEX (2016) Private correspondence with heathrow express safety & assurance advisor on 9th Feb 2016

Kirse K (2010) TriMet's WES commuter rail use of gauntlet track for level boarding. TriCounty metropolitan transportation District of Oregon. http://www.apta.com/mc/rail/previous/2010/Presentations/TriMet-WES-Commuter-Rail.pdf

Klohn H (1996) Retractable station platform extension. US Patent 5517921 A

Lee W (2006) Safety gap filler for railway platforms. US Application Patent no. US20100224097A1

Lemmerer H, Shibayama T, Rüger B, Seltenhammer K, Macoun T, Matsika E, Logueira DF, Battista U, Khosravi M, de Santiago Laporte A, Kuzmina D (2018) D2.1 User Needs and Expectations of the General Public and PRMs. EU FAIR Stations Project. NUMBER 777636—IP/ITD/CCA—IP1, IP3

Mass Transit (2016) Brightline Innovation: First Passenger Train in the U.S. to Utilize Retractable Gap Filler. https://www.masstransitmag.com/rail/press-release/12167187/florida-east-coast-railway-fec-brightline-innovation-first-passenger-train-in-the-us-to-utilize-retractable-gap-filler

Matsika E, Chirwa EC, Peng Q (2013) Influence of human related factors on railway vehicle wheelchair occupant crash safety. In: Joint Rail conference—ASME. Knoxville, Tennessee, USA, April 15–18, 2013. Paper No. JRC2013-2406, pp V001T06A002-; 11 pages. https://doi.org/10.1115/JRC2013-2406. ISBN: 978-0-7918-5530-0

Morlok EK (2001) The need for a new commuter car entranceway design for mixed high and low level platforms, Annual Meeting of the Transportation Research Board, Washington DC, Jan 2002

Muller RJ (1998) Railway platform gap filler. WO Patent 1998046465:A1

Palfinger (2020) Access lifts for wheelchair users in trains. The MBB TRAINLIFT. https://www.railway-technology.com/contractors/passenger/mbb-palfinger/

Network Rail (2015) Mind the gap! Volunteers test new Thameslink Programme platforms for accessibility. https://www.networkrailmediacentre.co.uk/news/mind-the-gap-volunteers-test-new-thameslink-programme-platforms-for-accessibility

Pipex px (2015) Platform Level Access (PLA) HUMP, July 2015

Pipex px (2019) [online] Available at: https://www.pipexpx.com/uploads/products/683.1363626138.pdf

Rail Insider (2020) Accessibility features on Greater Anglia's new Class 755 trains put through their paces. https://www.railinsider.co.uk/2019/08/09/accessibility-features-on-greater-anglias-new-class-755-trains-put-through-their-paces/

Rickert T (2010) Technical and operational challenges to inclusive Bus Rapid Transit: a guide for practitioners. World Bank and TFESSD—Disability Window, Norway and Finland

Rüger B, Petutschnig B, Simic G (2011) Improving railway vehicle accessibility for all persons with reduced mobility. PubTrans4All http://www.pubtrans4all.eu/en/

Rüger B, Tauschitz P (2010) Vienna University of Technology; Petutschnig B—BP International Consulting, Boarding Assistance System Evaluation Criteria Report, Deliverable 2.1, June 2010

RSSB (2015) Platform train interface strategy. Available from: https://www.rssb.co.uk/pticontent/pti041-2015-01-platform-train-interface-strategy.pdf

Shirres D (2019) Mind the gap: keeping trains and platforms level is a 'complex' issue. https://www.imeche.org/news/news-article/mind-the-gap-keeping-trains-and-platforms-level-is-a-'complex'-issue

Stadler (2015) Private correspondence with project engineer. 11th Nov 2015

STRAIL (2016) Information retrieved 15th Feb 2016. http://www.strail.de/index.php?id=186&L=1

Vincent-Genod B (2003) Apparatus for reducing gaps between a platform and a guided vehicle. US Patent 20030070576 A1

Winkelmann U, Hug R (2010) Gap-Bridging device for train platforms. US Application Patent no. US20100043664A1

Yamaguchi H, Maruyama A, Takahashi T, Mizobuchi T (2001) Step apparatus for platform. US Patent 20010042489 A1

Measuring the Impacts on the Economic Growth of Various Countries Using Key Performance Indictors—A Quantitative Approach Including Logistics Performance, Infrastructure, Rail Network and Volume of Goods Carried

Charles Lopes Ribeiro, João Paulo Albuquerque Kalil, Philipe Augusto de Paula Pacheco, Débora Magalhães Kirchmair, and Marin Marinov

Abstract This paper discusses the impacts of logistics development on logistics efficiency and economic growth of pre-selected countries. These impacts were analyzed in terms of: logistics infrastructure, rail network/1000km^2 and volume of goods carried by rail. Subsequently, the impact of 'Logistics Performance Indicator' on economic growth measured by current GDP was analyzed. For the analysis, a sample of eight countries (four developed and four developing) was used for a period of 10 years, 2007–2017. The research presentes theoretical framework that includes literature review, and analysis of Logistics Performance Indicator issues and economic growth, rail transport, and logistics performance of countries through indicators. The methodology, in general, includes both an explanatory and quantitative research. The data used were collected from the World Bank website. More specifically, multivariate statistical methodological approach was used with simple and multiple linear regression econometric models in panel data for this study. Based on this discussion, the results suggest that the quality of the infrastructure and the rail network/1000 km^2 have a positive impact on the logistics efficiency of the countries under study. Whereas Logistics Performance Indicators (like infrastructure quality and TKU/railways) indicate a positive impact on the countries' growth in terms of current GDP.

Keywords Impact assessment · Logistics performance indicator · Logistics infrastructure · Rail networks

M. Marinov
Engineering Systems and Supply Chain Management, Aston University, Aston Triangle, Birmingham, UK

C. L. Ribeiro · J. P. A. Kalil · P. A. de P. Pacheco (✉) · D. M. Kirchmair
Instituto Federal de Educação, Ciência e Tecnologia do Sudeste de Minas, Santos Dumont, Minas Gerais, Brazil
e-mail: philipe.pacheco@ifsudestemg.edu.br

M. Marinov and J. Piip (eds.), *Sustainable Rail Transport 4*, Lecture Notes in Mobility,
https://doi.org/10.1007/978-3-030-82095-4_3

1 Introduction

Countries with developed logistics systems tend to provide higher quality of life with better access to goods and services to consumers more efficiently at affordable costs. The people, companies and governments of these countries enjoy the benefits that an efficient logistics system offers by making the market and trade in these places more intense. Aspects such as infrastructure quality of existing modes, well-structured and localized ports, duplicate highways, extensive rail network, cargo terminals, quality of airports, bridges and viaducts, all this structure aims to bring together producers and consumers.

Railways are very important in our society and their operation and expansion are completely necessary. They are important because they allow passengers and cargo to be transported over long distances; In other words, they provide freedom of movement, which did not exist in the past. Rail networks are complex systems and consist of many different types of trains, corridors and services. Railways carry 21 billion passengers annually and 10 billion tons of cargo annually (IBM 2014 apud Bevrani 2015). Railroad history shows that it can be used as an efficient means of transport for moving goods and passengers. Building efficient transportation systems such as railways can improve local and global economies by allowing more goods to be transported, reducing greenhouse gas emissions, making roads safer, and reducing road congestion (IBM 2014 apud Bevrani 2015).

Studies indicate that infrastructure spending has a positive effect on growth and per capita growth. Signals of GDP and luminosity data coincide. If subnational authorities—the states—increased their spending by one percentage point, the regional GDP growth rate would increase by 0.11 percentage points per year, while per capita GDP growth rate would account for an increase of 0.072 percentage points per year (Amann et al. 2016).

Recent works demonstrate a need for sustainable expansion of the rail sector. This paper uses logistics indicators such as: rail network/1000 km^2, Logistics Performance Indicator, TKU (tonne-kilometer) and infrastructure quality, in addition to current GDP, using a sample of developed and developing countries comprising a series of 2007–2017 to define what are the impacts of logistics variables, especially rail transport and infrastructure quality, on the logistics efficiency of a country. Consequently, as a specific objective, we intend to analyze if the Logistics Performance Indicator impacts on the current GDP of the countries.

The results of this research contribute to the literature of the area and can be compared with the findings of other researchers, writers and logistics professionals.

The work in question is structured in 5 sections, with the introduction. The second section comprises the background and literature review that underpinned data analysis; The third section deals with the methodology used, followed by the results and the relevant discussions. Finally, the final considerations of the study are presented.

2 Background and Literature Review

According to Ballou (2015), the US, Japan, and members of the European economic community has a high standard of living and trade goods freely due to the efficiency of their logistics systems.

Marchetti and Ferreira (2012) studied the importance of logistics for the Brazilian economy. They addressed regulatory aspects, market performance and future challenges in the different modes. They add that in addition to the quality of the infrastructure, the transport network configuration itself influences logistics performance.

Logistics performance can also be measured not only at country level, but also at company level. Da Silva, Oliveira and Marinov (2020), analyzed the merger between two Brazilian companies. Using the DEA (Data Envelopment Analysis) methodology, the results indicated that the rail service after the merger was efficient, generating synergy and operational improvements.

Leans et al (2014) conducted a research using a dynamic structural model to test the relationship between logistics development and short-term and long-term economic growth in China. They concluded, using the Granger causality test, that economic development causes more demand for logistics services and therefore leads to the development of logistics. It was also noted that rail transport causes the development of road and waterway transport, which implies that rail played a key role in the transport network in China by 2014.

Boopen (2006) studied the relationship between investment and improvements in transport infrastructure with the development and economic progress of sub-Saharan African countries and 13 small developing states (Jamaica, Barbados, Mali, among others). Using cross-sectional and panel data analysis, he concluded that transportation capital contributed to the economic progress of these countries.

In his research Hayaloglu (2015) investigated the impact of the evolution of the logistics sector on economic growth in 32 OECD (Organization for Economic Development Cooperation) countries, covering the period 1994–2011. It used as methodology multiple linear regression with panel data analysis. One conclusion was that there was a positive and statistically significant relationship between investments in land transport infrastructure and GDP. In other words, infrastructure investments have a positive impact on economic growth.

Other results from this study, which used different indicators from the logistics sector, were that investments in the transport sector and development in the telecommunications sector contributed to economic growth in OECD countries.

Sturm, Jacobs and Groote (1999) found strong evidence of a positive impact of transport infrastructure investment on Dutch GDP in the second half of the nineteenth century.

From the above, it is clear that an efficient logistics system impacts on the economic performance of countries.

2.1 Transports

According to Arnold (2014), transport is essential for the economy of any area, as well as for social development. This statement is corroborated by Bowersox et al. (2014), who state that transportation represents more than 60% of a company's logistics expenses, explained by the great responsibility of moving stocks through the supply chain. Therefore, its performance is crucial for all areas of a company.

The emissions produced by the railways are amongst the lowest in the transport sector. If a greater mass of freights can be transported on one train, fewer trucks are needed to run, which have a positive effect on the decrease in the number of greenhouse gases produced. It would suggest that the solution to green transport is not to bring down car emissions, but to encourage use of the railways (Marinov et al. 2020).

According to Bowersox et al. (2014), the rail modal is historically known for its large capacity to transport goods (TKUs) efficiently and safely over long distances, which can be associated with the unitization usually with containers, often occurred in such modal. In this way, the high fixed costs associated with this type of transport are balanced with its low variable costs, which demonstrate its lower total cost.

The significant advantages of rail freight over its close competitors including road have been noted many times, some of which include benefits to the environment as rail creates less air pollution than road freight; the ability to create economies of scale and benefits for logistics operators. When completed efficiently, rail freight operations have been identified to benefit the economy through the proficient transfer of goods (woroniuk et al. 2013).

According to Da Silva, Oliveira and Marinov (2020), the strategy of using trucks increases export costs and decreases Brazilian sales, mainly due to the significant volume of cargo being transported by trucks over long distances.

In addition to the authors mentioned above, it can be noticed that in other countries with similar territorial and commercial structures, a substantial portion of transport is by rail.

Fraszczyk et al (2016) showed that extending the length of track, followed by offering more rail services stimulates economy as people are employed to build the railways, maintain it as well as more customers can use it. Besides that, the extending of the length of track to secluded villages and towns improve the quality of life for the individuals in these areas, allowing them to travel and exercise their mobility into towns and cities. And it could also potentially generate extra income for infrastructure managers as well as at journeys' destinations (employment, education, entertainment, etc.).

2.2 Analyzed Indicators

To support the importance of the logistics system, it was used indicators shown chronologically as follows: railway network/1000 km^2, Logistics Efficiency, Quality of Logistics Infrastructure, TKU and current GDP.

Current GDP was used in the second model of this research as an indicator of a country's growth. GDP is the main indicator of economic activity. It reflects the value of the final goods and services produced in an economy during a given period (Rogers et al. 2006).

The indicator Rail network/1000 km^2 was also used. For the calculation, a division was first made between the kilometers of railway network by the territorial extension in km^2 of each country of the sample within the chronological period. After that, this value was multiplied by 1000, thus acquiring a quantity of the rail network/1000 km^2 of each country. It is a more accurate indicator than just the length of each country's rail network. This calculation was performed due to large territorial differences of the sample countries.

According to World Bank (2019a, b, c, d, e, f), cargo traffic in any mode is usually measured in tons and tons-kilometers or TKU. One tonne-kilometer is equal to the weight of the cargo carried times the distance carried. In this research was used the term TKU/railroads (railroad freight), which is the volume of railroad freight, measured in metric tons times kilometers traveled. TKU values were logarithmized to facilitate calculations and analysis.

Logistics Efficiency is an indicator based on the logistics quality of a country according to the World Bank (2019a, b, c, d, e, f), using factors such as speed and efficiency of delivery and reception of products and raw materials, which is strongly involved in the predominant mode in the country.

The Logistics Performance Indicator indicator was created by the World Bank and considers the following dimensions: transport and communication infrastructure, consistency / reliability, cargo tracking, competence of public and private logistics services, availability of international transport, and customs procedures. The Logistics Performance Indicator ranges from 1 (worst) to 5 (best) and includes 155 countries. The higher the index, the better the logistics system (World Bank 2019a, b, c, d, e, f).

The logistics quality indicator was also used: the quality of the logistics infrastructure related to trade and transport. Details of the methodologies of these indicators are available on the World Bank website. According to the World Bank (2019a, b, c, d, e, f), Infrastructure Quality is an indicator that uses values based on existing infrastructures in a given country. Its time series was made available by the World Bank. The goal is to analyze whether infrastructure quality impacts logistics efficiency. The scale ranges from 1 (worst) to 5 (best). This indicator considers the quality of infrastructure related to trade and transportation of ports, railways, roads, information technology, among others.

3 Methodology

This section presents the methods that were used in the production of this study involving type of research, object of study and sampling, techniques and instruments for data collection, treatment and variables used in the regression model.

This research is explanatory and quantitative, as it aims to evaluate the logistics performance of countries through mathematical modeling applied to finance, economics and logistics. It is a bibliographical research, dealing with the relationship between economic growth and logistics development, rail transport system and logistics infrastructure. The secondary sources used come from books, articles and institutions endorsed as the World Bank, ANTF (National Railway Transport Agency) and IPEA (Institute of Applied Economic Research).

In the present work the object of study included a sample of developed and developing countries (BRIC) from 2007 to 2017, a total sample population of eight countries. The choice of countries was based on the highest GDPs for both developed and developing countries. To this end, it was necessary to identify and separate countries into two groups as well as collect their logistical information.

The sample population has as its starting point the year 2007 (beginning of the international financial crisis of the subprimes) until the year 2017 completing 11 years time series. Current GDP, Logistics Performance Indicator, infrastructure quality, TKU / rail, rail network and territorial extension of the sample countries were required. Table 1 shows the sample population divided into two groups.

The data collected through the World Bank (Logistics Performance Indicator, infrastructure quality, TKU / rail, current GDP and rail network) were processed in a spreadsheet editor software and in a graphical program for multivariate statistical analysis.

To treat the data, the statistical procedure with parametric hypothesis test was used. In this procedure a multivariate statistical technique was used, the multiple linear regression analysis using the generalized ordinary least squares method.

Table 1 Sample Population

Countries	Analysis group	Period	Continent
EUA	Developed	2007–2017	North America
Japan	Developed	2007–2017	Asia
Germany	Developed	2007–2017	Europe
Italy	Developed	2007–2017	Europe
China	Developing	2007–2017	Asia
India	Developing	2007–2017	Ásia
Brazil	Developing	2007–2017	South America
Russia	Developing	2007–2017	Asia

According to Hair et al. (2009), regression analysis is the most widely used and versatile dependency technique. Regression analysis is the foundation for business forecasting models ranging from econometric models that predict the national economy to business or country performance models.

The objective of regression analysis is to study the relationship between two or more explanatory variables that present linear form and a metric dependent variable (Gujarati and Porter 2011). Thus, a linear regression model can generally be written as follows:

$$\mathrm{Y} = \alpha + \beta_1 \mathrm{X}_1 + \alpha + \beta_2 \mathrm{X}_2 \cdots \beta_n \mathrm{X}_n u$$

where

- Y study phenomenon (metric dependent variable);
- α represents the intercept (constant);
- βk coefficients of each variable (angular coefficients);
- *u* disturbance or error term.

The disturbance also known as residual represents possible X variables that were not included in the model and that would be explanatory variables of the Y variable (Fávero et al. 2009).

In this research the panel data technique was used. Models involve data from various cross sections over time. Fávero et al. (2009) and Gujarati and Porter (2011) stated that, by combining time series with cross-sectional observations, panel data offer a reduction in multicollinearity problems of explanatory variables, generate more informative data, greater variability, more efficiency, better results and highest degree of freedom.

For panel data, two approaches were used: fixed effects model and random effects model. For Fávero et al. (2009), fixed models consider changes in cross sections over time, can be written as follows:

$$\mathrm{Y_{it}} = \alpha . D_\mathrm{i} + \beta_1 \mathrm{X_{it}} + u$$

Random effects models or error component models are composed of two or more errors. If the variables represent lack of knowledge about the model, this information can be expressed by the error term (Gujarati; Porter 2011). According to Fávero et al. (2009), the random effects model can be written as follows:

$$\mathrm{Y_{it}} = \alpha_i + \beta_1 \mathrm{X_{it}} + wi_t$$

According to Gujarati and Porter (2011), the compound error term w_{it} consists of two components: cross-sectional or country-specific component and error element combining time series and cross-section. The difference between the fixed and the random effect is the way the individual error component is handled.

The analysis of the regression model results is made in statistical inference based on the sample observations taking into consideration the level of significance.

3.1 Variables and Models

Information on the data (variables) used in the study was collected from the World Bank and is listed below.

A. Logistics Efficiency—$\text{Efilog}_{\text{it}}$
B. Infrastructure Quality—$\text{QualiInfra}_{\text{it}}$
C. TKU/Railways—TKUfe_{it}
D. Rail Mesh/km^2—$\text{MalhaFe}_{\text{it}}$
E. Current GDP in dollars (logarithm format)
F. $\text{Dummydesensubdesen}_{\text{it}}$

In this work, Stepwise method was used to evaluate the statistical significance of the parameters of each explanatory variable, considering for the final models only variables that did not present multicollinearity problems.

The binary variable Dummy was not included in the models to capture the impacts on the performance of variables separating countries into two distinct groups: developed (1) and developing (0) because GDPs of all these countries are among the largest in the world. Performing tests with Dummy on the models it was not significant.

For the following regression model, a hypothesis was elaborated based on the previous studies cited above and published books.

Hypothesis 1 With higher quality of logistics infrastructure in a country, added to TKU/railways plus density of rail network/1000km^2, countries improve their levels of logistics efficiency.

$$\text{Efilog}_{\text{it}} = a_0 + B1\text{QualiInfra}_{\text{it}} + B2\text{MalhaFe}_{\text{it}} + B3\text{TKU}_{\text{it}} + \text{u}_{it}$$

Logistics Performance Indicator represents the dependent variable. The control variables used in this model were the quality of infrastructure, TKU/railways and rail network/km^2.

Hypothesis 2 As logistics efficiency indicators increases, a country's growth measured by GDP increases.

$$\text{PIBcorrentLog}_{\text{it}} = a_0 + B1\text{QualiInfra}_{\text{it}} + B2\text{TKUfe}_{\text{it}} + B3\text{MalhaFe}_{\text{it}} + \text{u}_{it}$$

The control variables used in this model were the quality of infrastructure, tKU/railways and railway network/km^2. These variables represent logistic performance, the compound of logistic stake. Current GDP was treated as the dependent variable. The objective of this model is to capture only the impact of the variables of the Logistic Performance Indicator on the GDP of countries and not on groups.

4 Results and Discussion

After collecting the data on the indicators previously cited, an analysis of descriptive statistics and models used was carried out. These indicators were collected and separated into two groups: the BRIC (Brazil, Russia, India and China) and the developed countries group (USA, Germany, Japan and Italy).

4.1 Descriptive Statistics

Descriptive statistics analysis of the indicators was performed according to Table 2. It reports data on the maximum, minimum, mean and standard deviation from developed countries and the BRIC group. The indicators analyzed were: current GDP, Rail Mesh/1000 km^2, TKU, Logistics Performance Indicator, and Infrastructure Quality. There were 88 observations of each variable.

According to Table 2, the average GDP of BRIC was 12.412 while the average of the developed countries group was 12.701. This average of BRIC GDP indicator corresponds to 3.4 trillion dollars. It is influenced by the individual values of the India and China GDP time series. The logarithmized average of the developed countries GDP indicator is $6.8 trillion in current terms. The standard deviation of the GDP indicator of the BRIC group was 0.301 while in the developed countries group it was 0.333.

Regarding the Rail Network/1000 km^2, BRIC has a maximum value of 20.5/1000 km^2, belonging to 2017 India, but this value is well below the maximum

Table 2 Descriptive statistics of sample panel indicators

Country groups under review	Descriptive statistics	Current GDP (Log)	Rail Mesh/1000 km^2	TKU *railways*	Logistics Performance Indicator	Infrastructure Quality
BRIC	Mean	12.412	8.7	6.13	3.044	2.965
	Maximum	13.084	20.5	6.41	3.661	3.752
	Minimum	12.079	3.4	5.39	2.368	2.230
	Standard deviation	0.301	6.5	5.97	0.368	0.436
Developed Countries	Mean	12.701	54.4	5.82	3.911	4.063
	Maximum	13.290	94.9	6.43	4.226	4.439
	Minimum	12.263	16.1	4.00	3.575	3.519
	Standard deviation	0.333	27.3	6.03	0.176	0.240

Note Current GDP and TKU values are in logarithmic form. Efficiency and logistics infrastructure indicators are on a scale internationally predefined by the World Bank ranging from 0–5

value of the developed countries of 94.9/1000 km^2, which belong to Germany in 2007. That is, for every 1000 km^2 of the country, Germany has 94.9 km of railway network.

The minimum value of the Rail Network/1000 km^2 of BRIC belongs to Brazil (3468 km per 1000 km^2). It is a value considered too small. Already the minimum value of the group of developed countries was 16.110 km of rail network per 1000 km^2 for USA. Countries such as Japan, Germany and Italy have a rail network of 40 km per 1000 km^2 in the period under review.

By comparing the average of the rail networks of the two groups under analysis, a great difference is noticed. While developed countries enjoy a superiority of this indicator with an average value of 54.4 km per 1000 km^2, in developing countries the average 8,7 km per 1000 km^2.

The TKU variable has the values in logarithm format in the table. The maximum value in BRIC group was 6.41 while in developed countries it was 6.43. The average TKU was 5.82 for developed countries and 6.13 for BRIC. This last result represents, on average, the value of nearly one and a half trillion tons transported per kilometer via rail.

The Logistics Performance Indicator ranges from 0 to 5 according to the World Bank. The results of the descriptive statistics of these values show a superiority of the Logistics Performance Indicator of the developed countries. It is worth noting that the minimum value for logistics efficiency in developed countries was 3575, close to the maximum value in developing countries 3.661.

The results of the descriptive statistics for the Logistics Infrastructure Quality indicator show that BRIC has a maximum value of 3.752 reached by China in 2016–2017, which is lower than the maximum value of developed countries of 4.439 reached by Germany in 2016–2017. This superiority is also proved by the minimum value, in which BRIC had 2.23 from Russia in 2007–2009, considerably lower than the minimum value of developed countries, of 3.519 achieved by Italy in 2007–2009.

4.2 Logistics Efficiency Model Results

The data are in a strongly balanced panel, since there are an equal number of periods for each of the countries studied. Added to this information, the model has 88 observations for each variable considering 8 countries in a series from 2007 to 2017.

Table 3 shows the results of the Logistics Performance Indicator regression model as a dependent variable. This variable represents the logistics efficiency model of the countries. Based on the Hausman test, the random effects model was chosen. The significance of the Hausman test was 0.2982 accepting the null hypothesis of the test. The R^2 (overall) was 0.9846 indicating which model has high explanatory power in the variation of the dependent variable (Y). That is, 98.46% of the average variation of Logistics Performance Indicator performance can be explained by the model variables.

Table 3 Result of the logistics performance indicator model

EfiLog	Coefficient	Standard deviation	z	Z Valor
MalhaFe	0.0014	0.0008	1.77	0.077***
TKUfe	0.0075	0.0214	0.35	0.724
QualiInfra	0.7377	0.0624	11.82	0.000*
Const	07975	0.2411	3.31	0.001*

Note Significant values at 1%, 5% and 10% are indicated by *, ** and *** respectively

According to this model, infrastructure quality is significant at 1%, positively affecting a country's logistics efficiency. The infrastructure quality variable had a positive coefficient and z value test of 11.82 at a significance level of 1%, resulting in positive effects on the average Logistics Performance Indicator of the countries by 73.77%.

These results are similar to the literature in the area of logistics observed in publications by authors such as Ballou (2015), Bowersox et al. (2014), Hayaloglu (2015), in which the quality of transport infrastructure has a strong positive impact on a country's logistics efficiency. Also similar are the conclusions of the work of Marchetti and Ferreira (2012) in which they inferred that the quality of infrastructure and the configuration of the transportation network impact the logistics performance of a country.

These results show that improvements in the infrastructure quality indicator that includes the quality of transportation of ports, railways, roads, duplicate highways and information technology, increase the logistics efficiency of countries.

The Rail Mesh/1000 km^2 variable is also significant. This variable had a positive coefficient and z value test of 1.77 at a 10% significance level, resulting in positive effects on the average Logistics Performance Indicator of the countries by 0.14% for each kilometer of railway line built at 1000 km^2.

Based on the transport literature, the rail mode generates lower freight costs providing reductions in the cost of goods sold by companies in the market. A reduction in the cost of freight can lead to large scale reductions in product prices, moving the market more intensely.

Finally, the TKU variable was not statistically significant. One inference that can be made here is that the difference in population size of these countries not included in the calculation may have affected the indicator which, therefore, was not statistically significant.

4.3 Logistics Efficiency Growth Model Results

This model is a multiple linear regression which aims to assess whether a country's logistical efficiency variables affects GDP. In this model the current GDP in dollars with logarithmized values was used in order to facilitate the calculations. The control

Table 4 Result of the Logistics Performance Indicator development model

PibcorrentLogtmz	Coeficient	Standard deviation	t	*T* Value
MalhaFe	0,0032	0,0039	0.82	0.413
TKUfe	0,3387	0,1323	2.56	0.012**
QualiInfra	0,4594	0,0589	7.79	0.000*
Const	8,9904	0,7896	11.39	0.000*

Note Significant values at 1%, 5% and 10% are indicated by *, ** and *** respectively

variables used in this model were the quality of infrastructure, TKU/railways and railway network/km^2.

Based on the Hausman test, the fixed effects model was chosen. The significance of the Hausman test was 0.0147 not accepting the null hypothesis of the test.

According to Table 4 the infrastructure quality variable had a positive coefficient and t value test of 7.79 at a significance level of 1%, resulting in positive effects on the average current GDP indicator of the countries by 45.94%.

Also, according to Table 4, the TKU variable had a positive coefficient and t value test of 2.56 at a significance level of 5%, resulting in positive effects on the average current GDP indicator of the countries by 33.87%. The Rail Mesh/1000 km^2 variable is not significant in this model.

The result is similar to the results found by Lean et al. (2014) and Boopen (2006) in which they found a relationship between economic growth and logistics development and vice versa.

This result is also similar to the observed literature and related to the previous model of this research. The results of this model reveal that the Logistics Performance Indicator variables (infrastructure quality and TKU/railways) impacts the current GDP positively. And logistics efficiency is highly impacted by the quality of countries' transport and trade infrastructure as per the first regression model.

5 Conclusions

The objective of the this study evaluated the impact of logistics performances (including the quality of logistics infrastructure, rail network / 1000 km^2 and TKU / railways on the Logistics Performance Indicator) on a set of pre-selected countries (four developed and four developing). A peroid of 10 years was selected and it was analyzed whether 'Logistics Performance Indicator' in any way affects the country economic growth and development measured by current GDP.

The first results showed that the eight countries analyzed can be split into groups. According to descriptive statistics, the developed countries group has a higher Logistics Performance Indicator and average infrastructure quality than the developing countries of BRIC group. In addition, the average railroad/km^2 of developed countries in the sample is well above the BRIC average.

The results from the Logistics Performance Indicator model were statistically significant in relation to the variable infrastructure quality and rail network/1000 km^2. According to the regression analysis results, these two variables positively impact on the Logistics Performance Indicator of the sample countries.

Therefore, it can be concluded that investing in both logistics infrastructure and rail networks contributes to increases in logistics performance efficiency. In addition this study also revealed that increases in logistics infrastructure quality and TKU/railways impact on the economic growth of countries. It was shown that, the Logistics Performance Indicator positively affects the current GDP of the sample countries.

References

Amann E, Baer W, Trebat T, Lora JV (2016) Infrastructure and its role in Brazil's development process. Q Rev Econ Finance 62:66–73

Arnold JRT (2014) Administração de Materiais: Uma Introdução. São Paulo: Editora Atlas S.A, 2014

Brevani B (2015) Capacity determination and expansion models for rail networks. Thesis (Master of Science)—Faculty of Science and Engineering Queensland University of Technology

Ballou RH (2015) Logística Empresarial: Transportes, Administração de Materiais e Distribuição Física. Editora Atlas S.A, São Paulo

Boopen S (2006) Transport infrastructure and economic growth: evidence from Africa using dynamic panel estimates. Empirical Econ Lett 5(1):37–52

Bowersox DJ et al (2014) Gestão Logística Da Cadeia de Suprimentos. AMGH Editora Ltda, Porto Alegre

Da Silva FGF, de Oliveira RLM, Marinov M (2020) An analysis of the effects on rail operational efficiency due to a merger between brazilian rail companies: the case of RUMO-ALL. Sustainability 12(12):1–23

Fávero LP, Belfiore P, DA Silva FL, Chan, BL (2009) Análise de dados. Modelagem multivariada para tomada de decisões. Rio de Janeiro: Campus

Fraszczyk A, Lamb T, Marinov M (2016) Are railways really that bad? An evaluation of rail systems performance in Europe with a focus on passenger rail. Transp Res Part A 94:573–591

Gujarati DN, Porter DC (2011) Econometria Básica. 5. ed. Ed. AMGH, Porto Alegre

Hair JF et al (2009) Análise multivariada de dados. Bookman, Porto Alegre

Hayaloğlu P (2015) The impact of developments in the logistics sector on economic growth: the case of OECD countries. Int J Econ Financ Issues 5(2):523–530

IBM (2014) Smarter railroads. Retrieved from: http://www.ibm.com/smarterplanet/us/en/rail_transportation/ideas/

Lean HH; Huang W, Hong J (2014) Logistics and economic development: experience from China. Transp Policy 32:96–104

Marchetti DDS, Ferreira, Tiago Toledo. Situação atual e perspectivas da infraestrutura de transportes e da logística no Brasil. Ed. BNDES *Production*, 2012.

Marinov M, Mysore K, Wood B (2020) A study on greenhouse gas emissions, analyzing the potential of five rail transport passenger systems for meeting EU targets. IF Ingegneria Ferroviaria -Tecnologia e Innovazione per la Mobilità e il Transporto. Collegio Ingegneri Ferroviari Itallani Anno LXXV n.2 Febbrraio 2020

Marinov M, Woroniuk C, Lopes L, Oliveira A, Marujo L (2013) Railway freight research in Brazil and the EU. In: World conference of transport research (WCTR). Rio de Janeiro, Brazil

Rogers P, Ribeiro K, Securato J (2006) Governança corporativa. mercado de capitais e crescimento econômico no Brasil, Anais do Encontro Nacional da Associação Nacional de Pós-Graduação e Pesquisa em Administração, Salvador, BA, Brazil

Sturm J-E, Jacobs J, Groote P (1999) Output effects of infrastructure investment in the Netherlands, 1853–1913. J Macroecon 21(2):355–380

World Bank (2019) GDP (current US$). Disponible in: https://data.worldbank.org/indicator/ny.gdp.mktp.cd. Access in 23 Mar 2019

World Bank (2019) Logistics performance index: overall (1=low to 5=high). Disponible in: https://data.worldbank.org/indicator/LP.LPI.OVRL.XQ. Access in 23 Mar 2019

World Bank (2019) Logistics performance index: quality of trade and transport-related infrastructure (1=low to 5=high). Disponible in: https://data.worldbank.org/indicator/LP.LPI.INFR.XQ?locations=BR. Access in 23 Mar 2019

World Bank (2019) Railways, goods transported (million ton-km). Disponible in: https://data.worldbank.org/indicator/IS.RRS.GOOD.MT.K6?locations=BR. Access in 23 Mar 2019

World Bank. Rail lines (total route-km). 2019d. Disponible in: https://data.worldbank.org/indicator/IS.RRS.TOTL.KM?locations=BR. Access in 23 Mar 2019

World Bank (2019) Railways good transported. https://data.worldbank.org/indicator/IS.RRS.GOOD.MT.K6. Access in 15 Mar 2019

Woroniuk C, Marinov M, Zunder T, Mortimer P (2013) Time series analysis of rail freight services by the private sector in Europe. Transp Policy 25:81–93

Safety and Delays at Level Crossings in the United States: Addressing the Need for Multi-Objective Resource Allocation

Junayed Pasha, Maxim A. Dulebenets, Prashant Singh, Ren Moses, John Sobanjo, and Eren E. Ozguven

Abstract The level crossings in the United States experience a significant number of accidents every year. The accidents can be reduced with the application of various countermeasures (e.g., traffic signal preemption, flashing lights, barrier cubs, gates). However, the application of countermeasures for all the level crossings in the United States is not feasible due to monetary limitations. Moreover, each countermeasure has a unique level of effectiveness and installation cost (e.g., the most effective countermeasures are typically more expensive than the least effective ones). Hence, selection of potent and cost-effective countermeasures at the riskiest level crossings is imperative to improve safety. While improving safety at level crossings with the

J. Pasha · P. Singh
Department of Civil and Environmental Engineering, Florida A&M University-Florida State University (FAMU-FSU) College of Engineering, 2525 Pottsdamer Street, Building B, Suite B339, Tallahassee, FL 32310-6046, USA
e-mail: jp17j@my.fsu.edu

M. A. Dulebenets (✉)
Department of Civil and Environmental Engineering, Florida A&M University-Florida State University (FAMU-FSU) College of Engineering, 2525 Pottsdamer Street, Building A, Suite A124, Tallahassee, FL 32310-6046, USA
e-mail: mdulebenets@eng.famu.fsu.edu

R. Moses
Department of Civil and Environmental Engineering, Florida A&M University-Florida State University (FAMU-FSU) College of Engineering, 2525 Pottsdamer Street, Building A, Suite A129, Tallahassee, FL 32310-6046, USA
e-mail: moses@eng.famu.fsu.edu

J. Sobanjo
Department of Civil and Environmental Engineering, Florida A&M University-Florida State University (FAMU-FSU) College of Engineering, 2525 Pottsdamer Street, Building B, Suite B332, Tallahassee, FL 32310-6046, USA
e-mail: sobanjo@eng.famu.fsu.edu

E. E. Ozguven
Department of Civil and Environmental Engineering, Florida A&M University-Florida State University (FAMU-FSU) College of Engineering, 2525 Pottsdamer Street, Building B, Suite B313, Tallahassee, FL 32310-6046, USA
e-mail: eozguven@eng.famu.fsu.edu

M. Marinov and J. Piip (eds.), *Sustainable Rail Transport 4*, Lecture Notes in Mobility,
https://doi.org/10.1007/978-3-030-82095-4_4

application of countermeasures, there is a significant risk of waning highway vehicle flows, increasing delays, and negatively affecting the continuity of passenger and freight flows. In such a scenario, multi-objective resource allocation models could be instrumental, since such models can analyze the tradeoffs between conflicting objectives (e.g., minimizing the number of accidents vs. minimizing the total delay). Hence, this chapter presents a framework for multi-objective resource allocation to minimize the number of accidents and to minimize the total delay at level crossings. Furthermore, various methods for quantifying the number of accidents as well as delays due to the application of countermeasures at level crossings are reviewed. Solution methods for multi-objective resource allocation models, including exact and approximate optimization approaches, are also discussed. Finally, future research avenues for multi-objective resource allocation among level crossings are outlined.

Keywords Level crossings · Accident prediction · Accident prevention · Delays · Traffic queuing · Resource allocation · Multi-objective optimization

1 Introduction

Rail transportation plays an important role in the economy of the United States (U.S.). The impact of rail transportation has increased from 962 million rail vehicle revenue miles in 2004 to 1109 million rail vehicle revenue miles in 2014 (FHWA 2019). The magnitudes of both freight and passenger rail transportation have increased in the U.S. Freight rail intermodal traffic in the U.S. has increased from 765,883 in January 2000 to 1,105,989 in January 2020 (BTS 2020a). Meanwhile, passenger rail miles have increased from 445,873,041 passenger miles in January 2000 to 540,993,482 passenger miles in January 2020 (BTS 2020b). To accommodate this increase in rail transportation, several initiatives have been taken. For instance, the rail vehicle fleet in the U.S. (e.g., commuter rail locomotives, commuter rail self-propelled passenger coaches, commuter rail passenger coaches, light rail, heavy rail) has been increased from 19,623 vehicles in 2004 to 21,860 vehicles in 2014 (FHWA 2019).

Along with an increase in railway traffic, accidents at rail crossings, especially level crossings (where highways and railways meet at the same elevation), have increased as well. Figure 1 presents the frequency of level crossing accidents in the U.S. from 2010 to 2019, where it can be observed that a total of 2225 level crossing accidents occurred in the U.S. in 2019, as compared to 2052 level crossing accidents in 2010 (FRA 2020). These level crossing accidents resulted in numerous losses of lives, injuries to victims, and damages to property. For instance, the 2225 accidents at the U.S. level crossings led to 332 fatalities, 825 injuries, and a minimum of $17.3 million in reported highway vehicle property damages in 2019, as compared to 263 fatalities, 888 injuries, and a minimum of $14.8 million in reported highway vehicle property damages in 2010 (FRA 2020).

Safety issues at level crossings can be addressed with various strategies (e.g., traffic signal preemption, application of warning devices, grade separation). The

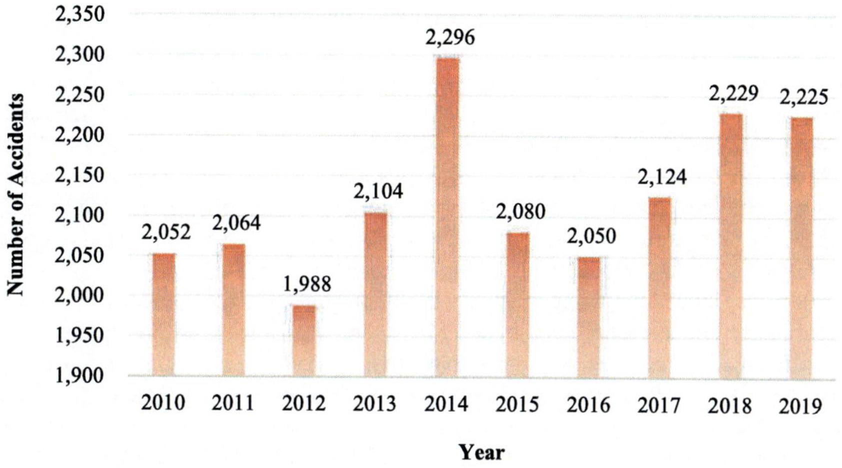

Fig. 1 Number of level crossing accidents in the U.S. from 2010 to 2019

application of warning devices, such as flashing lights, barrier cubs, gates, stop signs, wigwags, etc., is one of the most popular approaches for reducing the number of accidents at level crossings. However, the installation of such warning devices could be expensive. In fact, the installation of warning devices at each of the level crossings in the U.S. is not feasible with the annual budget that is set aside for the safety improvement projects at level crossings. Thus, only the riskiest level crossings can be improved with the installation of warning devices. Furthermore, each warning device has a unique level of effectiveness and a unique installation cost. The most effective warning devices are typically more expensive than the least effective ones. For instance, automatic gates are generally more effective in reducing the number of accidents at level crossings than stop signs. However, they are also substantially more expensive as well. If automatic gates are considered as the only countermeasure for all the riskiest level crossings, then, only a small number of level crossings can be upgraded. On the other hand, if stop signs are considered as the only countermeasure for all the riskiest level crossings, then, there might be significant safety concerns at many level crossings. Hence, a choice of countermeasures (e.g., warning devices) should be carefully examined for each of the level crossings under consideration. At the same time, a ranking of the riskiest level crossings is required, which can be achieved through the use of accident and hazard prediction models available in the literature, such as the U.S. DOT Accident Prediction Formula, the Coleman-Stewart Model, the Jaqua Formula, among others (Chadwick et al. 2014; Abioye et al. 2020).

The choice of countermeasures should also be incorporated with the consideration of highway traffic delays, as countermeasures generally lead to additional delays to highway traffic at/near level crossings. Note that highway traffic delays may vary from one countermeasure to another. For instance, automatic gates, which are generally more effective in reducing accidents at level crossings, also add more delays to highway traffic, as compared to stop signs. Even without the installation of additional countermeasures, highway traffic delays at/near the level crossings in the U.S. are already notable. Such delays have the potential to cause negative economic implications. For instance, in 2002, the Federal Highway Administration (FHWA) forecasted that highway traffic delays at/near the level crossings in the U.S. could lead to $7.9 billion in losses for the following 20 years (FHWA 2002). Extensive highway traffic delays could also lead to risky highway user behavior (e.g., driving through a level crossing while gates are moving, driving around closed gates in order to avoid delays) (Rilett and Appiah 2008; Khattak 2014; Khattak and Lee 2018).

Resource allocation or selection of the appropriate countermeasures for level crossings could be challenging, as several issues are associated with it. Some optimization models have been presented in the past for resource allocation among level crossings. However, most of them have focused on only one objective (e.g., minimizing the expected number of accidents at level crossings, minimizing the overall hazard at level crossings). Addressing only safety issues through the application of countermeasures might significantly affect highway traffic flows or lead to significant delays. Hence, there is a need for multi-objective resource allocation methodologies, which can analyze the tradeoffs between conflicting objectives. Such methodologies could establish a balance between the reduction in the number of accidents through the application of countermeasures and the highway traffic delays caused by those countermeasures. In particular, the decision makers would be able to select the appropriate countermeasures for the considered group of level crossings that will provide a desirable (or close to desirable) level of safety without causing a substantial increase in highway traffic delays. Hence, this study provides an outlook on multi-objective resource allocation models for level crossings as well as solution approaches for such models.

The remainder of this chapter is organized as follows. The following section contains an overview of the existing literature on safety issues at level crossings. The third section provides a multi-objective resource allocation modeling framework for level crossings and discusses some of the common approaches for quantifying the number of accidents and delays at level crossings. The fourth section highlights potential solution approaches for multi-objective resource allocation models. The fifth section highlights important implications of this work. The final section concludes this chapter and outlines a number of future research opportunities.

2 Literature Review

This section of the chapter presents a review of the literature particularly focusing on safety issues at level crossings and resource allocation among level crossings. Although the resource allocation studies are directly related to the theme of this research, the scope of the literature review also includes the studies that focus on safety issues at level crossings, since these studies evaluate the effects of different factors on safety at level crossings (hence, the findings from these studies can be directly used in resource allocation decisions). The existing gaps in the stat-of-the-art are also pointed out in this section as well.

2.1 Safety Issues at Level Crossings

A number of studies have addressed safety issues at level crossings. Hu et al. (2010) suggested that there was a lack of research efforts assessing the severity of accidents at level crossings. So, a logit model was developed for examining severity of level crossing accidents in Taiwan. Several factors were found to affect accident severity, including highway and railway traffic volumes, approaching crossing markings, highway separation, and obstacle detection devices. Yan et al. (2010) employed a hierarchical tree-based regression model in order to predict accidents at level crossings that were equipped with passive warning devices. The study highlighted that stop signs were effective passive warning devices to improve safety at level crossings. Evans (2011) analyzed the level crossing accidents in the U.K., which led to fatalities. The study considered three types of level crossings, including railway-controlled, automatic, and passive. It was found that automatic crossings were the least dangerous among the three types of level crossings. Lenné et al. (2011) analyzed driver behavior at two rural level crossings. Responses to active warning devices, including flashing lights and traffic signals, were compared to those of a passive warning device, which was a stop sign. The results indicated that the passive warning devices led to less compliance than the active warning devices.

Rudin-Brown et al. (2012) compared the use of boom barriers and traffic lights at the level crossings in Australia. It was suggested that traffic lights would not improve safety at level crossings that were already equipped with flashing lights and boom barriers. Tey et al. (2013) examined driver compliance towards different warning devices at level crossings. The study highlighted that socio-demographic characteristics (e.g., age, gender), speed, and warning device types significantly influenced driver compliance. Freeman and Rakotonirainy (2015) argued that pedestrian accidents at level crossings were prone to more fatalities and severe injuries, as compared to the other types of accidents at level crossings. The study analyzed pedestrian behavior at level crossings and indicated that pedestrians could violate traffic rules in spite of having adequate knowledge. Larue and Wullems (2015) conducted a driving simulator study in Australia, which highlighted that converting active level crossings into passive level crossings could lead to user errors. Borsos et al. (2016) assessed about 1700 level crossings in Hungary, and the study underlined that highway and railway traffic volumes significantly influenced safety at those level crossings. Laapotti (2016) indicated that more accidents occurred at the passive level crossings than the active level crossings in Finland. It was highlighted that active level crossings were more instrumental in preventing user errors.

Beanland et al. (2017) studied driver behavior at the rural level crossings in Australia. It was found that a significant portion of drivers did not acknowledge stop signs or made a complete stop. Beanland et al. (2018) conducted a driving simulator study assessing safety at level crossings. The findings suggested that users preferred standard level crossings as compared to the ones with new designs, which might be attributed to the fact that users were used to standard level crossings. Djordjević et al. (2018) developed a non-radial Data Envelopment Analysis model, which relied on desirable and undesirable inputs/outputs. The model was able to evaluate the railway safety performance of European nations. Khan et al (2018) highlighted that about half of the level crossing accidents occurred within 5 miles from an individual's residence. An accident prediction model, using binary logit regression, was proposed. Larue et al. (2018) highlighted that highway and railway traffic volumes as well as human errors could significantly change the level of risk at level crossings. Stefanova et al. (2018) analyzed pedestrians' risk perceptions and crossing intensions in case of level crossings. It was revealed that pedestrian gates were beneficial in promoting safer pedestrian behavior. Zhang et al. (2018) analyzed trespassing near-misses at level crossings. Using surveillance video footages, a computer vision algorithm was developed to trace trespassing near-misses.

Landry et al. (2019) examined the potential of in-vehicle auditory alerts for reducing the risk of accidents at level crossings. It was indicated that the in-vehicle auditory alerts reminded highway users that they were approaching level crossings and, thus, promoted safe driver behavior. Dent and Marinov (2019) discussed the automated obstacle detection implementation for the level crossings in Great Britain. It was highlighted that the automated obstacle detection would not just improve the level of safety at level crossings but also would reduce the costs that are associated with the level crossing signal systems. Keramati et al. (2020) employed a competing risk algorithm, which assessed the level crossing accident occurrence and severity. For a total of 3194 public level crossings in North Dakota, the main contributors of level crossing accidents were found to be geometric parameters, highway/railway characteristics, and traffic exposure. Russo et al. (2020) examined pedestrian and bicyclist behavior at level crossings using video footages. It was indicated that 27% of pedestrian accidents at the level crossings in the U.S. during 2008–2017 were related to disregarding gates. Zhou et al. (2020) introduced a Random Forest Algorithm for safety analysis of level crossings. The proposed methodology could improve the performance of unbalanced data forecasting, as it held a bootstrap characteristic. In addition to the attempts from researchers all over the world, different state Departments of Transportation (DOTs) in the U.S. have made a number of efforts to analyze and improve safety at level crossings (Faghri and Demetsky 1986; Elzohairy and Benekohal 2000; Weissmann et al. 2013; Hans et al. 2015; Sperry et al. 2017; Dulebenets et al. 2020a).

2.2 *Resource Allocation Amongst Level Crossings*

Farr (1981) developed the Transportation Systems Center Resource Allocation Model for level crossings, which accounted for the predicted accident rates at level crossings as well as the effectiveness of warning systems and cost parameters. For a number of allocation options, a funding priority ranking was provided as well. The proposed resource allocation model could maximize the total benefit by indicating a list of level crossings from the candidate level crossings and then selecting the appropriate warning systems for each of the level crossings. Berg (1986) argued that several studies at that time focused on the effectiveness ratios of countermeasures and integrated accident records into resource allocation procedures (as well as into accident and hazard prediction models). The study presented relevant data to reveal that such procedures could be biased, and, therefore, could misallocate resources. Hence, it was recommended to ignore the alteration of predicted accident rates based on accident records. The study also advocated for modeling approaches for the estimation of expected accident reductions. Forgionne (2002) developed a high-speed level crossing analysis tool for the evaluation of costs and benefits of potential projects. A system session was demonstrated to illustrate the potential benefits of the proposed decision support tool.

Cirovic and Pamucar (2013) stated that selection of level crossings for installation of safety equipment was subject to uncertainties of essential criteria in terms of decision making. In order to address such uncertainties, the study employed a fuzzy multi-criteria decision making approach along with fuzzy clustering techniques. The developed model was trained with expert knowledge and evaluated for a total of 88 level crossings in Belgrade (Serbia). The numerical experiments indicated that the proposed methodology provided straightforward results for selection of the most effective alternative from a given set of alternatives, as compared to different statistical models and mathematical transformations. Konur et al. (2013) proposed an optimization model for maximization of safety benefits through resource allocation among the level crossings in the State of Tennessee. A number of sorting methodologies were also developed for comparison purposes. The numerical experiments revealed that the optimization model, which was developed under the study, was more efficient than the sorting methodologies in improving safety at level crossings.

Rezvani et al. (2015) presented a cost/benefit framework to upgrade the most hazardous level crossings under safety improvement projects. The framework consisted of the following main steps: (i) measurement of accident cost; (ii) cost-based screening; (iii) benefit/cost analysis; (iv) prioritization of projects; and (v) funding. The study added that the presented framework was applicable to different modes and objectives (e.g., mobility, safety). Kavoosi et al. (2020a) proposed two mathematical models for resource allocation among the level crossings in the State of Florida. The objectives of the mathematical models were to minimize the overall hazard (i.e., vulnerability of level crossings to accidents) and to minimize the overall hazard severity (i.e., vulnerability of level crossings to fatalities, injuries, and property damage accidents). A set of heuristic algorithms were developed to solve the mathematical models. The heuristics were inspired by some sorting procedures that explicitly accounted for the total budget available for implementation of different countermeasures. A case study for the public level crossings in the State of Florida demonstrated the potency of the proposed methodology. A set of additional sensitivity analyses for the parameters of the developed model were conducted as well.

2.3 Literature Summary and Existing Gaps

From the review of the scientific literature, it was found that a number of research efforts have been carried out to improve safety at level crossings, while some studies have presented different methodologies for resource allocation among level crossings. Several mathematical models have been developed to minimize the predicted number of accidents, the overall level crossing hazard, or the overall hazard severity at level crossings. Yet, there is a lack of mathematical models, which can minimize the number of accidents with the application of countermeasures while considering the total delay caused by these countermeasures at level crossings. Hence, the multi-objective resource allocation models, which can not only minimize the number of accidents at level crossings but also minimize the total delay due to the application of countermeasures, are required to fill this gap in the state-of-the-art. Furthermore, the appropriate solution algorithms for such multi-objective resource allocation models with conflicting objectives have to be designed as well. Considering these shortcomings of the existing efforts on level crossing safety, this study provides an outlook on multi-objective resource allocation models for level crossings as well as solution approaches for such models.

3 Multi-Objective Resource Allocation Modeling Framework

A framework for the multi-objective resource allocation problem (**MORAP**) and safety improvement projects at level crossings is presented in this section. Under this framework, a set of level crossings $X = \{1, \ldots, n\}$ is considered for upgrades by the application of a set of candidate countermeasures $C = \{1, \ldots, m\}$. Due to monetary constraints, which are defined by the total available budget (TAB—USD), only a limited number of level crossings, based on the predicted number of accidents ($PA_x, x \in X$), can be upgraded. There are different methods for predicting the number of accidents at level crossings, which are discussed in Sect. 3.1 of this chapter. Each countermeasure has an effectiveness factor ($EF_{xc}, x \in X, c \in C$), which can reduce the potential of accidents at a level crossing to a certain degree. However, the countermeasures with high effectiveness factors may also have high implementation costs ($CA_{xc}, x \in X, c \in C$). Moreover, the application of a countermeasure could cause an additional delay ($TD_{xc}, x \in X, c \in C$) at a level crossing (see Sect. 3.2 for details). Therefore, the objectives of **MORAP** are to minimize the total predicted number of accidents as well as to minimize the total delay caused by countermeasures at the considered level crossings.

Sets

$X = \{1, \ldots, n\}$ set of level crossings (level crossings)
$C = \{1, \ldots, m\}$ set of countermeasures (countermeasures)

Decision Variable

$z_{xc} \in \mathbb{B} \forall x \in X, c \in C$	= 1 if countermeasure c is applied at level crossing x (=0 otherwise)

Parameters

$n \in \mathbb{N}$	number of level crossings (level crossings)
$m \in \mathbb{N}$	number of considered countermeasures (countermeasures)
$PA_x \in \mathbb{R}^+ \forall x \in X$	predicted number of accidents at level crossing x (accidents)
$p_{xc} \in \mathbb{B} \forall x \in X, c \in C$	=1 if countermeasure c can be potentially applied at level crossing x (=0 otherwise)
$EF_{xc} \in \mathbb{R}^+ \forall x \in X, c \in C$	effectiveness factor for countermeasure c when applied at level crossing x (effectiveness factor)
$\mathrm{CA}_{xc} \in \mathbb{R}^+ \forall x \in X, c \in c$	cost of implementing countermeasure c at level crossing x (USD)
$TD_{xc} \in \mathbb{R}^+ \forall x \in X, c \in C$	total delay caused by countermeasure c at level crossing x (seconds)
$TAB \in \mathbb{R}^+$	total available budget (USD)

Multi-Objective Resource Allocation Problem (MORAP):

$$\min \boldsymbol{F}_1 = \sum_{x \in X} \left[1 - \sum_{c \in C} (EF_{xc} \cdot z_{xc}) \right] \cdot PA_x \tag{1}$$

$$\min \boldsymbol{F}_2 = \sum_{x \in X} \sum_{c \in C} TD_{xc} \cdot z_{xc} \tag{2}$$

Subject to:

$$\sum_{c \in C} z_{xc} \leq 1 \forall x \in X \tag{3}$$

$$z_{xc} \leq p_{xc} \forall x \in X, c \in C \tag{4}$$

$$\sum_{x \in X} \sum_{c \in C} CA_{xc} \cdot z_{xc} \leq TAB \tag{5}$$

$$n, m \in \mathbb{N} \tag{6}$$

$$z_{xc}, p_{xc} \in \mathbb{B} \forall x \in X, c \in C \tag{7}$$

$$PA_x, EF_{xc}, CA_{xc}, TD_{xc}, TAB \in \mathbb{R}^+ \forall x \in X, c \in C \tag{8}$$

The objective function (1) aims to minimize the total predicted number of accidents at the considered level crossings, while the objective function (2) aims to minimize the total delay caused by countermeasures at the considered level crossings. Note that the objective function (1) and the objective function (2) will be further referred to as $\boldsymbol{F}_1$ and $\boldsymbol{F}_2$, respectively. Constraint set (3) implies that at most one countermeasure can be applied at a considered level crossing. Constraint set (4) ensures feasibility of applying countermeasures at level crossings (e.g., certain countermeasures cannot be applied at particular level crossings depending on certain physical and/or operational restrictions). Constraint set (5) guarantees that the total implementation cost of the adopted countermeasures at the selected level crossings is within the total available budget. Constraint sets (6) through (8) define the nature of the decision variable and different parameters of **MORAP**. A schematic illustration of the proposed **MORAP** framework is presented in Fig. 2. It can be observed that the estimation of predicted number of accidents at the considered level crossings and the estimation of highway traffic delays at the considered level crossings are the major steps. Different approaches for the estimation of predicted number of accidents at level crossings as well as different approaches for the estimation of highway traffic delays at level crossings are described in the following sections of the chapter.

3.1 Methods for quantifying the number of accidents at level crossings

Several methods can be adopted to predict the number of accidents at level crossings, including accident prediction models and advanced statistical models, which are highlighted in this section.

3.1.1 Accident prediction models

Different state DOTs in the U.S. have used a variety of accident prediction models, which include: (i) the Coleman-Stewart Model; (ii) the Iowa Accident Prediction Formula; (iii) the Jaqua Formula; (iv) the NCHRP Report 50 Accident Prediction Formula; (v) the Peabody-Dimmick Formula; and (vi) the U.S. DOT Accident Prediction Formula (Faghri and Demetsky 1986; Elzohairy and Benekohal 2000; U.S. DOT 2019; Chadwick et al. 2014; Pasha et al. 2020; Abioye et al. 2020). Among these accident prediction models, the Coleman-Stewart Model, the NCHRP Report 50 Accident Prediction Formula, the Peabody-Dimmick Formula, and the U.S. DOT Accident Prediction Formula have received national recognition in the U.S. On the other hand, the Iowa Accident Prediction Formula and the Jaqua Formula have been

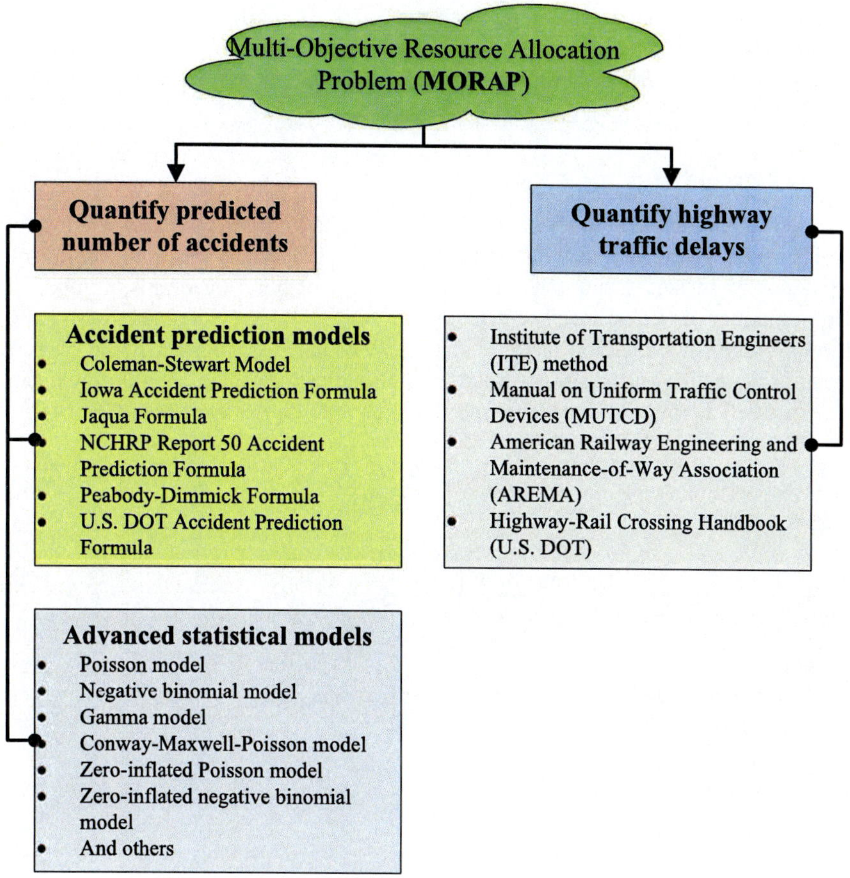

Fig. 2 A schematic illustration of the proposed **MORAP** framework

used by the Iowa DOT and the Oregon DOT, respectively. Some of the aforementioned accident prediction models are presented below.

Coleman-Stewart Model

The Coleman-Stewart Model considers level crossings with similar features (e.g., highway and railway traffic volumes, location, number of tracks, existing protection) to be in a group. The Coleman-Stewart Accident Prediction Model can be expressed as follows (Faghri and Demetsky 1986; Elzohairy and Benekohal 2000):

$$\log_{10} A = B_0 + B_1 \cdot \log_{10} C + B_2 \cdot \log_{10} T + B_3 \cdot \left(\log_{10} T\right)^2 \tag{9}$$

where:
A = average number of accidents per level crossing per year;

C = average number of vehicles per day (= 0.5 if $C = 0$);
T = average number of trains per day (= 0.5 if $T = 0$);
B_0, B_1, B_2, and B_3 = coefficients of the accident prediction equation (see Table 1).

Jaqua Formula

In order to predict the number of accidents, the Jaqua Formula uses a hazard rating, which depends on the angle of intersection, approach grade, curvature of the highway, existence of entrances and exits to streets, number of blind quadrants, number of lanes, number of tracks, speed of vehicles and trains, and street intersections near a level crossing. A protection factor is also used, which is influenced by the type of existing protection and location (urban/rural classification). The Jaqua Formula can be expressed as follows (Elzohairy and Benekohal 2000; Abioye et al. 2020):

$$ACC5 = \frac{A \cdot B \cdot C}{1610} \tag{10}$$

$$A = \sum_{i=1}^{n} T_i \left(\left(\frac{C_i \cdot V}{3 \cdot S_i} \right) + V \right) \tag{11}$$

where:
$ACC5$ = accident prediction for the next five years;
A = exposure factor;
n = number of train types;
T_i = number of trains of type i;
C_i = number of cars in a train of type i;
S_i = speed of a train of type i;
V = AADT;
B = hazard rating, which is estimated based on the number of blind quadrants, number of tracks, speed of vehicles and trains, angle of intersection, number of lanes, approach grade, curvature of the highway, existence of entrances and exits to streets and street intersections near the level crossing;
C = protection factor, which is estimated based on the location type (rural or urban) and the existing warning device type.

NCHRP Report 50 Accident Prediction Formula

The NCHRP Report 50 Accident Prediction Formula predicts the number of accidents at a level crossing based on a 10-year AADT, number of trains per day, existing protection, and location (urban/rural classification). The NCHRP Report 50 Accident Prediction Formula can be expressed as follows (Elzohairy and Benekohal 2000; Chadwick et al. 2014; Abioye et al. 2020):

$$\text{Number of accidents per year} = A \cdot B \cdot T \tag{12}$$

Table 1 The Coleman-Stewart Model coefficients and the associated R-squared values

Single-track urban						Multiple-track urban					
Item	B_0	B_1	B_2	B_3	R^2	Item	B_0	B_1	B_2	B_3	R^2
Automatic gates	−2.17	0.16	0.96	−0.35	0.186	Automatic gates	−2.58	0.23	1.30	−0.42	0.396
Flashing lights	−2.85	0.37	1.16	−0.42	0.729	Flashing lights	−2.50	0.36	0.68	−0.09	0.691
Crossbucks	−2.38	0.26	0.78	−0.18	0.684	Crossbucks	−2.49	0.32	0.63	−0.02	0.706
Other active	−2.13	0.30	0.72	−0.30	0.770	Other active	−2.16	0.36	0.19	0.08	0.65
Stop signs	−2.98	0.42	1.96	−1.13	0.590	Stop signs	−1.43	0.09	0.18	0.16	0.35
None	−2.46	0.16	1.24	−0.56	0.24	None	−3.00	0.41	0.63	−0.02	0.58
Single-track rural						Multiple-track rural					
Item	B_0	B_1	B_2	B_3	R^2	Item	B_0	B_1	B_2	B_3	R^2
Automatic gates	−1.42	0.08	−0.15	0.25	0.200	Automatic gates	−1.63	0.22	−0.17	0.05	0.142
Flashing lights	−3.56	0.62	0.92	−0.38	0.857	Flashing lights	−2.75	0.38	1.02	−0.36	0.674
Crossbucks	−2.77	0.40	0.89	−0.29	0.698	Crossbucks	−2.39	0.46	−0.50	0.53	0.780
Other active	−2.25	0.34	0.34	−0.01	0.533	Other active	−2.32	0.33	0.80	−0.35	0.31
Stop signs	−2.97	0.61	−0.02	0.29	0.689	Stop signs	−1.87	0.18	0.67	−0.34	0.32
None	−3.62	0.67	0.22	0.26	0.756	None	–	–	–	–	–

where:
A = factor based on the current amount of highway vehicles per day (see Fig. 3);
B = factor based on the existing protection and urban/rural classification (see Fig. 4);
T = current amount of trains per day.

Accident prediction models use a set of explanatory or independent variables to predict the number of accidents at a level crossing. Table 2 lists the explanatory variables of the accident prediction models that are commonly used by different state DOTs. Some of the common explanatory variables are the highway and railway traffic volumes that define the expose of a given level crossing to highway and railway traffic, existing protection (i.e., types of warning devices installed at a level

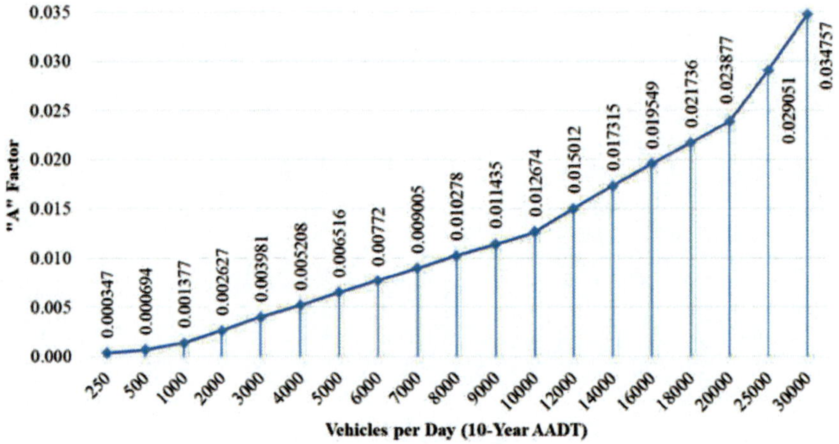

Fig. 3 The "A" factor values for highway vehicles per day based on a 10-year AADT

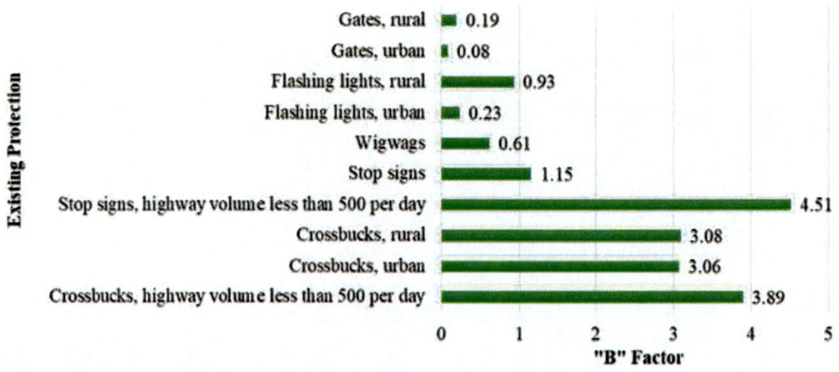

Fig. 4 The "B" factor values for existing protection

Table 2 Explanatory variables of different accident prediction models

Accident prediction model	Explanatory variables
Coleman-Stewart model	Existing protection; Location; Number of tracks; Trains per day; Vehicles per day
Iowa accident prediction formula	Accident history; Daylight thru trains per day; Existing protection; Highway pavement type; Number of tracks; Number of traffic lanes; Time of day; Train speed; Trains per day; Vehicles per day
Jaqua formula	Angle of crossing; Approach gradient; Existing protection; Highway vehicular speed; Location; Number of cars in a train; Number of tracks; Number of traffic lanes; Sight distance; Other highway geometrics; Train speed; Trains per day; Type of train; Vehicles per day
NCHRP Report 50 accident prediction formula	Existing protection; Location; Trains per day; Vehicles per day
Peabody-Dimmick formula	Existing protection; Trains per day; Vehicles per day
U.S. DOT accident prediction formula	Accident history; Daylight thru trains per day; Existing protection; Highway pavement type; Location; Number of tracks; Number of traffic lanes; Train speed; Trains per day; Vehicles per day

crossing), location (i.e., urban designation or rural designation), number of tracks, etc. Only the Iowa Accident Prediction Formula and the U.S. DOT Accident Prediction Formula directly account for the accident history. The number of traffic lanes is used by the Iowa Accident Prediction Formula, the Jaqua Formula, and the U.S. DOT Accident Prediction Formula in order to estimate the expected number of accidents at a given level crossing. Furthermore, the Jaqua Formula considers different highway geometric characteristics (e.g.., angle of crossing, approach gradient) and sight distance as well.

3.1.2 Advanced Statistical Models

In addition to the specific accident prediction models, different advanced statistical models can be used for accident prediction at level crossings. These models include: (1) the Poisson model; (2) the negative binomial model; (3) the gamma model; (4) the Conway-Maxwell-Poisson model; (5) the zero-inflated Poisson model; (6) the zero-inflated negative binomial model; and others (Lord et al. 2005; Shmueli et al. 2005; Oh et al. 2006; Cameron and Trivedi 2013; Lu and Tolliver 2016; Greene 2018; Washington et al. 2020). Such models can use the same explanatory variables, which are used by the accident prediction models employed by different state DOTs (i.e., highway and railway traffic volumes, existing protection, location, number of tracks, train speed, etc.). These advanced statistical models attempt to fit the observed accident data to the expected accident data by minimizing the difference between them (or the error term). When the average number of accidents is approximately equal to the accident variance over a given time period, the Poisson model can be used for quantifying the expected number of accidents.

When the variance is higher than the mean (i.e., over-dispersed accident data), which is more commonly observed in practice at level crossings, the negative binomial model is a popular advanced statistical model (Lord et al. 2005; Oh et al. 2006; Washington et al. 2020). In case of under-dispersed accident data (i.e., the mean number of accidents is higher than the variance in the number of accidents), the gamma model can be applied (Oh et al. 2006; Cameron and Trivedi 2013). The Conway-Maxwell-Poisson model can handle both over-dispersed and under-dispersed accident data (Shmueli et al. 2005; Lu and Tolliver 2016). Zero-inflated models (e.g., zero-inflated Poisson model, zero-inflated negative binomial model) can be particularly useful if the accident data has many zero values (Lord et al. 2005; Washington et al. 2020). Zero-inflated models offer better performance in such cases than the Poisson model or the negative binomial model, even though they raise some theoretical concerns. The theoretical concerns stem from the fact that the excess zeroes may result from a high level of heterogeneity in accident data, not from the presence of some virtually safe level crossings (Lord et al. 2005; Oh et al. 2006). Some of these advanced statistical models possess significant explanatory capabilities due to the potential of comprising a large number of explanatory variables. However, the state DOTs in the U.S. typically do not use these advanced statistical models for accident prediction at level crossings because of implementation difficulties.

3.2 Methods for Quantifying Delays at Level Crossings

The presence of a level crossing generally causes delays to the associated highway traffic. These delays can occur from stopping highway traffic until a train passes a level crossing, presence of different warning devices at a level crossing, and other reasons. Moreover, additional safety measures are sometimes taken at/near level crossings to prevent accidents, such as traffic signal preemption near level crossings. Preemption refers the situation when the traffic signal control equipment is interconnected with the adjacent level crossing signal control equipment, and the normal traffic signal control operations at the intersection should be preempted in order to operate in a special mode in case of an approaching train (i.e., prevent vehicles from driving in the direction of an approaching train). In the preemption mode, only traffic movements that do not conflict with the railway movements are allowed. The Institute of Transportation Engineers (ITE) has provided some recommendations regarding traffic signal preemption near level crossings (ITE 2006). According to ITE (2006), traffic signal preemption could be beneficial if a level crossing is in proximity to a highway intersection. In particular, traffic signals on highways should be interconnected with active warning devices, installed at level crossings, in case of the following scenarios: (1) traffic queues from a level crossing proceed toward a highway intersection; and (2) traffic queues from a highway intersection proceed toward a level crossing.

According to the Manual on Uniform Traffic Control Devices (MUTCD), the distance between signalized highway intersections and level crossings with active warning devices should be more than 200 ft, when some form of coordination (e.g., activation of variable message signs, omission of signal phases in case of queue detection) is existent (ITE 2006). In such a case, the 95th percentile queue length should not exceed a signalized intersection or a level crossing. If traffic queues from a level crossing proceed toward a highway intersection (i.e., the vehicles approach the signalized highway intersection from the railway tracks), the 95th percentile queue length (L—ft) can be estimated from the following equation when the volume/capacity ratio of a signalized intersection (v/c) is less than 0.9 (ITE 2006):

$$L = (2qr)(1 + p)(25) \tag{13}$$

where:
q = vehicle flow rate (vehicles/lane/second);
r = sum of red time and yellow time (i.e., effective red time) (seconds);
p = proportion of heavy vehicles in decimal points;
25 = sum of a passenger car's length and space between consecutive vehicles (i.e., effective length of a passenger car) (ft);
2 = a random arrival factor.

When the volume/capacity ratio of a signalized intersection is between 0.9 and 1.0, the following equations can be used to estimate the 95th percentile queue length (ITE 2006):

$$L = (2qr + \Delta x)(1 + p)(25) \tag{14}$$

$$\Delta x = 100(v/c - 0.9) \tag{15}$$

If traffic queues from a highway intersection proceed toward a level crossing (i.e., the vehicles approach the railway tracks from the signalized highway intersection), the 95th percentile queue length can also be estimated with Eqs. (13)-(15). Note that the vehicle flow rate (q) should then include both thru traffic and left/right turning traffic. Moreover, in case the traffic queues should be estimated from a highway intersection toward a level crossing, the symbol "r" would denote the effective time during which a train would block a level crossing. The effective time during which a train would block a level crossing (r) can be estimated as follows (ITE 2006):

$$r = 35 + \frac{l}{1.47S} \tag{16}$$

where:
l = train length (ft);
S = train speed (mph);
35 = the amount of time during which the level crossing is blocked by gates (about 25 s before a train enters the level crossing and 10 s after the train clears the level crossing) (seconds).

The queue lengths mentioned above would add delays at level crossings. The methodologies to quantify delays (in seconds or other time units) from queue lengths might vary for different regions. In the U.S., the recommendations that are provided by the U.S. Department of Transportation (U.S. DOT) could be applied. As highlighted in Eq. (16), warning devices could add delays at a level crossing. For instance, MUTCD recommends a minimum time of 20 s for activation of warning devices before arrival of a thru train at a level crossing (ITE 2006; U.S. DOT 2019). Furthermore, the American Railway Engineering and Maintenance-of-Way Association (AREMA) Communications and Signal Manual requires the minimum warning time for activation of warning devices before arrival of a thru train at a level crossing to be the sum of the minimum time before a train enters a level crossing (20 s) and the clearance time (AREMA 2004). The clearance time is 1 s for every 10 ft of additional level crossing length (i.e., the minimum track clearance distance) exceeding 35 ft plus the additional gate delay time, exit gate clearance time, and adjacent track clearance time (ITE 2006; U.S. DOT 2019). If required, AREMA (2004) recommends the addition of equipment response time (for variations in equipment response), buffer time (for variations in handling of trains), and advance preemption time (added by the public agency). Furthermore, the clearance time may differ from one level crossing to another and can be adjusted as needed by the appropriate agencies (ITE 2006).

The delays caused by different warning devices are generally expected to vary. In case of automatic gates, for example, a minimum of 3 s is required by gate arms to start moving toward their horizontal position after flashing lights start operation, indicating eventual train arrivals (ITE 2006; U.S. DOT 2019). The gate arms generally reach their horizontal position within 8 to 12 s, and that must be done at least 5 s before a train arrives (ITE 2006; U.S. DOT 2019). After a train clears the level crossing, a maximum of 12 s is required by gate arms to reach their vertical position (U.S. DOT 2019). Then, the flashing lights and the lights on the gate arms would stop operation. Furthermore, wayside horns are required to sound at least 15 s before a train's arrival at a level crossing. Moreover, wayside horns should account for the delay of 3–5 s before sounding after flashing lights start operation (U.S. DOT 2019).

4 Solution Approaches for Multi-Objective Resource Allocation

Unlike single-objective optimization models, which have only one optimal solution, multi-objective optimization models do not have a single solution that holds the best values for all the objective functions. In particular, the **MORAP** mathematical model has two objective functions, which are conflicting in nature. Obtaining the best value for one objective function will lead to a worse value for the other objective function. The application of a certain set of countermeasures may minimize the total predicted number of accidents at the considered level crossings (i.e., the best value of $\boldsymbol{F}_1$), but such a decision could result in significant traffic delays (i.e., the worst value of $\boldsymbol{F}_2$). On the contrary, the application of another set of countermeasures may minimize the total delay (i.e., the best value of $\boldsymbol{F}_2$), but these countermeasures would not be able to achieve the optimal safety performance (i.e., the worst value of $\boldsymbol{F}_1$). Hence, multi-objective optimization models with conflicting objectives, such as **MORAP**, have a set of optimal solutions (i.e., a Pareto Front [PF]), rather than a single optimal solution. Note that a PF denotes a set of non-dominant solutions, and so, improving an objective function without worsening the others is not possible in such a case (Dulebenets et al. 2020b). The analysis of tradeoffs between conflicting objectives using PFs could be very effective for multi-objective optimization.

A hypothetical example, which shows tradeoffs between the conflicting objectives using a PF, is illustrated in Fig. 5. In this example, the PF has a total of 5 optimal solutions for the **MORAP** mathematical model. Each of these optimal solutions, belonging to the considered PF, represents a resource allocation decision for the considered level crossings. Here, PF point "1" and PF point "5" can be referred to as corner PF points. Corner PF point "1" has the best value of $\boldsymbol{F}_2$ (total delay = 500 s) and the worst value of $\boldsymbol{F}_1$ (total predicted number of accidents = 75). On the contrary, the resource allocation decision, represented by corner PF point "5", has the best value of $\boldsymbol{F}_1$ (total predicted number of accidents = 45) and the worst value of $\boldsymbol{F}_2$ (total delay = 3500 s).

Fig. 5 Application of Pareto Fronts in multi-objective resource allocation

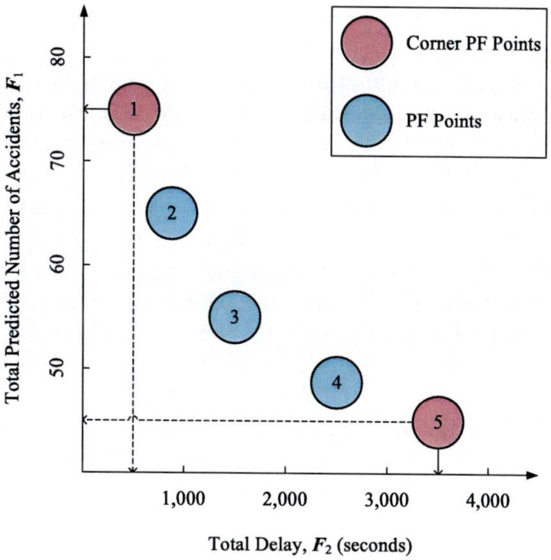

Thus, from this example, it is clear that the best value of $\boldsymbol{F}_1$ yields the worst value of $\boldsymbol{F}_2$ and vice versa. In other words, minimizing the predicted number of accidents only at the considered level crossings will increase the total delay, and minimizing the total delay only will increase the total predicted number of accidents. In such a case, the solution approaches that are specifically designed for the multi-objective optimization models with conflicting objectives have to be applied. Such solution approaches, which analyze the tradeoffs between conflicting objectives, will account for the fact that PF points "2", "3", and "4" do not have the worst values of either of the objective functions. For instance, the resource allocation decision, represented by PF point "3", reduces the total predicted number of accidents (i.e., the value of $\boldsymbol{F}_1$) by (75–55)/75 = 26.67%, as compared to corner PF point "1". At the same time, it reduces the total delay (i.e., the value of $\boldsymbol{F}_2$) by (3500–1500)/3500 = 57.14%, as compared to corner PF point "5". Such PF solutions that compromise the conflicting objectives can be viewed as more preferential by the relevant stakeholders rather than selecting the radical solutions, which are represented by PF point "1" and PF point "5". The multi-objective optimization solution approaches that can be applied for the multi-objective resource allocation problem, including exact optimization methods and approximate optimization methods, are discussed in the following sections.

4.1 Exact Optimization Methods

Multi-objective optimization models with PFs can be handled with several exact optimization methods, such as the ε-constraint method, augmented ε-constraint method, lexicographic ordering, multi-objective branch-and-bound, and others. The ε-constraint method, which is considered as one of the common exact optimization methods for multi-objective optimization models, is described in this section (Dulebenets 2018).

In case of multi-objective optimization models with two minimization objective functions, the ε-constraint method constructs PFs by minimizing only one objective function while setting an upper bound on the other. Let's consider a theoretical example, where the ε-constraint method tackles the **MORAP** mathematical model. This example is illustrated in Fig. 6. In step 1, the corner PF points, corresponding to the best values of the objective functions $\boldsymbol{F}_1$ and $\boldsymbol{F}_2$ are determined. Corner PF point $[\boldsymbol{F}_1^*; \boldsymbol{F}_2(\boldsymbol{F}_1^*)]$ represents the solution where $\boldsymbol{F}_1$ is minimized (i.e., $\boldsymbol{F}_1^*$) and an upper bound ε_2 is set on $\boldsymbol{F}_2$ (see the **MORAP-1** mathematical model for details). Similarly, corner PF point $[\boldsymbol{F}_1(\boldsymbol{F}_2^*); \boldsymbol{F}_2^*]$ represents the solution where $\boldsymbol{F}_2$ is minimized (i.e., $\boldsymbol{F}_2^*$) and an upper bound ε_1 is set on $\boldsymbol{F}_1$ (see the **MORAP-2** mathematical model for details). Note that **MORAP-1** and **MORAP-2** are mixed-integer linear programming models and can be solved to the global optimality with commercial mixed-integer linear programming solvers, such as CPLEX, GUROBI, MOSEK, etc.

MORAP-1:

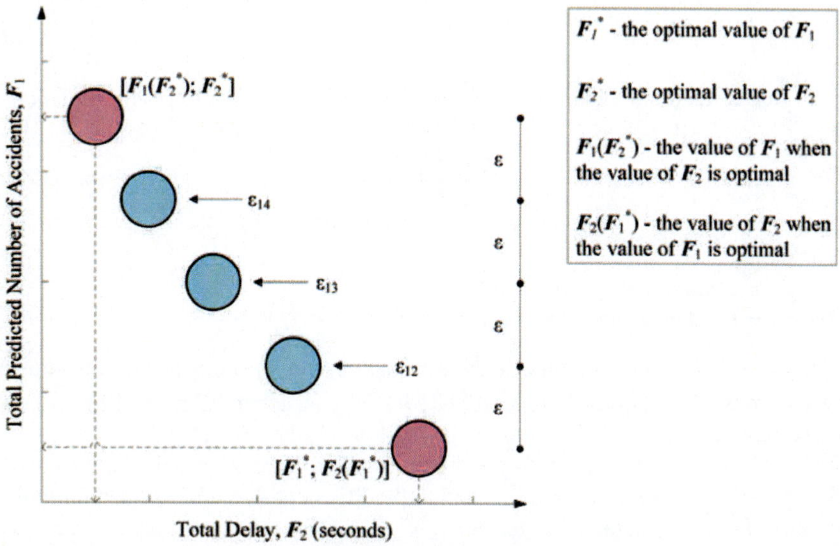

Fig. 6 Construction of Pareto Fronts with the ε-constraint method

$$\min \boldsymbol{F}_1 = \sum_{x \in X}\left[1 - \sum_{c \in C}(EF_{xc} \cdot z_{xc})\right] \cdot PA_x \tag{17}$$

Subject to:

Constraint sets (3)-(8)

$$\boldsymbol{F}_2 = \sum_{x \in X}\sum_{c \in C} TD_{xc} \cdot z_{xc} \tag{18}$$

$$\boldsymbol{F}_2 \leq \varepsilon_2 \tag{19}$$

MORAP-2:

$$\min \boldsymbol{F}_2 = \sum_{x \in X}\sum_{c \in C} TD_{xc} \cdot z_{xc} \tag{20}$$

Subject to:

Constraint sets (3)-(8)

$$\boldsymbol{F}_1 = \sum_{x \in X}\left[1 - \sum_{c \in C}(EF_{xc} \cdot z_{xc})\right] \cdot PA_x \tag{21}$$

$$F_1 \leq \varepsilon_1 \tag{22}$$

In step 2, the upper bound interval for $F_1(\varepsilon)$ is set using the following equation:

$$\varepsilon = \frac{F_1(F_2^*) - F_1^*}{N_{PF}^D - N_{PF}^C} \tag{23}$$

where:
N_{PF}^D = desired number of PF points;
N_{PF}^C = number of corner PF points.

In step 3, the initial upper bound on F_1 is set to F_1^*, and construction of the PF is started from F_1^*(i.e., corner PF point $[F_1^*; F_2(F_1^*)]$ is appended to the PF). In step 4, the upper bound on F_1 is increased by ε. In step 5, the **MORAP-2** mathematical model is solved with the updated upper bound on F_1, and a new PF point is generated. In step 6, the new PF point is appended to the PF. Steps 4–6 are repeated until the number of newly generated PF points is $N_{PF}^D - 1$. Once the repetitions are over, the desired number of PF points (N_{PF}^D) is reached, as corner PF point $[F_1(F_2^*); F_2^*]$ will be the last PF point. Note that a large number of PF points would improve the solution quality (i.e., will allow a more effective analysis of the tradeoffs between conflicting objectives) but would simultaneously increase the computational time required to solve the **MORAP** mathematical model.

4.2 Approximate Optimization Methods

Typical mathematical models, used for resource allocation among level crossings, can be reduced to the knapsack problem. The knapsack problem is one of the most-studied problems in operations research. The knapsack problem assumes that there is a knapsack with a restricted capacity (i.e., it cannot carry more weight than a particular threshold). In the meantime, there is a series of items with different weights that have different values. The objective of a decision maker is to select the items to be placed in the knapsack and to maximize the total value of the items placed, making sure that the knapsack capacity constraint will be satisfied (Dulebenets et al. 2020a). The resource allocation problems for level crossings have similarities with the knapsack problem. An example of a resource allocation problem for level crossings is illustrated in Fig. 7. It can be observed that there is a set of level crossings that are considered for upgrading. Furthermore, there is a set of candidate countermeasures with different attributes (i.e., these countermeasures have different effectiveness factors and implementation costs). The objective of a decision maker is to select the appropriate countermeasures to minimize the total predicted number of accidents at level crossings, considering a limited budget. Multi-objective resource allocation problems (e.g., **MORAP**) are particularly similar to the multi-objective knapsack

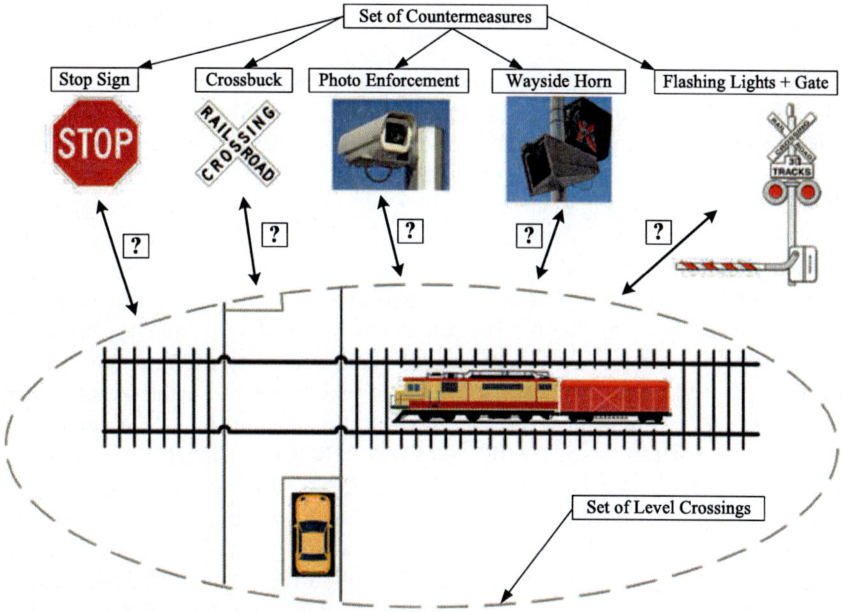

Fig. 7 An instance of the knapsack problem

problem. In such cases, the solution that minimizes only one objective function may not be the best solution for the overall problem. Thus, a set of solutions should be evaluated, considering the tradeoffs between different objective functions.

Resource allocation models for level crossings are generically non-deterministic polynomial time complete (NP-complete), like the knapsack problem. Note that NP-complete problems cannot be solved in polynomial time, but their solutions can be verified in polynomial time (Van Leeuwen 1990). In addition to exact optimization methods, which may require impractical computational times, approximate optimization methods (e.g., metaheuristics, heuristics) may be applied to obtain good-quality solutions for the knapsack problem or resource allocation models within a reasonable amount of time. Metaheuristic algorithms can be applied to a wide variety of problems. Moreover, they generally explore the search space more efficiently than heuristic algorithms, which are generally problem-specific (Eiben and Smith 2015). Several metaheuristic algorithms have been developed to solve the knapsack problem, and they might be adjusted to solve the resource allocation models for level crossings. For example, one of the pioneering studies by Drexl (1988) underscored that the multi-constraint 0–1 knapsack problem was non-deterministic polynomial time hard (NP-hard); therefore, obtaining optimal solutions was unlikely. Hence, a Simulated Annealing metaheuristic was proposed. NP-hard problems (like some variants of the knapsack problem) have higher computational complexity than NP-complete problems and require the development of efficient heuristics or metaheuristics (Kavoosi et al. 2019; Kavoosi et al. 2020b). Dueck and Scheuer (1990) proposed a Threshold

Accepting metaheuristic, which is a deterministic version of the Simulated Annealing metaheuristic.

Dammeyer and Voß (1993) presented a dynamic Tabu List management strategy, which was named as the reverse elimination method. The study also provided important directions to improve the computational effort of the proposed methodology. Khuri et al. (1994) was one of the early studies that first developed a Genetic Algorithm for the 0–1 multiple knapsack problem. In the proposed Genetic Algorithm, the fitness function penalized infeasible solutions in order to adequately capture the problem features. Battiti and Tecchiolli (1992) and Ohlsson et al. (1993) employed neural networks to solve the knapsack problem. It was revealed that neural networks had a tendency to return the final solutions that violated some of the constraints of the mathematical model. The aforementioned studies primarily addressed the single-objective knapsack problems. In case of multi-objective extensions (similar to the **MORAP** mathematical model that was formulated in Sect. 3), a number of multi-objective metaheuristic algorithms can be applied, including the following: (1) Multi-Objective Genetic Algorithm (MOGA); (2) Non-dominated Sorting Genetic Algorithm (NSGA); (3) Non-dominated Sorting Genetic Algorithm II (NSGA-II); (4) Niched-Pareto Genetic Algorithm (NPGA); (5) Pareto Archived Evolution Strategy (PAES); (6) Strength Pareto Evolutionary Algorithm (SPEA); (7) Multi-Objective Particle Swarm Optimization (MOPSO); (8) Multi-Objective Simulated Annealing (MOSA); and others.

5 Discussion

The state DOTs across the U.S. are responsible for safety of rail and highway travelers at level crossings. Implementation of various countermeasures is considered as one of the common approaches to improve level crossing safety and reduce the number of potential collisions between highway vehicles and trains. However, the state DOTs have to be also cognizant of potential delays that can be caused due to installation of new warning devices at level crossings. In fact, installation of new warning devices at level crossings may lead to disruption of passenger flows along with freight flows. In order to compromise safety goals and traffic continuity goals, this study proposed a novel optimization model for multi-objective resource allocation among level crossings. The developed optimization model not only minimizes the total predicted number of accidents but also minimizes the total delay caused by countermeasures at the considered level crossings. The proposed multi-objective resource allocation model can be easily implemented by the state DOTs. In particular, some of the well-established methods can be adopted to quantify the predicted number of accidents at level crossings (such as the Coleman-Stewart Model, the NCHRP Report 50 Accident Prediction Formula, and the Jaqua Formula).

Furthermore, the recommendations that are provided by the Institute of Transportation Engineers, the Manual on Uniform Traffic Control Devices, the U.S. Department of Transportation, and the American Railway Engineering and

Maintenance-of-Way Association can assist with quantification of traffic delays due to implementation of various countermeasures at level crossings. The scope of this study also included a complexity analysis of the proposed multi-objective resource allocation model. Since the developed optimization model has a lot of similarities with the knapsack problem, this study recommends application of heuristics or meta-heuristic algorithms to find solutions effectively in a reasonable computational time. In summary, the multi-objective resource allocation framework proposed herein can be instrumental for different stakeholders (not only for the state DOTs) that are concerned with operational challenges at level crossings in the U.S. and other countries as well. The proposed methodology would assist these stakeholders with selection of the appropriate countermeasures for the considered group of level crossings that will provide a desirable (or close to desirable) level of safety without causing a substantial increase in highway traffic delays.

6 Concluding Remarks

Rail transportation is a significant contributor to the U.S. economy. The magnitude of rail transportation has been increasing with time, so has the number of accidents at level crossings. Level crossing accidents result in numerous losses of lives, injuries to victims, and damages to property. The number of accidents at level crossings should be reduced, and the application of various countermeasures (e.g., warning devices) is one of the most popular approaches to do so. However, the installation of countermeasures at each of the level crossings in the U.S. is not feasible with the annual budget for safety improvement projects at level crossings. Since each countermeasure has a unique level of effectiveness and installation cost, a choice of the countermeasures should be examined for each of the level crossings under consideration. Furthermore, the application of countermeasures can cause additional delays to the associated highway traffic, which can lead to negative economic implications. Hence, consideration of highway traffic delays caused by countermeasures at level crossings should also be made. The aforementioned challenges in selection of the appropriate countermeasures for level crossings can be addressed by multi-objective resource allocation methodologies, which can analyze the trade-offs between conflicting objectives. Therefore, a framework for the multi-objective resource allocation problem (**MORAP**) and safety improvement projects at level crossings was presented in this chapter. The proposed model aimed to minimize the total predicted number of accidents as well as to minimize the total delay caused by countermeasures at the considered level crossings.

A number of different methods that could be used for accident prediction at level crossings were discussed, including the Coleman-Stewart Model, the Iowa Accident Prediction Formula, the Jaqua Formula, the NCHRP Report 50 Accident Prediction Formula, the Peabody-Dimmick Formula, and the U.S. DOT Accident Prediction Formula. Furthermore, some advanced statistical models for accident prediction at level crossings were discussed as well (such as the negative binomial model, the Poisson model, the Conway-Maxwell-Poisson model, the gamma model, the zero-inflated Poisson model, and the zero-inflated negative binomial model). The present study also provided a description of common methods that can be used to quantify the traffic queues at level crossings and measuring the associated traffic delays. Furthermore, this chapter reviewed a set of solution methods that could be applied for multi-objective resource allocation problems, including exact optimization methods (e.g., the ε-constraint method) and approximate optimization methods (different types of metaheuristics, including Multi-Objective Genetic Algorithm, Non-dominated Sorting Genetic Algorithm, Non-dominated Sorting Genetic Algorithm II, Strength Pareto Evolutionary Algorithm, Multi-Objective Particle Swarm Optimization, Multi-Objective Simulated Annealing, and others).

Future research on multi-objective resource allocation among level crossings could progress in the following directions:

- Develop region-specific accident and hazard prediction models to assess safety of level crossings;
- Design simulation models the can emulate collisions between trains and highway vehicles at level crossings;
- Consider highway user compliance for different countermeasures (some level crossing users might drive around the gates that are in a horizontal position due to an approaching train);
- Apply automated camera enforcements in human studies to accurately examine highway user compliance;
- Incorporate the expert and stakeholder opinions in the development of multi-objective resource allocation models;
- Allow the mathematical model to apply multiple countermeasures at a level crossing;
- Assess the effectiveness of countermeasures for different types of accidents;
- Integrate minimization of negative environmental impacts from applying different countermeasures (e.g., excessive harmful gas emissions due to long wait times) as an objective function;
- Develop advanced solution methodologies for multi-objective resource allocation models;
- Collect the required data and apply the proposed multi-objective resource allocation model for level crossings in a particular region.

Acknowledgements This study was supported by the Florida Department of Transportation (grant number BDV30-977-26). The opinions, findings, and conclusions expressed in this publication are those of the authors and not necessarily those of the Florida Department of Transportation or the U.S. Department of Transportation. The authors would like to thank Mr. Rickey Fitzgerald, Freight and Multimodal Operations Office Manager, for his involvement and valuable feedback throughout this study.

References

Abioye OF, Dulebenets MA, Pasha J, Kavoosi M, Moses R, Sobanjo J, Ozguven EE (2020) Accident and hazard prediction models for highway-rail grade crossings: a state-of-the-practice review for the USA. Railway Eng Sci 28(3):251–274

AREMA (2004) Highway-rail grade crossing warning systems. In: Communications and Signal Manual, Section 3. Landover, MD

Battiti R, Tecchiolli G (1992) Parallel biased search for combinatorial optimization: genetic Algorithms and Tabu Search. Microprocess Microsyst 16(7):351–367

Beanland V, Grant E, Read GJ, Stevens N, Thomas M, Lenné MG, Stanton NA, Salmon PM (2018) Challenging conventional rural rail level crossing design: evaluating three new systems thinking-Based designs in a driving simulator. Saf Sci 110:100–114

Beanland V, Salmon PM, Filtness AJ, Lenné MG, Stanton NA (2017) To stop or not to stop: contrasting compliant and non-Compliant driver behaviour at rural rail level crossings. Accid Anal Prev 108:209–219

Berg V (1986) Critique of rail-highway grade crossing effectiveness ratios and resource allocation procedures. Transp Res Rec 1069:88–100

Borsos A, Gabor M, Koren C (2016) Safety ranking of railway crossings in Hungary. Transp Res Procedia 14:2111–2120

BTS (2020a) Seasonally adjusted transportation data. In: Variable: rail freight intermodal. https://www.transtats.bts.gov/osea/seasonaladjustment/?PageVar=RAIL_FRT_INTERMODAL. Accessed 13 July 2020

BTS (2020b) Seasonally adjusted transportation data. In: Variable: rail passenger miles. https://www.transtats.bts.gov/osea/seasonaladjustment/?PageVar=RAIL_PM. Accessed 13 July 2020

Cameron AC, Trivedi PK (2013) Regression analysis of count data, 2nd edn. Econometric Society Monographs. Cambridge University Press, Cambridge, UK

Chadwick S, Zhou N, Saat MR (2014) Highway-rail grade crossing safety challenges for shared operations of high-speed passenger and heavy freight rail in the U.S. Saf Sci 68:128–137

Cirovic G, Pamucar D (2013) Decision support model for prioritizing railway level crossings for safety improvements: application of the adaptive neuro-fuzzy system. Expert Syst Appl 40:2208–2223

Dammeyer F, Voß S (1993) Dynamic tabu list management using the reverse elimination method. Ann Oper Res 41:29–46

Dent M, Marinov M (2019) Introducing automated obstacle detection to British level crossings. Sustainable rail transport. Springer, Cham, pp 37–80

Djordjević B, Krmac E, Mlinarić TJ (2018) Non-radial DEA model: a new approach to evaluation of safety at railway level crossings. Saf Sci 103:234–246

Drexl A (1988) A Simulated Annealing approach to the multiconstraint zero-one knapsack problem. Computing 40(1):1–8

Dueck G, Scheuer T (1990) Threshold accepting: a general purpose optimization algorithm appearing superior to ssimulated annealing. J Comput Phys 90(1):161–175

Dulebenets MA (2018) A comprehensive multi-objective optimization model for the vessel scheduling problem in liner shipping. Int J Prod Econ 196:293–318

Dulebenets MA, Moses R, Sobanjo J, Ozguven EE, Abioye OF, Kavoosi M, Pasha J, (2020a) Development of the Optimization Model for Improving Safety at Rail Crossings in Florida. A Technical Report Prepared for the Florida Department of Transportation. Florida Department of Transportation. Tallahassee, Florida, USA

Dulebenets MA, Pasha J, Kavoosi M, Abioye OF, Ozguven EE, Moses R, Boot WR, Sando T (2020) Multiobjective optimization model for emergency evacuation planning in geographical locations with vulnerable population groups. J Manag Eng 36(2):04019043

Eiben AE, Smith JE (2015) Introduction to Evolutionary Computing, 2nd edn. Springer-Verlag, Berlin Heidelberg, Berlin, Germany

Elzohairy Y, Benekohal R (2000) Evaluation of Expected Accident Frequency Formulas for Rail-Highway Highway-Rail Grade Crossings. Urbana, Illinois, USA. http://www.idot.illinois.gov/Assets/uploads/files/Transportation-System/Research/Illinois-Transportation-Research-Center/2000.09.01%20-%20Evaluation%20of%20Expected%20Accident%20Frequency%20Formulas%20for%20Rail-Highway%20Crossings%20-%20VC-HR1%20FY98.pdf. Accessed 24 June 2020

Evans AW (2011) Fatal accidents at railway level crossings in Great Britain 1946–2009. Accid Anal Prev 43:1837–1845

Faghri A, Demetsky M (1986) Evaluation of Methods for Predicting Rail-Highway Highway-Rail Grade Crossing Hazards. Charlottesville, Virginia, USA. http://www.virginiadot.org/vtrc/main/online_reports/pdf/86-r32.pdf. Accessed 24 June 2020

Farr EH (1981) Rail-Highway Crossing Resource Allocation Model. U.S. Department of Transportation. Research and Special Programs Administration. Transportation Systems Center. Cambridge Massachusetts, USA. https://rosap.ntl.bts.gov/view/dot/11160. Accessed 24 June 2020

FHWA (2002) Status of the Nation's Highways, Bridges, and Transit: 2002 Conditions and Performance Report. Chapter 26: Highway-Rail Grade Crossings. https://www.fhwa.dot.gov/policy/2002cpr/ch26.cfm. Accessed 24 June 2020

FHWA (2019) Status of the Nation's Highways, Bridges, and Transit: Conditions and Performance Report, 23rd edn. https://www.fhwa.dot.gov/policy/23cpr/pdfs/23cpr.pdf. Accessed 24 June 2020

Forgionne GA (2002) Selecting rail grade crossing investments with a decision support system. Inf Sci 144:75–90

FRA (2020) Accident/Incident Data. https://safetydata.fra.dot.gov/OfficeofSafety/publicsite/on_the_fly_download.aspx. Accessed 13 July 2020s

Freeman J, Rakotonirainy A (2015) Mistakes or deliberate violations? A study into the origins of rule breaking at pedestrian train crossings. Accid Anal Prev 77:45–50

Greene WH (2018) Econometric analysis, 8th edn. Pearson Education Inc., Upper Saddle River, New Jersey, USA

Hans Z, Albrecht C, Johnson P, Nlenanya I (2015) Development of Railroad Highway Grade Highway-Rail Grade Crossing Consolidation Rating Formula. A Technical Report Prepared for the Iowa Department of Transportation. Iowa Department of Transportation. Ames, Iowa, USA. https://lib.dr.iastate.edu/cgi/viewcontent.cgi?referer=&httpsredir=1&article=1061&context=intrans_techtransfer. Accessed 24 June 2020

Hu SR, Li CS, Lee CK (2010) Investigation of key factors for accident severity at railroad grade crossings by using a logit model. Saf Sci 48:186–194

ITE (2006) Preemption of Traffic Signals Near Railroad Crossings: An ITE Recommended Practice. Institute of Transportation Engineers. https://www.ite.org/pub/?id=e1dca8bc%2D2354%2Dd714%2D51cd%2Dbd0091e7d820. Accessed 24 June 2020

Kavoosi M, Dulebenets MA, Abioye OF, Pasha J, Wang H, Chi H (2019) An augmented self-adaptive parameter control in evolutionary computation: a case study for the berth scheduling problem. Adv Eng Inf 42:100972

Kavoosi M, Dulebenets MA, Pasha J, Abioye OF, Moses R, Sobanjo J, Ozguven EE (2020) Development of algorithms for effective resource allocation among highway-rail grade crossings: a case study for the State of Florida. Energies 13(6):1419

Kavoosi M, Dulebenets MA, Abioye O, Pasha J, Theophilus O, Wang H, Kampmann R, Mikijeljević M (2020) Berth scheduling at marine container terminals: a universal island-based metaheuristic approach. Marit Bus Rev 5(1):30–66

Keramati A, Lu P, Tolliver D, Wang X (2020) Geometric effect analysis of highway-rail grade crossing safety performance. Accid Anal Prev 138:105470

Khan IU, Lee E, Khan MA (2018) Developing a highway rail grade crossing accident probability prediction model: a North Dakota case study. Safety 4:22

Khattak A, Lee M (2018) Highway-Rail Crossing Safety Improvements by Diverting Motorist to Alternate Routes. https://www.utrgv.edu/railwaysafety/_files/documents/research/operations/

utcrs_khattak_highway-rail-crossing-safety-improvement_final-report.pdf. Accessed 24 June 2020

Khattak A (2014) Investigation of train warning times and gate violations. Transp Res Rec 2458:104–109

Khuri S, Bäck T, Heitkötter J (1994) The zero/one multiple knapsack problem and Genetic Algorithms. In: Proceedings of the 1994 ACM symposium on applied computing. Phoenix, Arizona, USA: ACM, pp. 188–193. http://citeseerx.ist.psu.edu/viewdoc/download?doi=10.1.1.51.2713&rep=rep1&type=pdf. Accessed 24 June 2020

Konur D, Golias MM, Darks B (2013) A mathematical modeling approach to resource allocation for railroad-highway crossing safety upgrades. Accid Anal Prev 51:192–201

Laapotti S (2016) Comparison of fatal motor vehicle accidents at passive and active railway level crossings in Finland. IATSS Research 40:1–6

Landry S, Jeon M, Lautala P, Nelson D (2019) Design and assessment of in-vehicle auditory alerts for highway-rail grade crossings. Transport Res f: Traffic Psychol Behav 62:228–245

Larue GS, Wullems C (2015) Human factors evaluation of a novel Australian approach for activating railway level crossings. Procedia Manuf 3:3293–3300

Larue GS, Naweed A, Rodwell D (2018) The road user, the pedestrian, and me: investigating the interactions, errors and escalating risks of users of fully protected level crossings. Saf Sci 110:80–88

Lenné MG, Rudin-Brown CM, Navarro J, Edquist J, Trotter M, Tomasevic N (2011) Driver behaviour at rail level crossings: responses to flashing lights, traffic signals and stop signs in simulated rural driving. Appl Ergon 42:548–554

Lord D, Washington SP, Ivan JN (2005) Poisson, Poisson-gamma and zero-inflated regression models of motor vehicle crashes: balancing statistical fit and theory. Accid Anal Prev 37(1):35–46

Lu P, Tolliver D (2016) Accident prediction model for public highway-rail grade crossings. Accid Anal Prev 90:73–81

Oh J, Washington SP, Nam D (2006) Accident prediction model for railway-highway interfaces. Accid Anal Prev 38(2):346–356

Ohlsson M, Peterson C, Söderberg B (1993) Neural networks for optimization problems with inequality constraints: the knapsack problem. Neural Comput 5(2):331–339

Pasha J, Dulebenets MA, Abioye OF, Kavoosi M, Moses R, Sobanjo J, Ozguven EE (2020) A comprehensive assessment of the existing accident and hazard prediction models for the highway-rail grade crossings in the State of Florida. Sustainability 12(10):4291

Rezvani AZ, Peach M, Thomas A, Cruz R, Kemmsies W (2015) Benefit-cost methodology for highway-railway grade crossing safety protocols as applied to transportation infrastructure project prioritization processes. Transp Res Procedia 8:89–102

Rilett L, Appiah J (2008) Microsimulation Analysis of Highway-Rail Grade Crossings: A Case Study in Lincoln, Nebraska. In: IEEE/ASME/ASCE Joint Rail Conference, Wilmington, Delaware, USA, USA. DOI: https://doi.org/10.1115/JRC2008-63063

Rudin-Brown CM, Lenné MG, Edquist J, Navarro J (2012) Effectiveness of traffic light vs. boom barrier controls at road–rail level crossings: a simulator study. Accid Anal Prev 45:187–194

Russo BJ, James E, Erdmann T, Smaglik EJ (2020) Pedestrian and bicyclist behavior at highway-rail grade crossings: an observational study of factors associated with violations, distraction, and crossing speeds during train crossing events. J Transp Saf Security 1–19. https://doi.org/10.1080/19439962.2020.1726545

Shmueli G, Minka TP, Kadane JB, Borle S, Boatwright P (2005) A useful distribution for fitting discrete data: revival of the Conway-Maxwell-Poisson distribution. Appl Stat 54(1):127–142

Sperry B, Naik B, Warner J (2017) Evaluation of grade crossing hazard ranking models. In: Ohio Transportation Engineering Conference, 2017. http://www.dot.state.oh.us/engineering/OTEC/2017Presentations/75/Sperry_75.pdf. Accessed 24 June 2020

Stefanova T, Oviedo-Trespalacios O, Freeman J, Wullems C, Rakotonirainy A, Burkhardt J, Delhomme P (2018) Contextual factors explaining risk-taking intentions at Australian level crossings. Saf Sci 110(2):145–161

Tey L, Wallis G, Cloete S, Ferreira L (2013) Modelling driver behaviour towards innovative warning devices at railway level crossings. Accid Anal Prev 51:104–111

US DOT (2019) Highway-Rail Crossing Handbook, 3rd Edn. https://safety.fhwa.dot.gov/hsip/xings/com_roaduser/fhwasa18040/fhwasa18040v2.pdf. Accessed 24 June 2020

Van Leeuwen J (1990) Algorithms and complexity. In: Handbook of theoretical computer science, 1st edn. Elsevier. Netherlands

Washington S, Karlaftis MG, Mannering F, Anastasopoulos P (2020) Statistical and econometric methods for transportation data analysis, 3rd edn. Chapman Hall/CRC, Boca Raton, Florida

Weissmann AJ, Weissmann J, Kunisetty JL, Warner J, Park E, Sunkari S, Protopapas A, Venglar S (2013) Integrated Prioritization Method for Active and Passive Highway-Rail Crossings. A Technical Report Prepared for the Texas Department of Transportation. Texas Department of Transportation. Austin, Texas, USA

Yan X, Richards S, Su X (2010) Using hierarchical tree-based regression model to predict train–vehicle crashes at passive highway-rail grade crossings. Accid Anal Prev 42:64–74

Zhang Z, Trivedi C, Liu X (2018) Automated detection of grade-crossing-trespassing near misses based on computer vision analysis of surveillance video data. Saf Sci 110:276–285

Zhou X, Lu P, Zheng Z, Tolliver D, Keramati A (2020) Accident prediction accuracy assessment for highway-rail grade crossings using random forest algorithm compared with decision tree. Reliab Eng Syst Saf 200:106931

Study of Sustainable Transport Enhancements Through Analysing Utilisation Levels of Rail Lines with Enhanced Passenger and Freight Services: A Case Study on a Local Rail Line Through Simulation Modelling and Scenarios

Joel Crannis, Jeremy Lee, and Marin Marinov

Abstract Due to climate change, more sustainable passenger and freight transport is in need. Rail is considered a more sustainable mode of transport compared to others such as road transport. Usage of rail lines that are currently under-utilised could help increase the sustainability of transport through enhanced utilisation of them. This study looks at the Marston Vale Line, identifying it as under-utilised, and through the use of event-based simulation modelling, observes the current utilisation level and develops ways in which the spare capacity can be utilised, with more passenger services, as well as local rail freight, which is identified as a potential use of spare capacity on the Marston Vale Line. Possible local freight that can be transported by rail is investigated and combined with current and possible additional passenger services in varying levels, in five different scenarios, which are evaluated to propose the best workable option for the Marston Vale Line, with journey time and reliability also considered. Particular theoretical attention is proposed to create a more environmentally friendly mode for the transportation of passengers and freight, and not just shifting more from road to rail using the idea that rail transport is more sustainable than road transport.

Keywords Rail lines · Rail routes · Utilisation levels · Passenger services · Freight services · Simulation modelling

J. Crannis · J. Lee · M. Marinov (✉)
Department of Engineering Systems & Supply Chain Management (ESSCM), College of Engineering and Physical Sciences (EPS), Aston University, Aston Triangle, Birmingham B4 7ET, UK
e-mail: m.marinov@aston.ac.uk

M. Marinov and J. Piip (eds.), *Sustainable Rail Transport 4*, Lecture Notes in Mobility,
https://doi.org/10.1007/978-3-030-82095-4_5

1 Introduction

1.1 Motivation

The need for more environmentally friendly transport; of both goods and passengers, is becoming increasingly relevant in recent times. Given that rail transport is seen as being more environmentally friendly than alternatives such as road transport, fully utilising rail lines that are currently under-utilised would help reduce the need for other forms of transport which are less environmentally friendly. At the same time, having an asset such as a rail line under-utilised leaves a gap for further revenue for the owner, in this case, being Network Rail (2020a, b, c). The Marston Vale Line being studied is a local rail line, which only has around 1 passenger train per hour (TPH) in each direction (Network Rail 2020a, b, c), with some freight services as well. Therefore, the line could be further utilised to transport more freight, as well as more passengers, in a more environmentally friendly way, particularly as 1TPH is inflexible for passengers using the line as an alternative to the car.

However, as illustrated by Givoni et al. (2009), the extent to how green rail transport is, as a whole, is not such a positive factor, as Givoni et al. (2009) identified that the least environmentally friendly trains polluted similar emissions to aeroplanes on the same routes. The Marston Vale Line is not electrified, and the passenger services on the Marston Vale Line are run by class 230 trains with diesel generators (Community Rail Network 2019), although they have traction motors (Rail magazine 2019). The class 230 trains are built from old train bodyshells and parts, being almost carbon neutral to build, and maintenance can be done trackside reducing necessary train movements for environmental benefit (Vivarai 2020). More of these old converted London Underground trains (Rail Magazine 2019) could also be converted to transport freight along the Marston Vale Line. This approach of converting old passenger trains to carry freight is being explored, with a test of roll pallets on a passenger train to carry parcels (Holden 2020). Some trains are actually being converted to carry parcels (Briginshaw 2019).

Having the same type of train on the line will help maintain the headway between trains on the Marston Vale Line as they would all be the same type of train with the same sectional running times, therefore maximising the potential capacity of the line. Additional passenger services could also be run by more of these class 230 trains. This could, therefore, be simply and accurately modelled using simulation software.

Using the class 230 with increased passenger and freight services would therefore consider and enhance the environmental benefits of rail, helping to provide 'greener' transport, especially when considering that presently a regular freight train would require a heavy diesel locomotive on the non-electrified Marston Vale Line, to pull the train.

1.2 Problem Formulation

Besides the approximate 1 passenger train per hour in each direction (Network Rail 2020a, b, c), the line is also used for rail freight, as it is a good connection between the West Coast Main Line at Bletchley and the Midland Main Line at Bedford. Data from Real Time Trains (2020) showed that the shortest time between two freight trains is 36 min apart, on a random weekday, being 27 July 2020. However, there is a clear problem that the line is being under-utilised. Combining this with the general problem of the need for more environmentally friendly transport, both of these problems could, at least, be partially mitigated by increasing the number of both freight and passenger trains running on the Marston Vale Line.

The current freight services using the line cannot be increased much, if at all, as they run on beyond either or both ends of the Marston Vale Line and onto the mainlines which the Marston Vale Line connects (Real Time Trains 2020). At the Bletchley end of the Marston Vale Line, the West Coast Main Line is at full capacity (RailTech.com 2020) and so cannot accommodate any more freight trains. According to Gleave (2013), the Midland Main Line also has a lot of services heading north out of London St Pancras, which is south of Bedford, and the title of the report itself; 'Capacity on North–South Main Lines', suggests there is a capacity problem on the line. Therefore, local freight could be transported along the Marston Vale Line itself where there is spare capacity, rather than the current way of mainly using the Marston Vale Line as a link route between the two main lines (Real Time Trains 2020).

The line only has a limited number of trains running on it at present, which is unlikely to experience much of a delay with a resulting increase in journey time, particularly concerning periods of disruption. However, further utilising the line with more passenger and freight services could result in journey times increasing due to congestion with the volume of traffic on the line, which could be exaggerated during periods of disruption, reducing the reliability of the line. Therefore, the journey times for all services using the line in this study will be observed when studying the possible increase in utilisation of the Marston Vale Line, to help maintain the reliability of the services.

As mentioned in Sect. 1.1, even local rail freight transported along the line would, at present, require a heavy diesel locomotive to pull the freight train, regardless of the number of wagons. From an environmental perspective, this would not be the best way to shift to a more environmentally friendly mode of transport, where rail's environmental credentials are less effective, as Givoni et al. (2009) have found. However, the class 230 train would offer a much more environmentally friendly way to transport local freight along the route than by alternatives such as road transport. This confirms that the focus is on a more environmentally friendly way to transport passengers and freight, and not just a shift from road to rail.

1.3 Objectives

The objectives of this study are to understand the current utilisation of the Marston Vale Line and be able to develop and evaluate ways to increase the utilisation, with additional passenger and freight services, that has a strong positive environmental benefit, while also being reliable and efficient, however prioritising the environmental benefit. The use of event-based simulation modelling software will aid in completing these objectives.

This study looks into simulation modelling to improve utilisation of rail line, however the particular attention given for the environmentally friendly aspects are only covered in the theoretical perspective of the use of rolling stock, and the increased utilisation of the line through more passengers and freight by rail than alternatives such as road transport.

1.4 Paper Organisation

Section 2 is a literature review on the history of rail operations of freight trains and current research into increased utilisation of rail lines with additional freight services using simulation software. Section 3, details of how the Marston Vale Line currently is simulated using the simulation software, and validated using the December 2019 Timetable. Section 4 then analyses the results of the simulation of the current line. This gives the actual utilisation of the line at present as data, rather than just observations of under-utilisation. Section 5 gives a background on freight by rail before investigating potential freight that could be transported by rail along the Marston Vale Line. Section 6 describes five different scenarios proposed to increase the utilisation of the Marston Vale Line with additional passenger and potential rail freight services, utilising the existing line infrastructure besides additional freight terminals along the route for the local rail freight. Detail is given on how these different scenarios are input into the simulation software. Section 7 evaluates the results of the 5 scenarios, considering the journey time and predicted reliability, as well as utilisation. Section 8 concludes the study and finally Sect. 9 proposes potential future work on this study.

Screenshots of the Simul8 event based simulation modelling software, presented as figures, are used throughout the study to show how the simulation model was created using Simul8.

1.5 Presentation and Analysis of the Marston Vale Line

The track layout of the Marston Vale Line is shown in Fig. 1. There are 10 intermediate stations along the line, with single-track sections at Fenny Stratford and at the end of the line at Bedford. There are only a few cross over sections along the

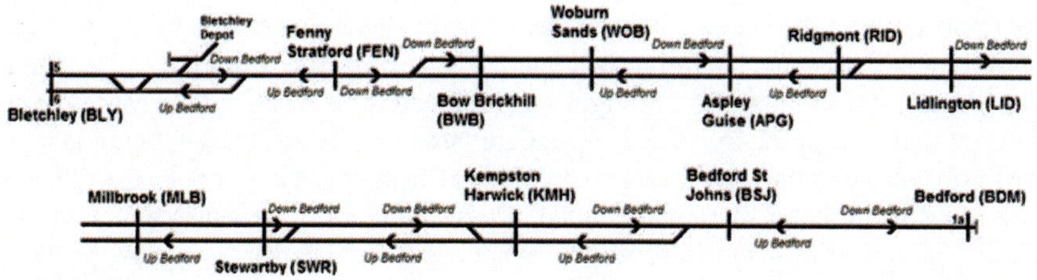

Fig. 1 The track layout of the Marston Vale Line (drawn using Google Maps (2020) and Train_PlaneHub (2015))

double-track section of line, so adding in a siding or loop to one side of the line as a freight terminal, with minimum investment in additional infrastructure while also minimising disruption during construction, would be very inflexible for operations. Therefore, this study will present the new additional freight terminals as being larger than a simple siding, because of the necessary additional track and point work required to have any sized terminal. Therefore, for simulation purposes, the terminals will have infinite capacity, however the exact size of terminals, with factors such as number of sidings, would be the size to support the freight train services planned, and possible additional space for future freight services. The modified track layout, in Fig. 11, shows these terminals, with question marks for possible future expansion and represent the unknown actual capacity due to having infinite capacity in the simulation model.

It is therefore clear from Fig. 1, that one passenger train per hour in each direction fits comfortably onto the Marston Vale Line, however the utilisation of the line as a whole is something that could be improved upon.

2 Literature Review

2.1 History of Rail Operations and Rail Freight Services

As railway transportation is operated by mainly passenger and only a few freight services in the UK, according to Haywood (1999), in the post-war era, both market share and volume of freight by rail has declined. Hence, in the 1960s, the decline has led to closure of many local freight depots and branch lines being cut by a third.

During the 1990s, the trend of rail freight services seems to continue to decline as, Stittle (2004) states that since the UK government privatised the railway industry in 1996, the overall result has been a further decline in rail freight traffic, hence fewer rail freight services. Their explanation is that the rail infrastructure has been transferred to a private sector infrastructure company, known as Railtrack plc (before being transferred to the public sector company Network Rail in 2002). This has led to the introduction of track access charges for privatised train and freight operators which, they only paid for the use of rail infrastructure and maintenance costs while services in operation. Consequently, rail freight operators are more likely to be penalised than rail passenger operators as, rail freight services tend to be heavier, thus cause more damage to the rail infrastructure. Hence, Crompton and Jupe (2003) both argue that the performance of the privatised system in the railway industry has been inefficient. They concluded that Railtrack attempted to reduce track maintenance, to maximise profit whilst, Railtrack's main source of revenue is from track access charge, paid by train and freight operators. As a result, privatisation has led to its rail infrastructure to deteriorate which may have discourage rail freight operators from using the rail infrastructure.

2.2 *Case Studies of Simulation in Rail Operations*

When investigating a potential implementation of more rail freight as well as potentially increasing passenger services, simulation models and scenarios can be used to investigate the current and potential utilisation levels of a local railway line. This has been shown in a study by Potti et al. (2019) as, the study uses simulation models and scenarios to investigate the utilisation levels of the Cross-City railway line in the UK, from Lichfield Trent Valley to Birmingham New Street Station. The railway line from their simulation models appear to be under-utilised, suggesting that there is not enough rail freight transport operated on the line. Thus, they set up scenarios such as, providing additional passenger services and implementing rail freight onto the line, to improve the line's utilisation levels.

Similarly, in a case study developed by Motraghi and Marinov (2012), they have developed simulation models and scenarios to analyse the performance of urban freight by rail, based around a Metro system in Newcastle upon Tyne. Their proposed ideas in using an existing Metro system, which offers passenger services, to operate urban freight is innovative. This is because it provides opportunities for businesses to utilise more the urban freight services by rail, rather than road freight, which would help to reduce road traffic in Newcastle. From their results, they discovered that urban freight by Metro is possible as, despite the increase of waiting times, the utilisation levels of the Metro system have increased, without making significant changes to the existing system in place. Hence, the benefits of implementing urban freight onto the existing Metro system is it minimises infrastructure costs. Also, this helps to improve the economy and the environment in Newcastle as, an additional option to transport urban freight by Metro, encourages more businesses to utilise rail freight services, which would evidently help to reduce their carbon footprint. Therefore, utilisation would help increase sustainable transportation around Newcastle. A common theme between these two studies is the location of the lines, as both are urban. The Marston Vale Line, although connecting two towns, is largely a rural line, and with the proposed local rail freight, appears to differentiate this study from current research in a more urban setting. It can be identified that there is a trend between additional utilisation, which increases the sustainability of transport, and increased journey time and reduced performance from a time perspective.

Furthermore, Singhania and Marinov (2017) have also used simulation models and scenarios to analyse the utilisation levels of a railway. Their focus has been based around a railway line section between Edinburgh Waverley and Glasgow Queen Street, to implement their proposals of increasing the number of freight trains. Thus, from their results, they revealed that more freight trains can be added along with the existing passenger trains. Therefore, this indicates there is an increase in rail utilisation as, their simulation model suggests there is adequate capacity to fit more freight trains onto its current railway line. This is the proposal with this study of the Marston Vale Line, that there is adequate capacity to add in additional freight and passenger services while not impacting the journey time and performance of the line. Again, this study is focussed in an urban area.

Additionally, in another case study developed by Potti and Marinov (2020), they have both looked at the utilisation level of the West Midlands Metro using simulation modelling. They found the line to be under-utilised and see potential research for additional services as well as urban freight. The research into this area has been focused around urban freight, again whereas this study differentiates itself by looking at an under-utilised, local, rural line. This study proposes particular theoretical attention in its proposals to ensure a more sustainable mode of transport for additional services rather than their current mode, which current research does not seem to focus so much on.

3 Simulation Modelling by Using SIMUL8

3.1 Studying the Rail Operations and Asset Utilisation on the Marston Vale Line

The purpose of the simulation modelling is to examine the rail operations and asset utilisation to measure and evaluate its efficiency and reliability throughout the model. In the initial studying of the Marston Vale Line, it is important to investigate the current passenger service first, before looking at different scenarios such as; increasing the passenger service patterns, and adding in freight services into the model.

3.2 Implementation of Simulation Modelling

Before implementing the simulation model to the study of the Marston Vale Line, it is essential to obtain actual data, such as a passenger timetable, which acts as a foundation to the model to ensure validity. Thus, the passenger timetable for this model is based around the December 2019 Timetable (Network Rail 2020a, b, c). The simulation model can be created, after obtaining information about the timings and stations throughout the Marston Vale Line.

3.3 Understanding of SIMUL8

The SIMUL8 software has been used to create the simulation model because it allows to set a specific time for the model to be simulated from start to end, whilst the work items act as passenger and freight entities to be served throughout the system. After running the model, results will be generated to determine the utilisation and reliability of the rail operations on the line at present.

3.4 Clock Properties

In Fig. 2, the simulation model is set to be simulated for 9 h/day from 5:00 to 14:00, through weekdays from Monday to Friday, as shown in the results collection period which is set at 1 week. The clock is set at minutes to allow all results to display the same unit, to ensure consistency to make comparisons with other results. There is no warm-up period for this model as, the rail operations on the Marston Vale Line for the passenger and freight services operate after the start time at 5:00 and before the end time at 14:00 each day on a set timetable.

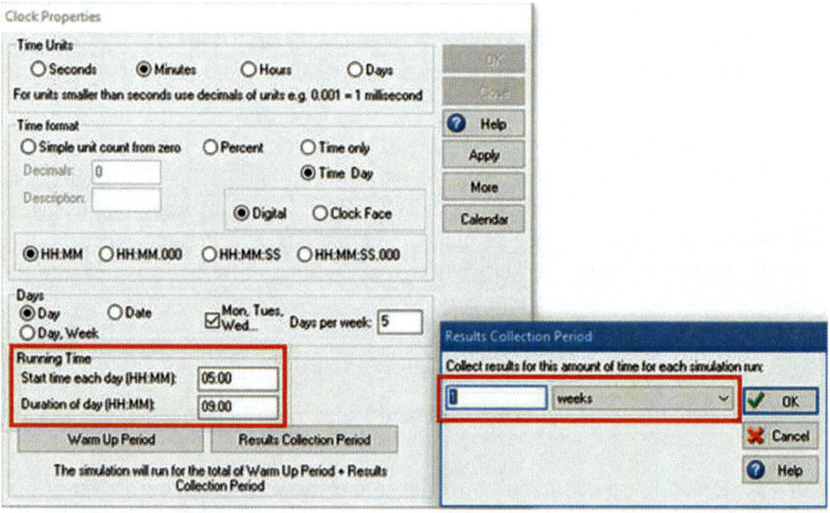

Fig. 2 Clock set up timings, with results collection period

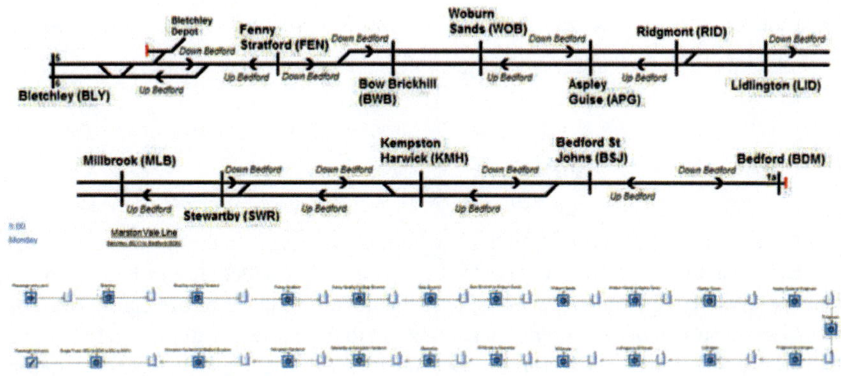

Fig. 3 Simulation model (left), showing the activities and queues, with the track layout design (right)

3.5 Activities

From Fig. 3, the 'Activity' modules in SIMUL8 are used across the simulation model. These activities have two types: to simulate the timings for the train to travel and station dwell times.

The first activity is the journey time between stations, with the accordance of the December 2019 timetable. The timings are fixed, sectional running times, and they are calculated by using timetabled time between two departures from the December 2019 Timetable and subtracting 0.5 min (or 30 s) for the station stop.

The second activity is station stops/dwell times. All station stop activities are set at a fixed dwell time of 0.5 min which allows for trains at each intermediate station to stop, and allow passengers to board on or off the train, prior to their departure time.

The exception is at Bletchley, where the station stop activity fixed time at Bletchley is set at 0 min. The reason for this is because the start point, which is before Bletchley, is the departure time from Bletchley, however having the activity Bletchley station stop after the start point has no effect on the timings as it just passes through the activity, with a working time of 0 min. This is just to show Bletchley as a station activity in the simulation model.

3.6 Queues

There are queues in between each activity, using the 'Queue' modules (see Fig. 3). The maximum queue capacity is set at one because this allows only one passenger or freight train to be in one section of track, or block at any one time. The purpose is to prevent collisions if there is a service blockage in the system, focusing the trains to wait at stations. It is therefore similar to a block signalling system, as used in the UK (The Railway Technical Website 2019).

3.7 Start and End Points

Using the 'Start Point' modules. This is where the trains enter the simulation model and the arrival times for this model are set by using a fixed schedule, see Fig. 4. This is done by creating a spreadsheet, which uses the departure times at Bletchley from the timetable, as shown in Fig. 5. The trains then run to the same arrival pattern following the activity times, the sectional running and dwell times at each line section/station activity.

Using the 'End Point' modules. This is where the trains then leave the simulation model.

Fig. 4 Start point input before the activity point at Bletchley

Fig. 5 Passenger timetable at the start point, showing the departure times from Bletchley

Sheet: Passenger entry into the simulation timetable

	A	B	C	D
1	Time	Items		
2	16	1		
3	84	1		
4	152	1		
5	181	1		
6	246	1		
7	318	1		
8	361	1		
9	421	1		
10	481	1		

3.8 *Using Time Dependent Distribution at the Single-Track Activity*

Looking at the track diagram and using the timetable of the Marston Vale Line, the sections between Kempston Hardwick, Bedford St Johns and Bedford stations are situated on single-track, which additional services might use the single track in different directions at the same time, which cannot happen, as they would conflict. Even though this model is only looking at a single direction, departing from Bletchley, time distributions need to be created and put into a time dependent distribution to consider the other direction.

As shown on Fig. 6, the timings are inputted in the activity 'Single Track (BSJ to BDM to BSJ to KMH)' which uses the time dependent distribution. This type of distribution combines all the time distributions from each passenger service time using the single line from the timetable, and the timings are all run in a fixed schedule. Thus, this is calculated by using the timetabled departure times between Kempston Hardwick (towards Bedford) and return to Kempston Hardwick (towards Bletchley). This is necessary because it accounts for the trains having to travel through the single-track section, as well as the dwell time spent at Bedford station, and the return journey back towards Bletchley through the same single-track section.

4 Analysis of Results on the Current Passenger Service

Based on the December 2019 Timetable of the Marston Vale Line, just looking at the passenger service, it is expected that the utilisation level is low for the whole model however, the single-track section is likely to have a higher utilisation level than the double-track sections as, trains need to pass through the same single-track section

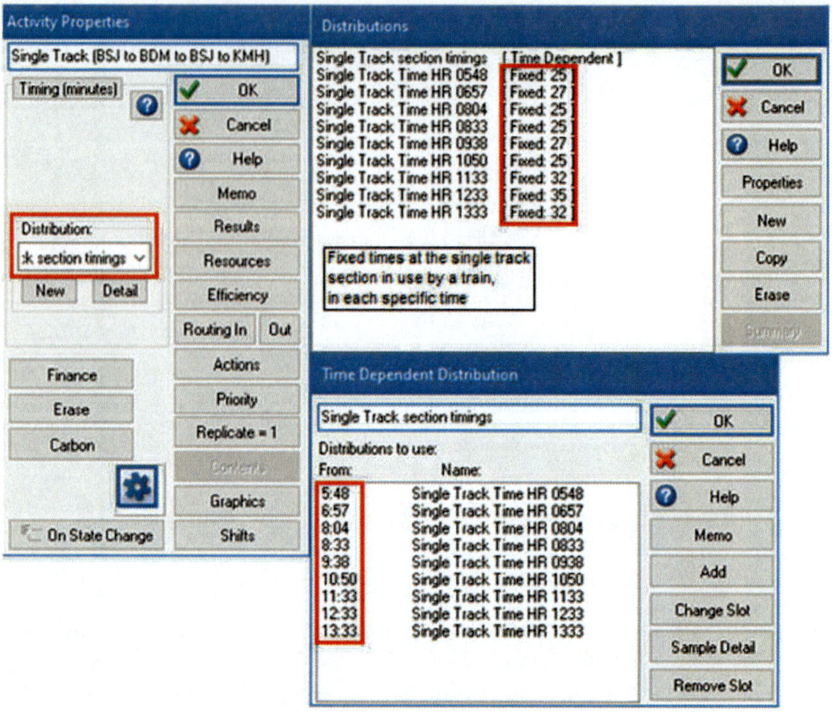

Fig. 6 Time dependent distribution which combines all fixed distribution timings

from each direction. While the return journey towards Bletchley from Bedford is being modelled until Kempston Hardwick in the Up Bedford direction, on the double track sections, the utilisation would not increase from Bletchley bound services, even if they were modelled in the simulation, as they would use the Up Bedford track.

The results from Table 1 (also shown in a bar chart, see Fig. 7) show the passenger journey time is consistent, with the average of 65.6 min and the difference of 10 min between minimum (62.5 min) and maximum (72.5 min). However, despite the journey time's consistency, the utilisation levels of the line between stations, see Table 2, varied from 0.83 to 1.83%, whilst the single-track utilisation level is 9.37%. This indicates that the line is poorly served with only 1 TPH in each direction, which explains the low utilisation levels in the double-track sections in between stations.

Table 1 Results of current passenger service time spent in the simulation

Statistics	Passenger (min)
Minimum	62.50
Average	65.61
Maximum	72.50
Standard deviation	3.86

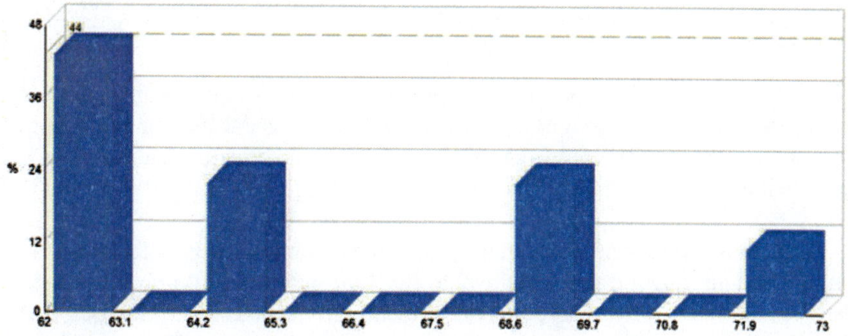

Fig. 7 Bar chart showing the percentage of different passenger journey times in the simulation

Table 2 Utilisation levels on the current passenger service of each activity point

Activity point	Utilisation levels (%)
Woburn sands	0.17
Woburn sands to Aspley Guise	0.83
Aspley Guise	0.17
Aspley Guise to Ridgmont	1.17
Lidlington	0.17
Lidlington to Millbrook	0.83
Stewartby	0.17
Stewartby to Kempston Hardwick	1.17
Single Track (BSJ to BDM to BSJ to KMH)	9.37

Hence, improvements to the utilisation levels can be implemented through use of simulation and setting up scenarios, refer to Sect. 6.

5 Rail Freight

5.1 Background of the Rail Freight in the UK

According to Woodburn (2003), the reasoning behind the low utilisation level on rail is that between the Second World War into the mid-1990s, the UK has experienced a long period of decline in transporting freight by rail. As a result, this has led to an increase of road freight because of its flexibility in terms of the routes and timings, which does not involve following the fixed rail freight timetable. However, with the concerns of increasing road freight traffic and the environment, rail freight in the UK could potentially be revived, perhaps implementing rail freight on local rail lines.

5.2 Potential Local Rail Freight

Since the utilisation level is low on the Marston Vale Line, it provides an opportunity to implement local rail freight across the line, which would help to improve the utilisation of the line. Therefore, new freight depots could be constructed, along with remodelling the line with new junctions to enable freight to enter and exit from the line.

The potential freight depot locations in this study are chosen to be at Ridgmont and Millbrook, and freight trains will travel towards the outskirts of Bedford at a Retail Park. As this study only looks at the one direction from Bletchley to Bedford, additional freight trains will only be considered towards Bedford, and will not consider the return journeys of the freight trains back to Millbrook and Ridgmont, where further alterations to the timetable for freight and passenger services may be required to fit in the return journeys.

The explanation for the freight transported towards Bedford Retail Park Depot is that the rail freight service avoids using the single-track sections as, it prevents blockage in the system to ensure the passenger service is not impacted by the implementation of the rail freight service.

The reason for having a freight depot at Ridgmont is that there is an Amazon Logistics Fulfillment Centre located in Ridgmont (Courier Locations UK 2018). Freight trains could transport goods from the fulfilment centre to the village stations along the line for local collection, reducing the need for road transport for local delivery, and also into Bedford Retail Park Depot, for serving the town of Bedford, for local collection, meaning in terms of the logistics from the fulfilment centre to Amazon's customer, there would only be the use of rail freight. Meanwhile, before goods arrive at the Amazon Logistics Fulfillment Centre in Ridgmont, as the site is very close to the M1, allows for efficient road transportation of goods to the fulfilment centre in place of where improvements in rail freight utilisation from further afield are still to come, and so adds potential for future rail freight using this depot for areas further afield.

For Millbrook, the reason for a freight depot is that there is an energy-from-waste plant being built, named as 'Rookery South Energy Recovery Facility' which uses solid waste from landfill sites and converts it into electricity (Covanta Holding Corporation 2019). Freight trains could stop at each station along the Marston Vale Line, before terminating at Bedford Retail Park Depot, to pick up waste to transport to the Energy Recovery Facility. The energy-from-waste plant can benefit from rail freight as, it provides much higher capacity to transport solid waste through connected freight carriers whereas, road freight requires separate freight carriers which are usually transported by individual tipper trucks. Therefore, this terminal also adds potential for future rail freight to serve the facility from further afield with improvements in rail freight.

6 Development of Scenarios to Improve the Utilisation of the Rail Line

After observing the results of the current passenger service patterns on the Marston Vale Line, as opposed to just observations, the utilisation level across the model is low. Therefore, creating different scenarios to improve the utilisation of the line can be achieved by adapting the current model, such as the use of labels to separate the passenger and freight entities, new freight depots for the freight to enter and exit the system, routing out, and changes to the timetable.

6.1 Labelling and Routing Out

For the scenarios that require freight to be entered into the system, it is important to assign labels to passenger and freight trains (see Fig. 8), because it allows to control routing out from an activity, for instance, a passenger label is assigned to continue travelling on the Marston Vale Line, while a freight label is assigned to branch off the line and travel towards a freight depot, ready to be unloaded. The label value is created at the start point for both passenger and freight, with a value of 1 given to passenger, and a value of 2 given to freight. Also, labels allow activities to input different timings for both passenger and freight trains. However, in this simulation, freight trains will follow the same timings as passenger trains, due to being the same

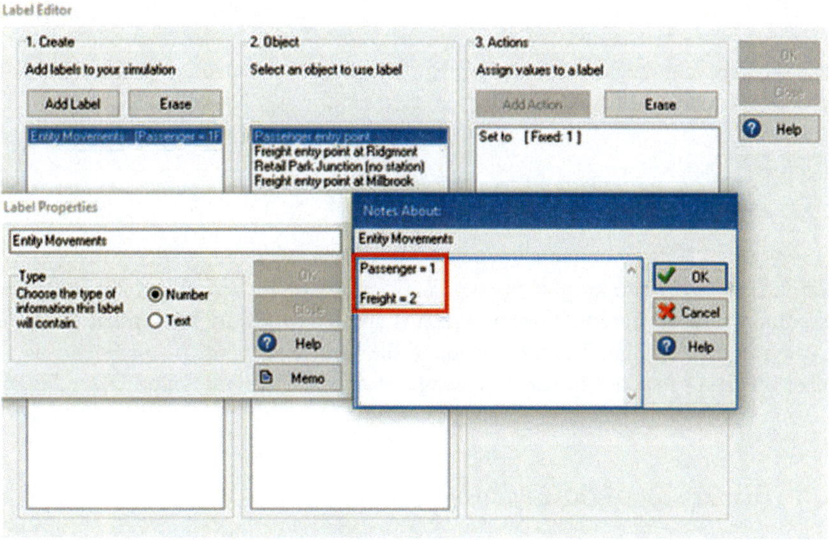

Fig. 8 Labels assigned to passenger and freight

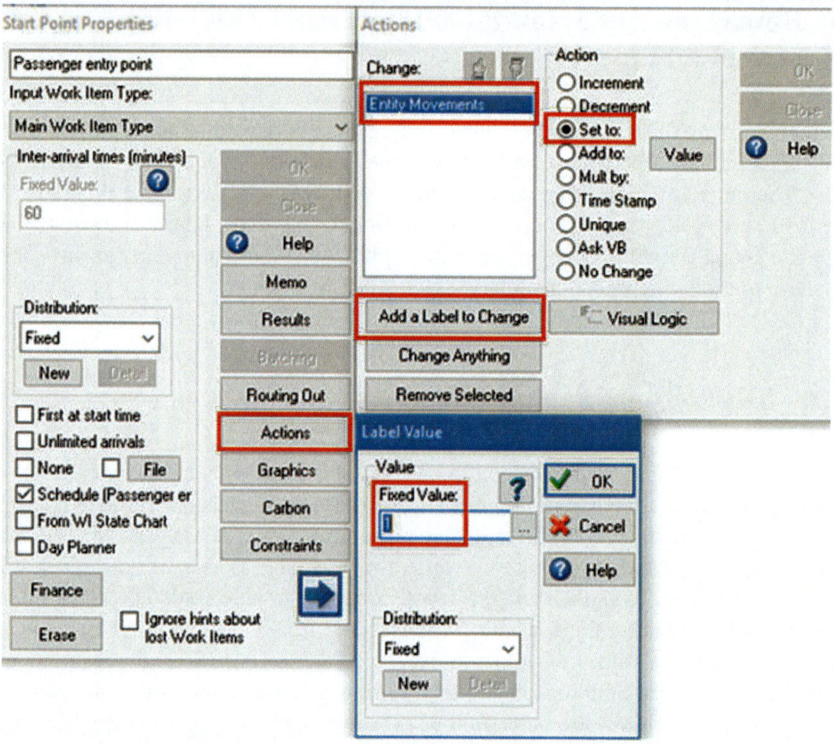

Fig. 9 Inputting actions data at the passenger start point

type of train, and the need for both to stop at all stations along the route, for either passengers or freight parcels/waste.

6.2 *Actions*

For labels to work properly, at the separate start points for passenger and freight, it essential to 'add a label to change' which is applied to 'Entity Movements', where it enters at a start point. Then, the value of the 'Entity Movements' is set to a fixed value which corresponds to either passenger (value 1) or freight (value 2), see Fig. 9.

6.3 *Discussion of the Proposed Scenarios*

The low utilisation presents an opportunity to create scenarios to look at ways to increase the utilisation levels from its current passenger service model, whilst keeping the journey times consistent. Hence, 5 scenarios (named as Scenario A, B, C, D and E) were proposed, in order to improve the utilisation levels. Each scenario encompasses a varying number of passenger and freight services, which are added as additional services within the current passenger model.

Firstly, in Scenario A, to improve the utilisation level, increasing the number of passenger service patterns can be achievable as, it does not require much infrastructure change to the line but, still relying on the single-track section. The only change from the current model is by altering the fixed timetable schedule for passenger service. This is done by increasing the passenger service from 1 to 2TPH, and these additional train times are determined by calculating the median average between two currently timetabled departure times, see Fig. 10.

However, in the next four scenarios, there will be an introduction of freight services into the model, which departs from two depots, at Millbrook and Ridgmont, and both terminate at the Bedford Retail Park Depot.

In Scenario B, retaining the same passenger service as present of 1TPH, freight services (1TPH from both Millbrook and Ridgmont) have been allocated into the model. From Fig. 11, at stations in Ridgmont and Millbrook, new freight depots are added as activities, as well as freight start points. These freight trains are transported towards Bedford Retail Park Depot, requiring the construction of a junction on the Marston Vale Line which acts as an activity named as 'Retail Park Junction (no station)', allowing freight trains to switch to a branch line towards the freight depot.

Fig. 10 Increased passenger timetable, showing the departure times from Bletchley

Sheet: Passenger entry into the simulation timetable

	A	B	Time Depart from BLY
1	Time	Items	
2	16	1	05:16
3	50	1	05:50
4	84	1	06:24
5	118	1	06:58
6	152	1	07:32
7	167	1	07:47
8	181	1	08:01
9	214	1	08:34
10	246	1	09:06
11	282	1	09:42
12	318	1	10:18
13	340	1	10:40
14	361	1	11:01
15	391	1	11:31
16	421	1	12:01
17	451	1	12:31
18	481	1	13:01

Times highlighted are added from the existing passenger service timetable, making the passenger service operate from 1TPH to 2TPH.

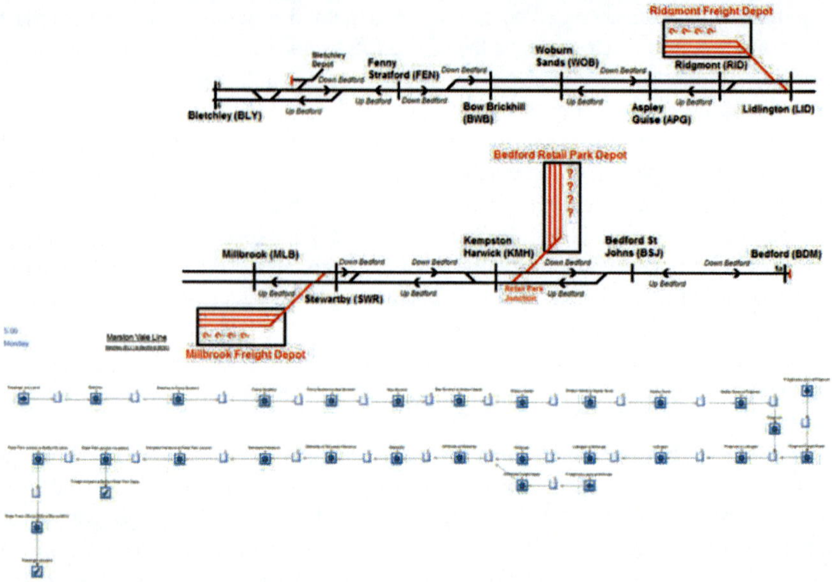

Fig. 11 Revised model (on top), featuring new freight depots into the system, with the proposed track layout (next). Question marks at the freight depots represent possible future expansion and represent the unknown actual capacity due to having infinite capacity in the simulation model

As mentioned before about the labels in Sect. 6.1, the timings at the junction are slightly different between passenger and freight, with timings for sectional running times obtained and estimated from The Chiltern Trainspotters (2019). An estimated additional 1 min is required for freight services to slow down when approaching the junction, to diverge onto the branch line towards Bedford Retail Park Depot. For passenger services, no additional time is required as the passenger train is ahead of the freight train and it carries on the line towards Bedford, with the junction having no effect on it, see Fig. 12. If the other direction was also being modelled, activities for the junctions to Millbrook and Ridgmont freight depots would also be included,

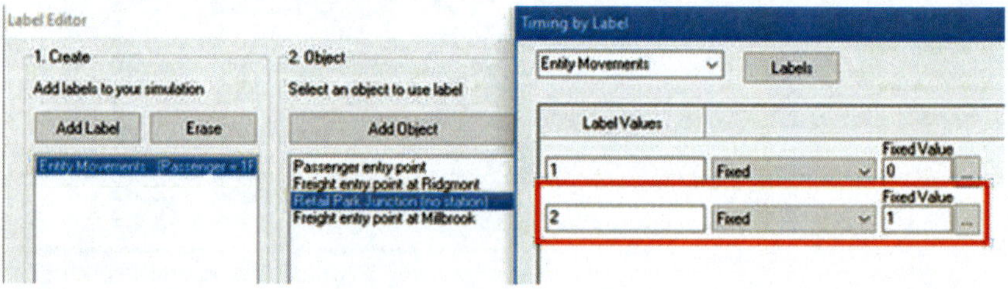

Fig. 12 Inputting times for each label, where freight (value 2) is set at a fixed value of 1, representing 1 min

as the freight trains would need to slow down to navigate the junction whereas the passenger services would continue, resulting in different sectional running times between passenger and freight trains.

The freight service timetable depends on the passenger's departure time at the same station where the depot is located at so, for example, Ridgmont freight service departure time uses the passenger's departure time at Ridgmont, and follows the passenger service with a headway maintained through block/queue and activity occupation.

Scenario C is similarly to Scenario B however, it focuses on having 2TPH for passenger service while, retaining the 1TPH for freight service. This means that this scenario prioritises passenger services as there is currently only 1TPH, with the freight services in this scenario less focused.

Scenario D aims to increase freight service to 2TPH (from each freight depot) while, retaining 1TPH for passenger service.

Lastly, Scenario E is the same as Scenario B except, both the passenger and freight services are increased from 1 to 2 TPH.

7 Results and Evaluation of the Scenarios

7.1 Journey Time

Firstly, the time spent in the simulation will be evaluated to consider the potential reliability of the additional services, before looking at actual utilisation levels. Freight train services' journey times will be less of a priority than passenger trains, as passenger trains have priority on the UK railway. The time factor is less crucial to within minutes for freight services, although dependability would still be required.

It seems that the journey times across the 5 scenarios are somewhat consistent, which is important for the passenger services to operate reliably. However, in some scenarios, namely in Scenarios A, C and E, the passenger service has a longer time spent in the simulation, which suggests there is some delays or blockages in the system, see Table 3 for the detailed results.

From the scenarios, it seems that the freight service does not impact the passenger service despite having to slow down at the junction to change towards Bedford Retail Park Depot. However, it suggests that the increase from 1 to 2 TPH for passenger services, combined with any freight services, has an impact on passenger services' time spent in the simulation.

When comparing Scenarios B and E, Scenario E has a higher standard deviation of passenger and freight service time spent in the system than Scenario B. This means that the more passenger and freight services on the line, there is an increase in chance of service delays hence, making the service less reliable.

In terms of its time spent in the simulation, including running freight services, Scenario D would be the most desirable because since the Marston Vale Line is not

Table 3 Results of scenarios, on passenger and freight services, time spent in the simulation

	Passenger (min)					Freight (min)				
	A	B	C	D	E	A	B	C	D	E
Minimum	62.50	62.50	62.50	62.50	62.50	–	13.00	13.00	13.00	13.00
Average	72.26	65.61	72.26	65.61	72.26	–	19.14	19.14	17.54	19.07
Maximum	90.50	72.50	90.50	72.50	90.50	–	24.50	24.50	24.50	24.50
Standard deviation	9.77	3.86	9.77	3.86	9.77	–	5.41	5.41	4.42	5.34

fully a double-track railway, the single-track sections limit the number of passenger trains, and increasing the passenger service is likely to cause delays (as seen in Scenarios A, C and E in Table 3), especially when there are service disruptions, such as signalling problem and train faults. Therefore, the single-track at Bedford is a bottleneck on the Marston Vale Line from significantly increasing capacity which is possible on most of the rest of the line which is double track. This also suggests that in the other direction, from Bletchley to Bedford, the single-track at Fenny Stratford may also be a bottleneck. However, for freight trains, increasing the freight service from 1 to 2 TPH (for each depot) is likely to not have an impact on the passenger service. This is because from the simulation model, freight trains terminate at Bedford Retail Park Depot. This means freight trains do not cross through the single-track sections as they branch off towards the depot, which allow passenger trains to travel through the single-track sections without potentially getting blocked by freight trains. Doubling the whole of the Marston Vale Line would therefore significantly increase capacity.

In terms of the freight services, as can be seen from Table 3, they are much more consistent than passenger service times, and with freight service times less of a priority, would work reliably and dependably in any of the scenarios with freight services.

7.2 *Utilisation*

The next perspective is the utilisation of the Marston Vale Line, see Table 4 for the detailed results.

Table 4 Utilisation levels, combining with both passenger and freight services, in each scenario of each activity point

	Utilisation levels (%)				
Activity point	A	B	C	D	E
Woburn Sands	0.31	0.17	0.31	0.17	0.31
Woburn Sands to Aspley Guise	1.57	0.83	1.57	0.83	1.57
Aspley Guise	0.31	0.17	0.31	0.17	0.31
Aspley Guise to Ridgmont	2.20	1.17	2.20	1.17	2.20
Lidlington	0.31	0.33	0.48	0.48	0.63
Lidlington to Millbrook	1.57	1.67	2.41	2.41	3.15
Stewartby	0.31	0.50	0.65	0.80	0.94
Stewartby to Kempston Hardwick	2.20	3.50	4.54	5.58	6.61
Retail Park Junction (no station)	–	0.67	0.67	1.26	1.26
Single Track (BSJ to BDM to BSJ to KMH)	17.56	9.37	17.56	9.37	17.56

From the evaluation in Sect. 7.1, Scenario D would be the most desirable for the shortest possible journey time, and reliability. However, looking at the utilisation in Scenario D compared to the other scenarios and current services, there is an increase in utilisation of the line when freight trains use the line from Ridgmont onwards, whereas before Ridgmont Freight Depot, the utilisation of the line is the same as it is currently at the moment, as the number of passenger trains in Scenario D is still the same. Scenario E, with both 2TPH for passenger and freight (from each depot), utilises the line the most, with the line further utilised before Ridgmont Freight Depot and after, with additional passenger, and freight and passenger services respectively, and would be the most desirable scenario from the perspective of utilisation and therefore sustainable transportation.

7.3 Utilisation Versus Journey Time

As identified in Sect. 2.2 of the Literature Review by Motraghi and Marinov (2012), and confirmed by the results of this study, increased utilisation can lead to increased journey times and congestion. Therefore, there needs to be a compromise between the increased utilisation of the line, journey time and reliability. To facilitate increased utilisation, changes in priorities for journey time would need to be considered, with a possible change in policy to implement increased utilisation and more sustainable transportation.

8 Conclusions

The study of the current utilisation levels on Marston Vale Line has found that the current utilisation level is low as, it is limited to passenger service of 1TPH (Network Rail 2020a, b, c), although it is consistent in terms of journey time. Through using SIMUL8, a simulation model of the rail line in study has been created. It is based around the December 2019 Timetable (Network Rail 2020a, b, c), ensuring validity in the model.

This gave an opportunity to implement freight services into the model, as well as increase the number of passenger services from 1 to 2TPH. Therefore, the simulation model developed could be modified by creating new scenarios to improve the line utilisation levels, combined with keeping the journey time consistent for both passenger and freight services.

Two passenger trains per hour utilise the Single Track (BSJ to BDM to BSJ to KMH) considerably more than one passenger train per hour. It was found that despite the increase in maximum journey time and likelihood of delays, Scenario D is the best scenario to increase the utilisation of the Marston Vale Line in an effective, reliable way. To increase passenger services which would be more reliable, the single-track section at Bedford would need to be doubled, as it is a bottleneck for more than one passenger train per hour, with consideration for the same work at Fenny Stratford for services running in the opposite direction.

However, from a theoretical point of view, just shifting more people and goods from road to rail by increasing the utilisation levels of a local railway line to maximise the passenger and freight services could bring other issues, such as by just using existing heavy diesel locomotives for freight services (one of the potential issues is sustainability in the environment). The concern is that increasing the amount of passenger and freight services mean that the railway's carbon footprint would be increased from additional passenger services, and especially freight services, where it is currently required to use heavy diesel locomotive to pull the freight train, irrespective of the number of wagons. Hence, alternative freight vehicles to transport the freight, for example, by converting older passenger trains to carry freight is a solution to reduce its carbon footprint and, as well as, potentially reducing rail infrastructure damages by not using heavy diesel locomotive to shift freight. Also, another concern is, due to mainly the limitations of the single-track sections, increasing the rail line utilisation at very high levels, combined with passenger and freight services, can cause delays. This would then lead to blockages in the system, which could impact the journey time, therefore suggesting a change in priority between journey time and utilisation is necessary to benefit the environment.

9 Future Work

The methodology applied to study the Marston Vale Line utilisation levels could be applied to other rail lines which are under-utilised, both in the UK and further afield. Increasing the utilisation of rail lines which are already electrified would enhance more sustainable transport solutions. This study looked at improving the utilisation levels of a non-electrified line.

Future work around the Marston Vale Line itself, with additional services, of both passenger and freight, could be considered. Removing the bottlenecks of the single-track sections along the Marston Vale Line in particular and remodelling would add potential for further utilisation of the line. As discussed in Sect. 5.2, additional rail freight from further afield could also be modelled, particularly with HS2's capacity benefits on the horizon, to further enhance the utilisation of the Marston Vale Line. As well as this, the East West Rail (2020) will enhance and change the line, with the section of the project between Bletchley and Bedford given approval (Railway Technology 2020), so would be an opportunity for further work in the future.

Also, another aspect to consider for future work, is increasing utilisation levels of the line through digital rail applications. As, Network Rail (2020a, b, c) indicates that the implementations of digital technologies on the railways could provide significant additional capacity. This is because instead of using conventional methods to increase capacity, i.e. through building new railway infrastructure, 'digital railway' can be applied onto conventional rail lines by utilising modern signalling and train control technology. Hence, the benefits of digital rail on Marston Vale Line could allow more passenger and freight services into the system as, the train control technology uses the European Train Control System (ETCS), comprised by the European Rail Traffic Management System. This system allows trains to operate narrower headway distances together and travelling at optimised speeds, while maintaining safe braking distances (European Union Agency for Railways 2020). Therefore, digital rail can help to improve the utilisation levels of the line, as well as safety, which would be an opportunity for a future work.

References

Briginshaw D (2019) Bi-mode trains ordered to develop parcels traffic. Available at https://www.railjournal.com/freight/bi-mode-parcels-trains-ordered/. Accessed 3 Oct 2020

Community Rail Network (2019) New trains arrive for the Marston Vale Line. Available at https://communityrail.org.uk/marston-vale-july19/. Accessed 29 July 2020

Courier Locations UK (2018) Amazon Logistics, Marston Gate Fulfilment Centre (LTN1) Ridgmont Bedfordshire MK43 0ZA United Kingdom. Available at https://courierlocations.uk/amazon-logistics-marston-gate-fulfilment-centre-ltn1-ridgmont-bedfordshire-mk43-0za-united-kingdom/. Accessed 3 Aug 2020

Covanta Holding Corporation (2019) Project overview. Available at https://www.rookerysouth.co.uk/about-the-facility/project-overview. Accessed 27 July 2020

Crompton G, Jupe R (2003) Such a silly scheme: the privatisation of britain's railways 1992–2002. Crit Perspect Account 14(6):617–645. https://doi.org/10.1016/S1045-2354(02)00187-9

East west Rail (2020) East West Rail will connect communities from Oxford through to Cambridge. Available at https://eastwestrail.co.uk/the-project. Accessed 4 Aug 2020

European Union Agency for Railways (2020) European Rail Traffic Management System (ERTMS). Available at https://www.era.europa.eu/activities/european-rail-traffic-management-system-ertms_en. Accessed 12 Oct 2020

Givoni M, Brand C, Watkiss P (2009) Are railways 'climate friendly'? Built Environ 35(1):70–86. https://doi.org/10.2148/benv.35.1.70

Google Maps (2020) Google Maps. Available at https://www.google.com/maps/. Accessed 30 July 2020

Haywood R (1999) Land development implications of the British rail freight renaissance. J Transp Geogr 7(4):263–275. https://doi.org/10.1016/S0966-6923(99)00006-X

Holden M (2020) Class 319 train used in GB Railfreight parcel test at London Euston. Available at https://www.railadvent.co.uk/2020/04/class-319-train-used-in-gb-railfreight-parcel-test-at-london-euston.html. Accessed 3 Oct 2020

Motraghi A, Marinov MV (2012) Analysis of urban freight by rail using event based simulation. Simul Model Pract Theory 25:73–89. https://doi.org/10.1016/j.simpat.2012.02.009

Network Rail (2020a) About us. Available at https://www.networkrail.co.uk/who-we-are/about-us/. Accessed 26 July 2020

Network Rail (2020b) Digital Railway. Available at https://www.networkrail.co.uk/running-the-railway/railway-upgrade-plan/digital-railway/. Accessed 12 Oct 2020

Network Rail (2020c) Electronic national rail timetable (Table 64). Available at https://www.networkrail.co.uk/running-the-railway/the-timetable/electronic-national-rail-timetable/. Accessed 8 May 2020

Potti P, Marinov M (2020) Evaluation of actual timetables and utilization levels of the west midlands metro using event-based simulations. Urban Rail Transit 6(1):28–41. https://doi.org/10.1007/s40864-019-00120-4

Potti P, Marinov M, Sweeney E (2019) A simulation study on the potential of moving urban freight by a cross-city railway line. Sustainability 11(21):1–19. https://doi.org/10.3390/su11216088

Rail Magazine (2019) From the archives: Class 230 train test. Available at https://www.railmagazine.com/news/rail-features/from-the-archives-class-230-train-test. Accessed 29 July 2020

RailTech.com (2020) HS2 will unlock capacity of three mainlines in the UK. Available at https://www.railtech.com/infrastructure/2020/07/21/hs2-will-unlock-capacity-of-three-mainlines-in-uk/. Accessed 27 July 2020

Railway Technology (2020) Second phase of East West Rail receives approval. Available at https://www.railway-technology.com/news/second-phase-east-west-rail/. Accessed 4 Aug 2020

Real Time Trains (2020) Woburn sands around 10:33. Available at https://www.realtimetrains.co.uk/search/detailed/WOB/2020-07-27/1033. Accessed 27 July 2020

Singhania VR, Marinov M (2017) An event-based simulation model for analysing the utilization levels of a railway line in urban area. Promet Traffic Trans 29(5):521–528. https://doi.org/10.7307/ptt.v29i5.2306

Gleave SD (2013) Capacity on North-South main lines. Available at https://assets.publishing.service.gov.uk/government/uploads/system/uploads/attachment_data/file/286617/capacity-on-north-south-main-lines.pdf. Accessed 29 July 2020

Stittle J (2004) Accounting for UK rail freight track charges: privatisation, politics and the pursuit of private sector vested interests. Acc Forum 28(4):403–425. https://doi.org/10.1016/j.accfor.2004.05.002

The Chiltern Trainspotters (2019) *Exclusive* | Class 230 Cabride | Bletchley—Bedford. 5 May. Available at https://www.youtube.com/watch?v=lxqHpBSNock. Accessed 29 July 2020

The Railway Technical Website (2019) Signalling. Available at http://www.railway-technical.com/signalling/. Accessed 3 Aug 2020

Train_PlaneHub (2015) London Midland Class 150 Bedford to Bletchley *Drivers Eye View*. 10 March. Available at https://www.youtube.com/watch?v=SFJd44OIhf8. Accessed 30 July 2020

Vivarail (2020) Vivarail Green credentials. Available at https://vivarail.co.uk/vivarail-green-credentials/. Accessed 30 July 2020

Woodburn AG (2003) A logistical perspective on the potential for modal shift of freight from road to rail in Great Britain. Int J Transp Manag 1(4):237–245. https://doi.org/10.1016/j.ijtm.2004.05.001

Analysis and Characterization of Noise Generated in Heavy-Haul Railways in Brazil

Thiago Antonio Fiorentin, Yesid Asaff, and Laura Dacoreggio Volpato Braz

Abstract The noise generated by transportation systems is one of the most important causes of noise-induced annoyance, since exposure to high level noises for long periods of time can be detrimental to health. Amongst the different types of noise that originate from railroads, the squeal noise generated at curves, the impact noise generated at rail joints and the rolling noise stand out. To propose mitigation measures to this problem, noise generation must be understood as well as by which parameters it is influenced. To that end, analytical models have been developed to predict the generated Sound Pressure Level (SPL) and to verify which factors are related to noise generation. This paper aims to validate analytical models for impact noise, squeal noise and rolling noise, which were originally designed for urban railways, used for passenger transportation. In this work, the validation is made using SPL measurements taken at a Brazilian heavy-haul railway, and a railway which carries bulk cargo in general. The models are implemented on a numerical computation software considering vehicle and track characteristics, train speed and a reference SPL. The results obtained from the models were compared with SPL measurements performed at the railroad during regular operation for the three cases (rolling, squeal and impact), which, despite the measurement limitations imposed by practical constraints during regular traffic operation, showed good agreement.

Keywords Railway noise · Noise measurement · Freight railway · Squeal noise · Impact noise

1 Introduction

Despite the convenience and advances provided by technological development, environmental degradation is accelerating as other sources of pollution arise. Amongst

---

T. A. Fiorentin (✉) · Y. Asaff · L. D. V. Braz
Joinville Technological Centre—Federal University of Santa Catarina, Joinville, Brazil
e-mail: Thiago.antonio@ufsc.br

Y. Asaff
e-mail: Yesid.a@ufsc.br

M. Marinov and J. Piip (eds.), *Sustainable Rail Transport 4*, Lecture Notes in Mobility,
https://doi.org/10.1007/978-3-030-82095-4_6

them, the environmental noise caused by sirens, automotive and domestic alarms, religious temples, horns, transportation systems and other sources, has become a cause of incidents in urban areas (Zannin and Bunn 2014; Fiorentin et al. 2013, 2016, 2017). As a result of urbanization, economic growth and motorized traffic, noise pollution is one of the main sources of complaints in urban areas since it can be detrimental to health, being associated to hearing damage, cardiovascular diseases, irritability, sleep disturbance, headaches, high blood pressure and nervous stress reactions (Stassen et al. 2008; Sánchez et al. 2017). Traffic noise, generated by transportation systems, is one of the most important causes of noise-induced annoyance (Kasess et al. 2013).

Brazil has a railway network, which is mainly used for freight transportation, that covers approximately 30,000 km. Although its density is low when compared to other countries with continental dimensions, railways are fundamental for Brazilian logistics, as they are widely used to move goods that will be exported through seaports. On their way to the ports, the trains run through several cities—there are approximately 12,000 level crossings in the country—crossing residential neighbourhoods and inconveniencing those who live nearby (Zannin and Bunn 2014; Confederação Nacional dos Transportes 2020).

A railway system has many noise sources, and the dominant one depends on a series of factors such as speed, loading, track and wheel conditions. The noise can come from the traction system, the brake system, the aerodynamic flow, the wheels rolling over the rails, among others. Freight railways in Brazil operate at speeds around 70 km/h. Some research indicates that the main noise source near such an operating speed are the wheels rolling over the rails. Traction-system noise is important for speeds up to 20 km/h and aerodynamic noise for speeds above 250 km/h (Failed 2008). Amongst the different types of noise originating from a railroad, the squeal noise produced in curves, the impact noise at rail joints and the rolling noise stand out.

Noise-control measures implemented at source are usually more effective than transmission-path or receiver treatments, as they offer greater noise reduction and better cost-effectiveness (Jiang et al. 2015). At-source noise control is also more complex than the alternatives since it requires identifying and understanding each noise source and its generation mechanisms. First, the dominant source must be identified and then the paths or contributions quantified. The next step looks at which parameters influence each source, which can be performed through a sensitivity analysis using analytical models, for example. Finally, possible solutions are developed and tested, first through theoretical models, then in laboratory and lastly through practical tests. Also, when it comes to railway noise, the solutions must not compromise operational safety (Thompson 2009).

The literature offers solutions for cars, aircraft, electrical systems and others (Fiorentin et al. 2013, 2016; Oliveira et al. 2008). There are, also, some solutions that can be applied to railways, but they are expensive and need to be customized for each railroad. Using models is, then, a necessary tool to verify such solutions before they are implemented. The objective of this work is to validate analytical models (Thompson 2009; Rudd 1976; Vér et al. 1976) used to predict noise levels from Brazilian freight railways using SPL measurements taken during the railroad regular operation. Three noise types are studied—squeal, impact and rolling noise. The models are implemented separately on a numerical computation software, and the results are compared against the experimental measurements of SPL.

Following this introduction, an explanation of each studied noise type is presented. Next, there is a brief description of the three analytical models used, as well as the methodology used to validate them. At last, results are presented and discussed.

2 Wheel/Rail Noise

The noise from railways arises primarily from the action of the steel wheels rolling on the steel rails, defined as wheel/rail noise. As train speeds increase, so does the importance of wheel/rail noise and, at typical freight railway speeds, it dominates environmental noise emission. Much research has been conducted to understand the mechanisms through which wheel/rail noise is generated (Dittrich and Janssens 2000; Kondo and Yamazaki 2013). The interaction between wheel and rail results in a force at their interface, which induces mechanical vibration that is transformed into sound radiation. The wheel/rail noise is the combination of the sound generated by the wheels and the rails.

The wheel/rail noise can be divided into three categories: squeal, impact and rolling. Squeal noise occurs when the wheels of the vehicle alternately stick and slip, as they run through short-radius curves. Impact noise is the noise generated by the wheels encountering discontinuities such as rail joints or flat wheels. Rolling noise is caused by rail roughness when in contact with the wheels (Failed 1975).

2.1 Squeal Noise

Freight railway systems have great noise-emission potential, since they are designed to meet safety requirements, rather than comfort. Also, they are subject to rougher operation and cruder maintenance procedures than passenger cars, factors that lead to noise and vibration generation (Stanworth 1983). Railway noise originates from many different sources, whose contribution to overall noise levels may vary under different situations (Hanson et al. 2014). One of the most important noises produced by railways is a high-pitched and tonal sound generated when a train traverses a curve, known as squeal noise.

The squeal nose is tonal and associated with wheel vibration, which is caused by a sliding mechanism, as a force on the wheel tends to make it slide and the static friction tends to “stick” it to the track. The bogie of freight trains is usually formed by wheelsets with rigid parallel axles. So, when a train negotiates a curve, the wheels do not align to the rail: those of the first wheelset tend to run away from the curve, while the ones on the trailing axle have a tendency to run into it.

Figure 1 shows the behaviour of a bogie and the longitudinal and lateral forces between the wheels and the rails at a curve. The sliding of the wheels over the rail induces dry friction forces that excite them and hence generate squeal noise (Rudd 1976; Heckl and Abrahams 2000).

Due to the influence of many parameters, squeal is described as a random event, whose occurrence is difficult to predict (Hanson et al. 2014). It is not necessarily constant while the train runs through the curve. It can have an intermittent character and the noise-level amplitude may vary (Thompson 2009). However, some authors associate it with tight curves, stating that its occurrence probability increases in curves with radii smaller than 500 m, that it is only sporadic for radii between 200

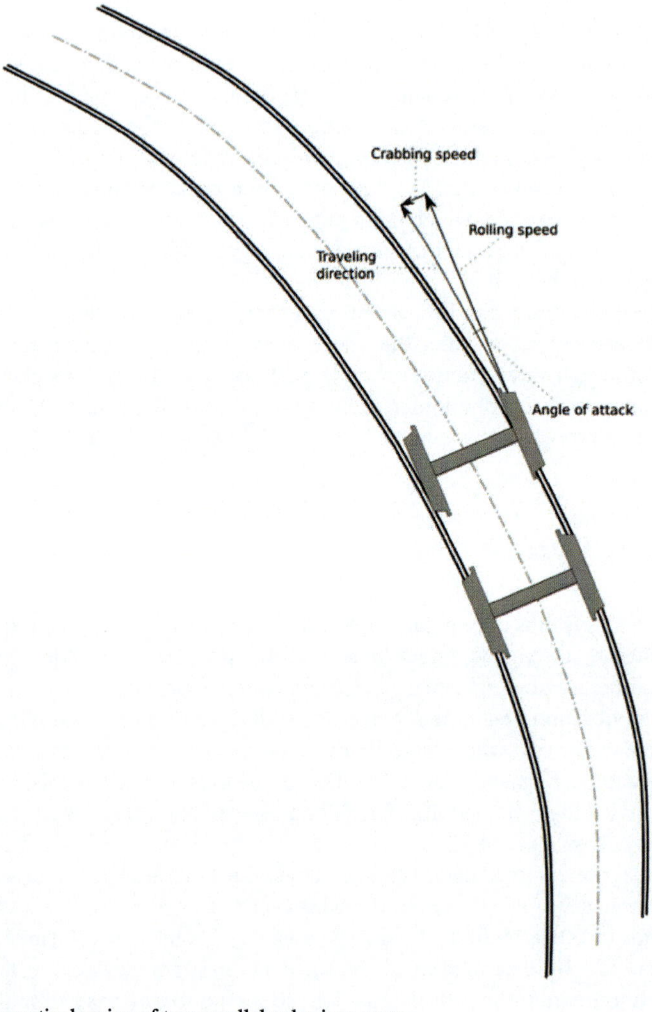

Fig. 1 Schematic drawing of two parallel axles in a curve

and 500 m, and that it becomes common in curves with radii smaller than 200 m (Eadie and Santoro 2006).

The squeal noise can exceed regular rolling noise in up to 30 dB, and it usually occurs in the frequency range in which human hearing is the most sensitive, thus really inconveniencing those who live near railroads (Hanson et al. 2014; Eadie and Santoro 2006). According to some researchers, the squeal noise has a fundamental frequency of approximately 4 kHz (Glocker et al. 2009). For others, there is a frequency for each type of squeal noise, lower than 5 kHz for wheel squeal and between 5 and 10 kHz for trim noise (Eadie and Santoro 2006).

Analytical models have been developed over the years to help understand and predict squeal noise. Therefore, they are useful to assist the development of noise control strategies. One of the first developed models to predict squeal noise was designed for subways and urban trains. It predicts squeal sound pressure level (SPL) as a function of train speed, curve radius and vehicle characteristics (Rudd 1976). Since then, more complex models have been elaborated, but they are usually not purely analytical and consist of many sub models (Heckl and Abrahams 2000; Schneider et al. 1988; Fingberg 1990; Beer et al. 2000; Pieringer 2013).

2.2 *Impact Noise*

According to some researchers, the impact noise occurs when a wheel and/or rail is excited by some discontinuity on the rolling surfaces (Failed 1975). Railway tracks are not perfectly smooth and flat. The most common wheel discontinuities are the flattening caused by the brake system during the contact between the wheel and the shoe pad. Typical rail discontinuities are the joints, characterized by the gap and the height difference between the rails (Wu and Thompson 2003).

Joining tracks with fishplates is one of the methods used in railroads to connect rails. Joints also absorb variations in track length caused by temperature changes. Using long welded rails is an alternative to eliminate rail joints (Silva Nabais 2015), however, it is more expensive and usually used at railways that operate at higher speeds.

When a train wheel rolls over a rail joint, great impact forces are generated by the relative displacement between them. The impulsive sound produced by such forces is known as impact noise, and is one of the main noise sources in railway operations (Wu and Thompson 2003). The impact noise generation mechanism at railway joints is shown in Fig. 2. Studies suggest that the main impact noise tones have frequencies of 250 Hz to 2 kHz, approximately, close to the frequency range in which human hearing is more sensible (Sueki et al. 2017).

The dynamic behaviour of the wheel and rail at the presence of discontinuities has been studied over the years, and equations have been elaborated relating train speed and sound pressure level (SPL) for different types of discontinuities (Vér et al. 1976; Failed 1975; Wu and Thompson 2002; Xiao et al. 2008; Zhen et al. 2018; Mazilu 2007; Yang and Thompson 2014). These models help understand the impact of each

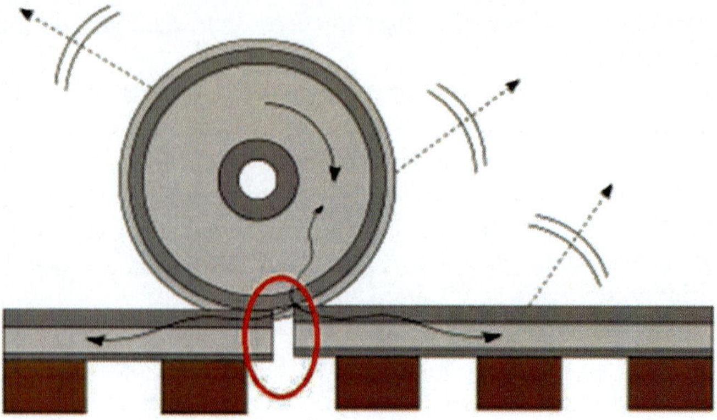

Fig. 2 Impact noise generation at a rail discontinuity

noise generation mechanism and predict its sound pressure levels. Therefore, they are also useful to assist the development of noise control strategies.

The first analytical model to identify the factors that influence impact noise was created based on wheel and rail geometry, kinematics and dynamics. The concept of critical speed is established as the train speed above which the wheel and rail lose contact when crossing a joint, for example. It also establishes equations to estimate the impulse and peak SPL for different discontinuities (Vér et al. 1976). It was later improved by other researchers who estimated a roughness spectrum equivalent for railway joints, making it possible to determine the noise generation potential at a discontinuity (Failed 1975).

This approach was also used in another study, but the equivalent roughness spectrum was used as input data to predict railway noise, and a piece of software was developed (Wu and Thompson 2003). Later, a hybrid model was created, consisting of a vehicle/track dynamic interaction model working in the time domain coupled with an FE-BE vibroacoustic model for the wheel working in the frequency domain (Xiao et al. 2008). To simulate the vibration and noise generated by insulated joins, some authors also employed the Finite Element Method at high frequencies (Zhen et al. 2018).

2.3 Rolling Noise

Considering a railway operation speed of 70 km/h, the main noise source is the rolling noise, caused by the vibration of the wheels and rails on their contact area due to surface irregularities, as represented in Fig. 3.

Surface irregularities are commonly referred to as "roughness" and are defined as non-uniform or non-periodic undulations that form on the surface of the rail or the tread of the wheel. Although micro-roughness is essential for adhesion, larger

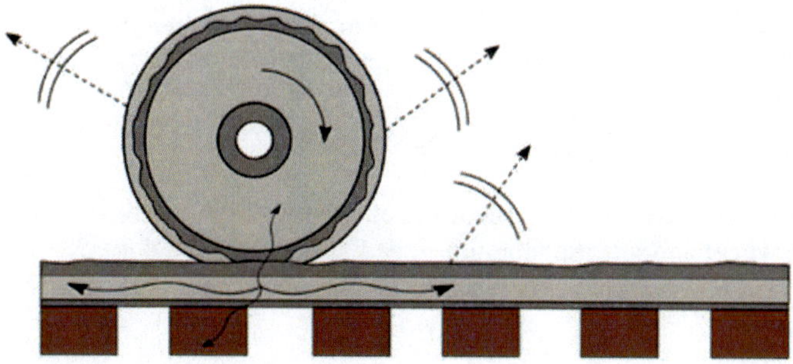

Fig. 3 Rolling noise generation mechanism

wavelength irregularities are undesirable because of the noise. The relevant roughness for rolling noise generation on both wheels and rails has a wavelength in the range of 5–200 mm, with a peak-to-peak amplitude of 0.3–120 μm (Failed 2004).

The frequency of vibration brought about by roughness depends on the train speed, following an $f = V/\lambda$ ratio. Thus, there is a frequency range for each speed range and roughness wavelength. Therefore, setting a specific frequency for rolling noise is difficult. Some authors claim that the wheel is the dominant source of rolling noise at frequencies from 200 to 400 Hz because, according to them, the irradiation efficiency of the rail is low within that frequency range. The rail becomes the dominant source between 400 and 1000 Hz because the wheel impedance is much higher than that of the rail and, within that frequency range, the irradiation efficiency is similar for the wheel and the rail. Above 1000 Hz, the dominant source depends on the distance between the receiver and the railway (Failed 1975).

3 Mathematical Models

Models for wheel/rail interaction can be formulated in the frequency or time domain. Frequency domain models use frequency–response functions to represent the dynamic behaviour of the vehicle, track, and their contact area. Time domain models are used to solve the system of differential and algebraic equations that describe the vehicle, the track and the contact area in a time scheduling procedure. Specific models are commonly found for each type of railway noise.

3.1 Model for Curve Squeal Noise

There are some analytical models for squeal noise generated in curves, which is caused by a sliding mechanism. There is a limited range of conditions under which vibration is unstable and, therefore, noise can occur. According to the literature (Rudd

1976), to calculate the sound pressure level of the squeal generated from a single wheel 50 ft away, the expression below can be used:

$$\mathrm{SPL}_{\mathrm{squeal}} = 10\log\log\left(\sigma_W A v^2\right) + 114 \tag{1}$$

In these expressions, σ_W is the acoustic efficiency of the wheel, A is the radiation area of the wheel and v is the vibration speed, which is defined as:

$$v = \varepsilon' V \tag{2}$$

where V is rail speed in m/s and ε' is the creep parameter, defined as:

$$\varepsilon' = 1.15\left(\underline{\varepsilon} - \varepsilon_0\right)^{(1/2)} \tag{3}$$

The creep on rounding curve $\underline{\varepsilon}$ is approximately $0.7(L/R)$, where L is the wheelbase of the truck, and R is the radius of the curve. The creep for maximum friction ε_0 is defined as 0.7% for common situations. Replacing the variables on Eq. 8:

$$\varepsilon' = 1.15\left[\left(0.7\left(\frac{L}{R}\right) - \left(\frac{0.7}{100}\right)\right)V\right]^{1/2} \tag{4}$$

Rearranging Eq. 1, 2, 3 and 4, the squeal noise can be predicted by the expression:

$$\mathrm{SPL}_{\mathrm{squeal}} = 10\log(\sigma_W A) + 10\log\left(1.3225\left(0.7\frac{L}{R} - 0.007\right)V^2\right) + 114 \tag{5}$$

The maximum curve radius on which wheel squeal will occur has also been found to be about 100 times the length of the truck's wheelbase.

The equation to predict the squeal-generated sound pressure level can also be expressed in another way:

$$\mathrm{SPL}_{\mathrm{squeal}} = 10\log\left(\sigma_W \mathrm{AV}^2\right) + 10\log\left[\left(\frac{L}{R} - \frac{1}{100}\right)/\left(\frac{3}{100} - \frac{L}{R}\right)\right] + 93 \tag{6}$$

According to Eq. 6, the squeal noise is governed by the acoustic efficiency and the radiation area of the wheel (σ_W and A, respectively), the train speed (V), the bogie's wheelbase (L), and the curve radius (R). In Eq. 6 it is possible to separate the speed term and the geometric terms. In Eq. 5 they are together. The SPL of the squeal tends to increase with train speed and in curves with small radius.

3.2 Analytical Model for Impact Noise

The research developed by Vér et al. (1976) is focused on estimating impact noise levels caused by discontinuities in wheels and rails (Vér et al. 1976). Its approach to the impact dynamics depends on the rail mounting, whether it is rigid or resilient, which is the case for Brazilian railways. It also influences the critical speed calculation, the speed above which the wheel separates from the rail when rolling over a joint. This model establishes different equations to estimate the critical speed, depending on the joint's geometry. Furthermore, it shows how the peak SPL of the impact relates to the train speed. Finally, equations are elaborated to relate the speed of the compositions and the sound pressure level for three types of rail joints.

Three joint geometries are studied in this paper: step-up, step-down and level joints. Step joints are characterized by a gap and a height difference between rails, while level joints are vertically aligned but still have a horizontal gap. The behaviour of a step joint as a step-up or step-down joint depends on the train's moving direction.

Step-up joints generate the highest peak SPL of all geometries, since it grows as the speed increases according to the expression $20*\log_{10}V$ and does not reach a maximum value when the train goes over the critical speed. For step-down joints, the peak SPL increases likewise until the train reaches the critical speed, above which the peak SPL stabilizes. Finally, peak SPL also grows with train speed for level joints, but its behaviour above critical speed is unknown (Vér et al. 1976).

The critical speed (V_{CE}) for step-down and level joints depends on the critical train speed for wheel/rail separation in rigid railways (V_{CR}). This parameter is a function of the wheel radius (a), the gravitational acceleration (g), the spring supported mass of car per wheel (M), and the wheel mass (m). This calculation is already considered in Eq. 7. Therefore, the critical speed of wheel/rail separation on resilient railways is also governed by the rail mass per length unit (ρ) and a parameter β, described in Eq. 8. This parameter depends on the foundation stiffness per length unit of rail (K), also known as track modulus, the Young's modulus of the rail material (E) and the moment of inertia of the rail's cross section (I).

$$V_{CE} = \left[\mathrm{ag}\left(1 + \frac{M}{m}\right)\right]^{\frac{1}{2}} x \left[1 + \frac{m}{\rho}\frac{\beta}{2}\right]^{\frac{1}{2}} \tag{7}$$

$$\beta = \left(\frac{K}{4\mathrm{EI}}\right)^{\frac{1}{4}} \tag{8}$$

Although the relation between an increasing speed and the peak SPL is described, the equation to predict the peak impact noise SPL is not specified by the model. To that end, an expression was chosen from another study that relates SPL and train speed, based on reference data for those variables (Thompson 2009). Equation 9 shows a relation between speed and SPL, defined by a factor N, which will be considered $N = 20$, since the analytical model estimates that the slope of peak SPL vs. train speed curve is $20*\log_{10}V$. In the expression below, V_0 and SPL_0 are the reference data from measurements performed at the railway, and V is the train speed range.

$$\mathrm{SPL_{peak}} = \mathrm{SPL_0} + N \log_{10} \frac{V}{V_0} \tag{9}$$

3.3 *Analytical Model for Rolling Noise*

An equation to predict rolling noise was developed by Thompson (2009). The expression, similar to that used to predict impact noise, is given by Eq. (10), and establishes that the SPL of rolling noise ($\mathrm{SPL_{roll}}$) is proportional to the logarithm of the train speed (V).

$$\mathrm{SPL_{roll}} = \mathrm{SPL_{ref}} + N \log_{10} \frac{V}{V_{ref}} \tag{10}$$

The terms $\mathrm{SPL_{ref}}$ and V_{ref} refer to speed values and corresponding SPL that should be used as a reference for the calculations. According to Thompson (2009), the value of factor N is determined from measurements based on linear regression and usually varies from 25 to 35, with a peak value of 30.

4 Experimental Analysis

Many countries adopt specific standards to measure rail noise. For example, the standard ISO 3095:2013 (International Organization for Standardization 2005) specifies measurement methods and conditions to obtain reproducible and comparable exterior noise emission levels and spectra for all kinds of vehicles operating on rails or other types of fixed track.

In Brazil, however, there are no specific rules to quantify the noise generated by railway systems. Hence, the analyses presented on this work are based on standard NBR 10,151, which deals with noise assessment in inhabited areas (Associação Brasileira and de Normas Técnicas 2000). According to the standard, the measurements should be taken at points approximately 1.2 m above the floor and at least 2 m from any other reflecting surfaces, such as walls, constructions, etc. Measurements should not be taken when there is audible interference from natural phenomena (thunder, heavy rain, etc.). The length time of measurement must be chosen to allow characterization of the noise in question. The measurement may involve a single sample or a sequence of them.

In this section, the results of the sound pressure level measured on two railways in Brazil are presented. The first railway is used for iron ore transportation and the second carries bulk cargo in general.

Table 1 Values of parameters used in the analytical model

Wheel radiation efficiency	1.00
Wheel radius	0.5080 m
Truck wheelbase	2.4380 m
Range of curve radius	73.14–243.80 m

4.1 Case 1—Squeal Noise

4.1.1 Values of Parameters Used in the Analytical Model

To make the predictions, Eq. 6 is implemented on a numerical calculation software. The squeal SPL was predicted considering each train speed recorded during in-field measurements, and curve radii ranging from 30 to 100 times the truck wheelbase. The values considered as input data on the squeal SPL estimate are shown in Table 1. The wheel radius and the truck wheelbase values are typical of a train locomotive, since it was the vehicle from which the squeal originated on most in-field measurements. As mentioned before, the wheel radiation efficiency is given by the author of the analytical model (Rudd 1976).

Since Eq. 1 is valid for curve radii between 30 and 100 times the truck wheelbase ($30L < R < 100L$), the value of the curve radius (R) was limited, in the calculations, to a range between 73.14 and 243.80 m.

4.1.2 In-Field Measurements

The values obtained by the model are compared with the ones recorded during in-field measurements. These measurements were performed at different curves of a Brazilian freight railroad, during regular operation. A sound-level meter was placed at the centre of each curve, around 15 m away, and 1.20 m above the ground.

The measurements were taken at four different curves that are presented on Fig. 4, with radii of 156, 200, 206 and 230 m, respectively. The red dots indicate the position of the sound-level meter, the black dashed lines indicate the centre of the railway and the yellow arrows indicate the radius of the curve. Two measurements were taken of trains passing by in different directions at each curve. Data was recorded in the time domain, using A-weighting. The squeal was heard throughout the measurements and the noise peak was later identified during data analysis.

4.1.3 Results and Discussion—Squeal Noise

As stated in the previous section, during each measurement, the noise was recorded and the noise peak corresponding to squeal was later identified during data analysis. The predictions were made considering each train's characteristics and speed. In Fig. 5, continuous lines show the squeal values predicted by the model at three

Fig. 4 Curve radius

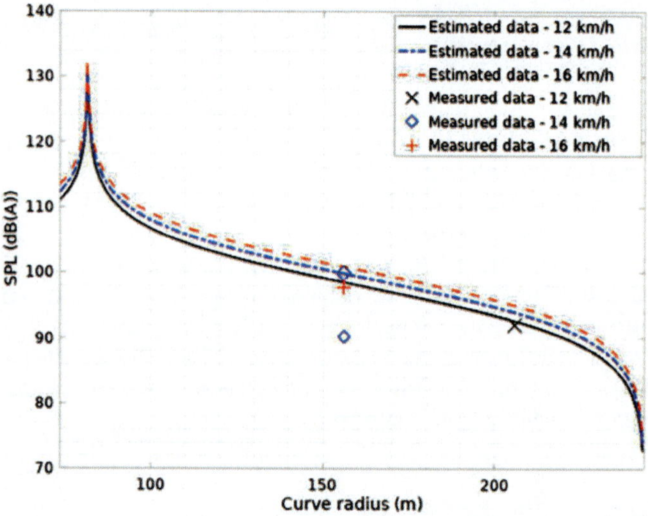

Fig. 5 Estimated and measured data of SPL verses curve radius for three different train speeds (12, 14 and 16 km/h)

Table 2 Results of estimated and predicted data

Train speed (km/h)	Curve radius (m)	Estimated data dB (A)	Measured data dB (A)
12	206	92.59	92.00
14	156	99.81	99.90
16	156	100.97	97.80

different speeds, 12, 14 and 16 km/h, and the corresponding data recorded at each in-field measurement.

Squeal was generated only in Curves 1 and 3. In Curves 2 and 4, with a 200 m and a 230 m radius, respectively, no such noise was heard. The absence of squeal noise during some of the measurements can be explained by different track and vehicle conditions, or even environmental conditions, such as temperature and wind. This result agrees with the description of squeal noise as a random phenomenon due to the interplay of many variables in its generation mechanism (Hanson et al. 2014).

To help understand the results, they are also presented in Table 2, along with train speed, curve radius and sound pressure level.

The values presented in the table show small differences between estimated and predicted data. Even with a small amount of data, the results suggest that this analytical model can be a useful tool to predict and understand the squeal SPL variation as a function of speed and curve radius.

However, more measurements should be taken at curves with radii smaller than 150 m to determine if the model is accurately adequate within the full range of curve radii for which it is valid. Also, noise measurements with others speeds must be performed. This was not performed on this study since the measurements were taken at a real freight railroad, which does not usually have many tight curves. The only curves with radii smaller than 150 m were in hard-to-access locations. It is important verify the relation between a significative velocity variation and the difference between sound pressure level measured and predicted, in others words, the relative error.

4.2 Case 2—Impact Noise

The tests were carried out on a metric gauge railway used exclusively for cargo transportation located in southern Brazil. The rail is of the TR-37 type, fixed on wooden sleepers seated on the ballast, with a mostly exclusive HFD, HFE and FHD freight wagons circulation. The results estimated by the analytical model were obtained considering a speed range from 2 to 30 m/s.

Table 3 Values of parameters used in the analytical model

Wheel radius (a)	0.3625 m
Gravitational acceleration (g)	9.81 m/s^2
wagon mass	73 kg
Spring supported mass of car per wheel (M)	9,137.5 kg
Wheel mass (m)	308 kg
Rail mass per length unit (ρ)	37.10 kg/m
Foundation stiffness per length unit of rail (K)	10 MPa
Young's modulus of rail material (E)	2.10 MPa
Moment of inertia of the rail cross section (I)	951.10×10^{-8} m^4

4.2.1 Values of Parameters Used in the Analytical Model

Table 3 presents all the values considered on the critical speed estimate. The wheel mass and radius values are typical of a multiple wear freight car wheel CD-29, after its first reprofiling. The spring-supported mass of car per wheel (*M*) was obtained by dividing the mean mass of the wagons, 73.1 t, by the number of wheels, which resulted in 9125 kg. Rail-characteristic parameters were tabulated values according to the rail type, in this case, an ASCE 75. The value considered for the track modulus was typical of a permanent railway in bad conditions (Failed 2015).

Those values, along with speed and SPL reference data, were used as input for the analytical model, which was implemented on a numerical computation software. The implemented code analysed the joint geometries separately—step joints, up or down, and level joints, with or without a horizontal gap. Finally, it returned the critical speed for a resilient railway, as well as the peak SPL variation according to train speed for each joint.

4.2.2 In-Field Measurements

The analytical model estimate was compared to the SPL measurements. In-field measurements were taken during railway regular traffic. A sound-level meter was placed 1.8 m away from the joint, at the same height as the top of the rail, recording the SPL during the passage of the trains.

For step joints, one specific joint was selected for this study and the data of three trains passing by, two in one direction, one in the other were collected. The gap between the rails was 23.6 mm long with a 5 mm height difference. For level joints, two were chosen, one with an 11 mm gap and the other one an ideal joint, without any misalignment. In both cases, only one train passed by each joint, so only one measurement was taken for each one. Figure 6 shows the geometry of the three joints studied on this work.

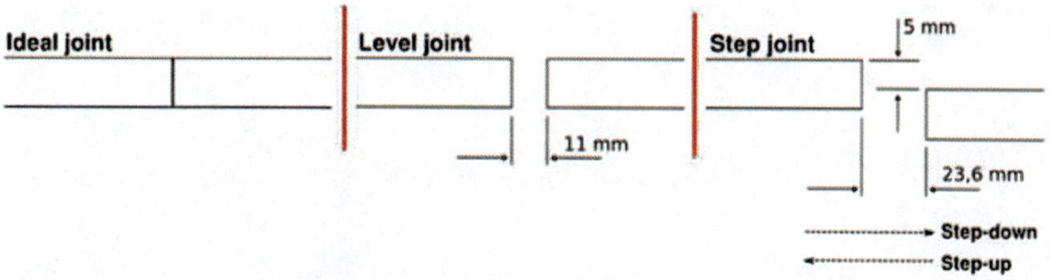

Fig. 6 Geometry of the studied joints

It is important to elucidate that the expression "ideal joint" is, in this work, in terms of impact noise emission, once smaller gaps between rails are expected to generate lower noise levels. The ideal gap, in terms of permanent railway construction, depends on the weather at the railway site, operational speed, among other factors.

The studied joints are presented in Fig. 7 and 8. A side view of the step joint is presented on Fig. 7. Figure 8 shows a top view of the level joint, with an 11 mm gap between the vertically aligned rails, and a side view of the ideal joint.

Performing the measurements at a real railroad during regular operation may impose some limitations to the study. Achieving significant speed variations along a specific section of the rail line can be difficult, as there are speed limits imposed for safety reasons. That is why gathering a large amount of data for each studied joint, to verify whether the analytical model would correctly predict the SPL at higher speeds, is impossible.

Fig. 7 Side view of the step joint

Fig. 8 Top view and side view of the level joint with gap between rails

4.2.3 Results and Discussion—Impact Noise

In this section, the results obtained for each case—step and level joint—will be presented and discussed separately.

(a) **Step joint**

Figure 9 shows the SPL results obtained for a specific joint. The SPL generated

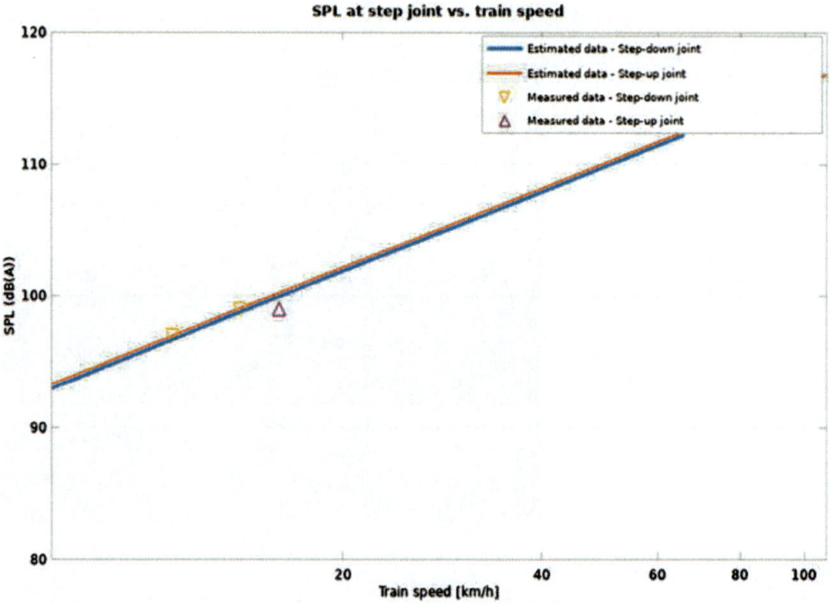

Fig. 9 Estimated and measured data of SPL verses train speed for a step joint

from the impact on a joint was recorded during the passage of three trains traveling at speeds of 11, 14 and 16 km/h.

The direction of the first two compositions gives the joint a step-down configuration, while the last composition, moving in the opposite direction, does otherwise (a step-up configuration). In the image, the SPL estimated by the analytical model is observed considering step-up and step-down joints with the same geometry (red and blue overlapping lines). The values measured in the field are marked with orange triangles for step-down joints and a purple one for the step-up joint.

Since one pair of speed and SPL data are used as reference, the estimated and recorded values, in this case, are the same. For the two other trains passing by the joint on its step-down direction, the relative difference between estimated and measured data reaches 1.20%. It is a small difference that may have been caused by differences in the structure of the wagons and the permanent way, factors that influence noise generation. Even with a small amount of data, the SPL variation according to train speed seems to be well predicted by the model. The analytical model validation made by the authors had a similar result (Failed 1975).

(b) **Level Joint**

The critical speed for a level joint is calculated the same way as for a step-down joint. Therefore, for the analysed level joints, the critical speed is 334 km/h. The previously stated information about the operational speed in Brazilian railways also applies in this case.

Figure 10 shows the level joints values of SPL vs. train speed estimated by the model. The estimate is based on speed and SPL reference data of an ideal joint—an 11 km/h speed and a corresponding SPL of 86.00 dB (A). The continuous line indicates the estimated data. The data collected on the railway are marked with a blue square for the ideal joint and a red cross for the non-ideal one.

In this case, the data collected from the ideal joint was used as the reference for the analytical model, which is why the recorded data matches the estimated data line.

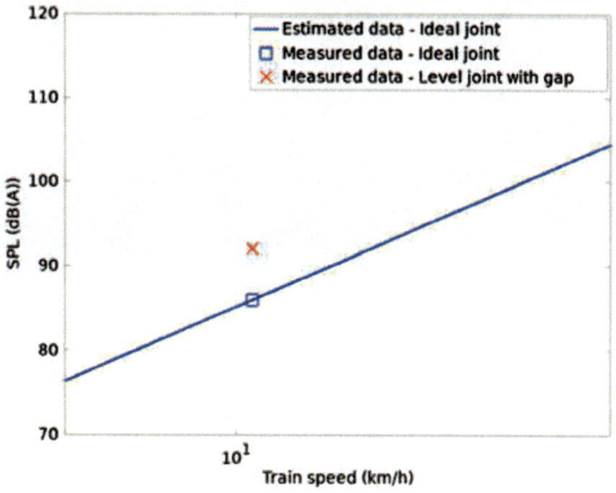

Fig. 10 Estimated and measured data of SPL verses train speed for level joints

Here, no speed variations were registered. Instead, the influence of the gap between rails on impact noise generation was investigated.

The SPL registered at the ideal joint was 86.00 dB (A) for a train passing by at 11 km/h. At the non-ideal level joint, with an 11 mm horizontal gap, the SPL recorded was 92.00 dB (A), for the same train speed. Studies show that the impact noise generated at joints is 5 to 8 dB greater than regular rolling noise (Sueki et al. 2017). So, the results obtained in this case well agree with the ones found in the literature.

4.3 Case 3—Rolling Noise

4.3.1 Typical Rolling Noise

The measurements were carried out on a single metric gauge railway located in southern Brazil, used for bulk cargo transportation in general. The composition consisted of 75 type HFD and FHD empty wagons. When empty, each wagon weighs about 21.0 tons.

The analysis was based on standard NBR 10,151, the sound pressure level (SPL) was measured 1.2 m above the ground, 10 m away from the centre of the track. A 22,270 B&K sound-level meter was used. Figure 11 presents the results of the SPL measured during the passage of an empty wagon and the background noise. The

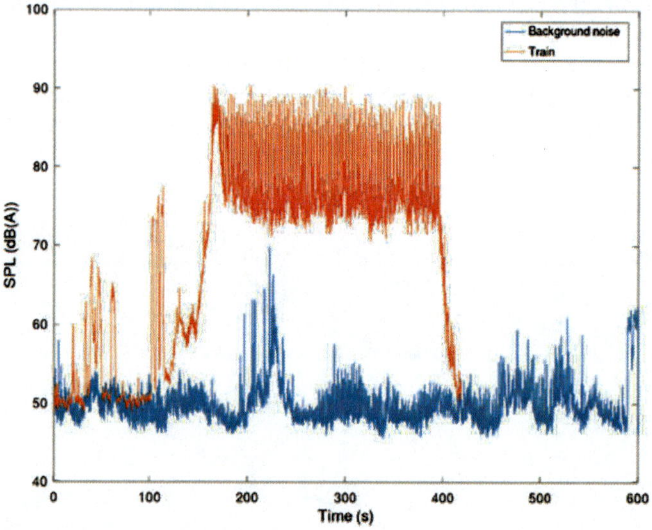

Fig. 11 Sound pressure level at time domain

Table 4 Speed and sound pressure level measured during the passage of the test composition

Velocity [km/h]	SPL [dB (A)]	Velocity [km/h]	SPL [dB (A)]
10	69	30	83
10	70	32	84
15	72	35	85
15	73	43	86
20	76	45	88
20	77	52	90
23	80	55	91
25	80	60	92
30	82	60	93

measurements yield overall values of 80.5 dB (A) for the empty composition and 48.4 dB (A) for the background noise.

Measurements presented in Fig. 11 are on the time domain, clearly showing when the noise increased due to railway traffic. The background noise is around thirty dB quieter than the train noise.

4.3.2 In-Field Measurements

Speed and SPL values were recorded during the passage of a test train at different speeds through the same stretch of the railroad, and the values are presented in Table 4.

The lowest speed and measured SPL were 10 km/h and 69 dB (A), then used as reference values in Eq. (10). This expression was implemented in a computer routine, considering three values for the factor N—25, 30 and 35. Figure 12 presents the predicted values and those obtained experimentally.

The black continuous line indicates the values predicted by Eq. (10) considering N equal to 25, while the blue dashed line and the red dotted line indicate the predicted values with N equal to 30 and 35, respectively. The values registered on the railroad are indicated by the " + " symbol.

4.3.3 Results and Discussion—Rolling Noise

According to Thompson (2009), every time the speed doubles, a rolling noise SPL increase of 8 to 10 dB (A) is expected, which is corroborated by the values presented in Table 4. With the train travelling at 15 km/h, for example, the rolling noise reaches 73 dB (A), while at 30 km/h, the noise level reaches 83 dB (A), an increase of 10 dB (A).

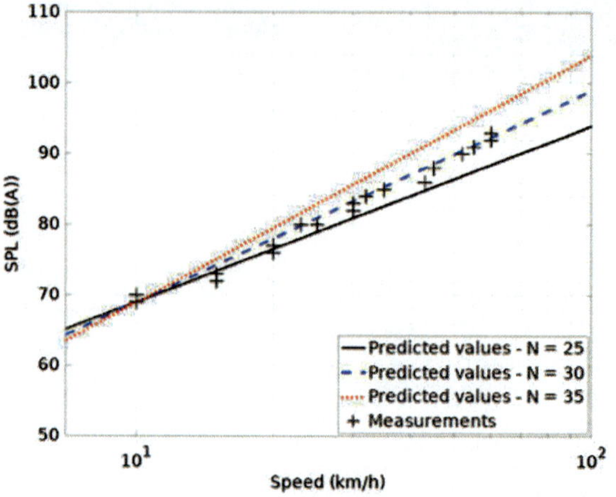

Fig. 12 Sound pressure level versus speed

The data recorded in the field agree with the estimated values, showing the expected behaviour due to the increase in the composition speed. The factor $N = 30$ is the most suitable to predict rolling noise, as suggested by Thompson (2009), especially for speeds above 20 km/h.

5 Conclusions

The present research discusses the sources of noise in a railway, introduces models to predict some types of noise and evaluates noise emission in Brazilian railways. The main goal was to validate analytical models for squeal, impact and rolling noise predictions. Analytical models available in the literature allow predicting these types of noise. Each one was implemented on numerical computation software, considering the required parameters in each case. The results obtained from the models were compared with experimental measurements of SPL taken at the railway during regular traffic. Generally, the calculated results showed good agreement.

For squeal noise, the used variables are typical of a railway and are available in the design phase, which enables to predict the noise generated by the railway during design steps. The model may be a useful tool to predict and understand the relationship between train speed, curve radius and squeal noise.

In Case 2, for impact noise, the equations considered vehicle and track characteristics, train speed and a reference data of speed and corresponding SPL. The SPL was measured at different joints during regular railway operation and compared with the predictions made by the model, showing good agreement.

The main limitations found during the validation of squeal and impact noise prediction models were related to the fact that experimental measurements were taken at the railroad, during its regular operation. Hence, it was not possible to achieve greater speed and curve radius variations, or even gather a large amount of data, since the region where these measurements took place has low traffic density. Also, some tests were cancelled due to train delays, cancelations or even accidents on the route.

In the case of rolling noise, a simpler analytical model was used. However, the measurements taken for the model experimental validation used a test composition, which simplified the collection of a larger volume of data across the desired speed range. The obtained results converge with the literature, and the model used can be considered a useful tool to predict rolling noise in freight trains.

Future work may aim to study the accuracy of the squeal and impact noise prediction model over a wider speed range. The use of a test track is suggested since it allows greater variation of the parameters studied.

References

ABNT (2000) NBR 10151: Acústica – Avaliação do ruído em áreas habitadas, visando o conforto da comunidade – Procedimento. Rio de Janeiro

Confederação Nacional dos Transportes (2020) Boletim unificado—Fevereiro 2020. [S.l.]

De Beer F, Janssens M, Koojiman PP, Van Vliet WJ (2000) Curve squeal of railbound vehicles (part 1): frequency domain calculation model. Proceedings of Internoise 2000, Nice, 27–30 Aug 2000

de Oliveira A, Fiorentin T, da Silva M (2008) Design of system and components—NVH view—test and simulation. SAE Technical Paper 2008-36-02139. https://doi.org/10.4271/2008-36-0239

da Silva Nabais RJ (2015) Manual básico de engenharia ferroviária. (Oficina de Textos, 2015). ISBN: 9788579752087

Dittrich MG, Janssens MHA (2000) Improved measurement methods for railway rolling noise. J Sound Vib 231(3):595–609

Eadie DT, Santoro M (2006) Top-of-rail friction control for curve noise mitigation and corrugation rate reduction. J Sound Vib 293(3–5):747–757. https://doi.org/10.1016/j.jsv.2005.12.007

Fingberg U (1990) A model of wheel-rail squealing noise. J Sound Vib 143(3):365–377. https://doi.org/10.1016/0022-460X(90)90729-J

Fiorentin TA, Ferguson NS, Renno JM, Lenzi A (2013) Structural response of an aircraft fuselage to hydraulic system—a wave and mobility approach. Noise Control Eng J 61:87–99

Fiorentin TA, Lopes LF, Silva Junior OM, Lenzi A (2016) Vibroacoustic models of air-core reactors. Int J Acoust Vib 21:453–461

Fiorentin TA, Mikowski A, Silva Junior OM, Lenzi A (2017) Noise and vibration analysis of a heat exchanger: a case study. Int J Acoust Vib 22:270–275

Glocker C, Cataldi-Spinola E, Leine RI (2009) Curve squealing of trains: measurement, modelling and simulation. J Sound Vib 324(1–2):365–386. https://doi.org/10.1016/j.jsv.2009.01.048

Hanson D, Jiang J, Dowdell B, Dwight R (2014) Curve squeal: causes, treatments and results. Proceedings of Internoise 2014, Melbourne, 16–19 Nov 2014

Hardy AEJ, Jones RRK (2004) Rail and wheel roughness—implications for noise mapping based on the calculation of railway noise procedure. A report produced for Defra

Heckl MA, Abrahams ID (2000) Curve squeal of train wheels, part 1: mathematical model for its generation. J Sound Vib 229(3):669–693. https://doi.org/10.1006/jsvi.1999.2510

Hemsworth B (2008) Noise consultant LLP: environmental noise directive development of action plans for railways. Prepared for International Union of Railways (UIC)

International Organization for Standardization (2005) 3095: railway applications: acoustics: measurement of noise emitted by railbound vehicles. Switzerland

Jiang J, Hanson D, Dowdell B (2015) At-source control of freight rail noise: a case study. Acoust Aust 43(3):233–243. https://doi.org/10.1007/s40857-015-0026-3
Kasess C, Noll A, Majdak P, Waubke H (2013) Effect of train type on annoyance and acoustic features of the rolling noise. J Acoust Soc Am 134(2):1071–1081. https://doi.org/10.1121/1.4812771
Kondo O, Yamazaki Y (2013) Simulation technology for railway vehicle dynamics. Nippon Steel Sumitomo Metal Tech Rep 105(12):77–83
Mazilu T (2007) A dynamic model for the impact between the wheel flat and rail. UPB Sci Bull, Ser D: Mech Eng 69
Pieringer A (2013) On the modelling of wheel/rail noise. Presented at AIA-DAGA 2013, Merano, 18–21 Mar 2013
Rangel GWA, Aragão FTS, Motta LMG (2015) Proposta de uma metodologia para estimativa da deflexão do pavimento ferroviário. Presented at XXIX Congresso Nacional de Pesquisa em Transporte da ANPET, Ouro Preto, 9–13 Nov 2015
Remington PJ, Rudd MJ, Vér IL (1975) Wheel/Rail noise and vibration—volume 1: mechanics of wheel/rail noise generation. U.S. Department of Transportation
Rudd MJ (1976) Wheel/rail noise—Part II: wheel squeal. J Sound Vib 46(3):381–394. https://doi.org/10.1016/0022-460X(76)90862-2
Schneider E, Popp K, Irretier H (1988) Noise generation in railway wheels due to rail-wheel contact forces. J Sound Vib 120(2):227–244. https://doi.org/10.1016/0022-460X(88)90431-2
Stanworth CG (1983) Consideration of some noise sources due to railway operation. J Sound Vib 87(2):233–239. https://doi.org/10.1016/0022-460X(83)90558-8
Stassen KR, Collier P, Torfs R (2008) Environmental burden of disease due to transportation noise in Flanders (Belgium). Transp Res Part d: Transp Environ 13(6):355–358. https://doi.org/10.1016/j.trd.2008.04.003
Sueki T, Kitagawa T, Kawaguchi T (2017) Evaluation of acoustic and vibratory characteristics of impact noise due to rail joints. Q Rep RTRI 58(2):119–125. https://doi.org/10.2219/rtriqr.58.2_119
Sánchez M, López-Mosquera N, Lera-López F, Faulin J (2017) An extended planned behavior model to explain the willingness to pay to reduce noise pollution in road transportation. J Cleaner Prod 177. https://doi.org/10.1016/j.jclepro.2017.12.210
Thompson DJ (2009) Railway noise and vibration: Mechanisms, modelling and means of control, 1st Edition. Oxford, Elsevier, pp 1–10. ISBN: 9780080451473
Vér IL, Ventres CS, Myles MM (1976) Wheel/rail noise—part III: impact noise generation by wheel and rail discontinuities. J Sound Vib 46(3):394–417. https://doi.org/10.1016/0022-460X(76)90863-4
Wu T, Thompson D (2002) A hybrid model for the noise generation due to railway wheel flats. J Sound Vib 251(1):115–139
Wu TX, Thompson DJ (2003) On the impact noise generation due to a wheel passing over rail joints. J Sound Vib 267(3):485–496. https://doi.org/10.1016/S0022-460X(03)00709-0
Xiao, Xinbiao & Jin, Xuesong & Sheng, X.. (2008). A Hybrid Model for Noise Generation from a Railway Wheel Due to Wheel/Rail Impact. https://doi.org/10.1007/978-3-540-74893-9_39
Yang J, Thompson DJ (2014) Time-domain prediction of impact noise from wheel flats based on measured profiles. J Sound Vib 333(17):3981–3995. https://doi.org/10.1016/j.jsv.2014.04.026
Zannin PT, Bunn F (2014) Noise annoyance through railway traffic—a case study. J Environ Health Sci Eng 12
Zhen Y, Boogaard A, Chen R, Dollevoet R et al (2018) Numerical and experimental study of wheel-rail impact vibration and noise generated at an insulated rail joint. Int J Impact Eng 113:29–39. https://doi.org/10.1016/j.ijimpeng.2017.11.008

Solving Some of the Issues of Luggage Storing on Passenger Trains

Zhiyuan Zeng, Marin Marinov, and Bernhard Rüger

Abstract The aim of this research is to elaborate, describe and propose solutions for some of the issues of luggage storage on passenger trains, using a systematic approach. Specifically, the paper discusses and works out different technical solutions for better baggage handling, which include alternative transformed luggage racks (oven, trench and adjustable table). Pet compartment is also designed to solve the inconvenience of traveling with pets on passenger trains. These solutions are evaluated from different perspectives, taking different views into account. To obtain the optimal result when applying a particular solution, the characteristics, interior design and specific equipment of different types of passenger trains should be considered.

Keywords Luggage storing · Passenger trains · Rail transport · Interior design

1 Introduction

Since the end of twentieth century, the velocity of railway development has reached a higher level than ever before, and can be divided into two periods; mature period and new development period. The technological transformation of the railways has also made significant progress. Almost all railways around the world use traction power network for their passenger train services now. The advantage of this new mode is in the level of energy efficiency, resulting in less pollution. With the changes in the

---

Z. Zeng
School of Engineering, Newcastle University, Newcastle upon Tyne NE1 7RU, UK

M. Marinov (✉)
Department of Engineering Systems & Supply Chain Management (ESSCM), College of Engineering and Physical Sciences (EPS), Aston University, Aston Triangle, Birmingham B4 7ET, UK
e-mail: m.marinov@aston.ac.uk

B. Rüger
Institut Für Verkehrswissenschaften, Forschungsbereich Eisenbahnwesen, Verkehrswirtschaft Und Seilbahnen, Technische Universität Wien, Karlsplatz 13/230-2, 1040 Wien, Austria
e-mail: bernhard.rueger@tuwien.ac.at

M. Marinov and J. Piip (eds.), *Sustainable Rail Transport 4*, Lecture Notes in Mobility,
https://doi.org/10.1007/978-3-030-82095-4_7

railway energy supply and the adoption of new technologies, the economic benefits of the railways have seen reasonable improvement.

More recently, the high-speed rail has been developed in many countries including Germany, Italy, Japan, China and France. Germany was one of the first countries to start developing high-speed rail and has some of the most mature technology. The ICE (InterCity Express) is a high-speed train that connects all major cities in Germany (ICE High Speed Train 2017) (SIEMENS 2017). With speeds up to 300 km/h, this is one of the fastest ways to travel between cities such as Berlin, Hamburg and Cologne. The ICE has international connections to Denmark, the Netherlands, Belgium, France, Switzerland and Austria, with 5.5 million passengers using it in 2015. The development speed of China's high-speed rail is the fastest in the world (Duddu 2013), and in certain aspects like the maximum operating speed (430 km/h) and single operating mileage (2298 km), it is seen by many as the best in the world. As Nie Ligao's report explains (Ligao 2015), it is possible to balance a coin on the window sill of a train traveling at 300 km/h (186mph) for nearly 8 min from Shanghai to Beijing. Hence the high-speed rail technology in terms of speed, stability and comfort has reached a satisfactory level.

However, improvements are still required when it comes to luggage storing and handling in order to satisfy the growing requirements of passengers. A recent survey titled 'Carrying luggage on train, is it a problem?' was created and put forward to the public, (Tripadvisor. 2017) show that people feel comfortable taking one piece of luggage on their journey. However, problems may occur when that number goes above one piece per person, meaning that the existing luggage area may not accommodate all passenger's luggage during different time periods (peak hours, holiday and/or weekends). Hence the question is: why many railway companies release policies to limit the size, amount and number of luggage units like Cross-country who strongly recommend carrying one item of luggage (90 by 70 by 30 cm) (Cross-country. 2017)?. The Virgin Trains East Coast (VTEC) limits the amount of luggage to three pieces per person (Virgin rail company 2015), but the problem is still not resolved. The passengers who have no place to store their luggage take up space near the train entrance, passenger wagon door and aisles, which makes it inconvenient for other passengers to embark and disembark, walk through the aisle, use the toilet etc., causing major disruption. This also has a serious impact on individuals who are disabled, elderly people, children, pregnant women and pushchairs/baby prams. It is a difficult task to take care of luggage storage whilst maintaining the well-being and safety of passengers onboard. The situation with passenger behavior on the train is usually difficult to predict and describe, and often there are high risks for accidents to happen. For example, when people need to move their luggage for whatever reason and children are sitting or moving nearby, sometimes the heavy luggage can pose a danger as it might hurt the children if passengers moving it do not see them. It can also be difficult for parents to take care of their children and look after their luggage when their train arrives at the station and they need to get off the train.

It is apparent that measures should be taken to effectively deal with the luggage issue on passenger trains whilst maintaining the comfort and safety of all passengers and staff onboard.

1.1 Problem Definition

With the government's advocacy, more people would consider choosing train as their first choice of travelling. However, luggage will always be an issue that affects passengers, regardless of the purpose of travelling. People have to take luggage onboard. Train operating companies, though, need to maximize revenue by filling up their trains with passengers. And here the conflict between the profit of a train operating company and passenger's luggage should be resolved with an appropriate method.

1.2 Aims

The aims of this study are trifold: 1) to come up with solutions that improve the rate of space utilization onboard, 2) to discuss different situations and interior designs to accommodate more luggage and 3) to propose suitable solutions for every situation and design. High speed rail is the primary future of the train industry development; therefore it will be the primary focus here. An aim of this study is also to elaborate relevant solutions for storing special luggage securely, especially for the transport of pets.

1.3 Methodology

The methodology of this research contains the following. Firstly, identifying the problem, such as the conflict between railway companies' profit and luggage storing space (the space of luggage storing for passengers cannot be a gradual change as it is a current issue). Next work out the alternative methods, where these methods are used to solve the corresponding problems (i.e. lack of storing space, inefficient ways to storing pets onboard), and alter the design if need be. The next step is based on the characteristics of types of train, and the analysis of those alternative methods to find out which can apply to the various train designs.

The expected outcomes include:

(1) Additional space of luggage storing for each passenger;
(2) An improved level of space utilization onboard;
(3) Ensure the profit of railway companies are maintained whenever possible;
(4) Solving the problem of safely transporting pets onboard, and provide suggestions for alternative solutions.

2 Existing Problem of Storing Luggage

Dr. Rüger (Rüger, N.D) (Rüger 2008; Rüger N.D.; Rüger N.D.) has researched the issue of luggage storage, and part of the discussion deployed here has been drawn from achievements reported in (LOK PAN LO et al. 2020/21) and (Brice et al. 2015). According to recent analyses, a passenger will take 0.28–0.41 medium size luggage and 0.66 large size luggage onboard for one trip. In addition, studies also show that 90% of passengers take more than one bag for a long trip. This has led to the issue of passengers having no space to store their luggage.

Luggage issues are considered one of the most important factors that affect passenger's decision making when it comes to travelling by train. The expectations of passengers and train operators concerning the number of seats and train efficiency might defer though. The wants and needs of railway companies and passengers appear to be totally opposite. The train operating companies expect to maximize their revenue by increasing the number of seats onboard (however, more seats mean less luggage space). On the other hand, passengers would prefer to have the best travelling experience with high efficiency, fewer seats and more luggage room. Therefore, a compromise needs to be made to find a balance point that will satisfy both passengers and railway companies alike.

There is another issue though, regarding passengers' willingness to lift luggage items of different size and weight up to various heights. It was discovered that passengers are not really willing to lift medium or heavy luggage to the overhead rack. What a surprise.

Next would it be of interest to know what is the importance for passengers to keep visual contact with their luggage onboard; there were 45% and 43% who think it is very important and important respectively. Only 2% of passengers said it is unimportant. Most large luggage racks are located near the entrances, which means it is an unsafe solution of luggage storage in many passengers' opinion.

2.1 Types of Baggage and Luggage

Before dealing with the baggage issue, the size of luggage needs to be clearly identified. To improve revenue for the train operator, railway companies tend to maximize the number of seats, meaning there are fewer spaces for passenger luggage. As a result, train operators have restricted the number and size of luggage passengers can take onboard.

The size of luggage can be divided into four types (LOK PAN LO et al. 2020/21): large size (78×48×29 cm), medium size (67×45×29 cm), compact size (63×38×21 cm) and cabin size (48×32×18 cm). Different public transport modes can accept luggage items of different sizes. For example, airline companies only allow passengers carry cabin size luggage onboard but not the other three types of luggage. Other luggage needs to be checked in due to its large size. The rules of luggage storage with railway companies contain the same principles as the airline companies. The overhead luggage racks are designed for the cabin size luggage. The other three types of luggage are needed to be stored in large luggage racks.

Now let us consider the size of backpacks compared to the size of suitcases. Needless to spend a massive amount of time to conclude that the sizes are different, meaning these items have different dimensions. The current problem with that is that the UIC code, which is used for dimensioning the baggage racks of passenger trains, now specifies dimensions that are way too small for passenger's baggage. As a result, baggage racks in today's trains are not sufficiently dimensioned and needs revision.

If passengers are hiking, going cross-country or are doing similar activities which could last over a week, many decide to choose large travel bags with capacity of 65 L or more. This version suit for long-distance travel.

Medium-sized carpetbags typically have a capacity between 30 and 50 L. These bags are more versatile than others, it is suited for passengers who are going on 2–4 days of field trips or inter-city travelling.

If passengers prefer to do short leisure travelling, daypacks are a good alternative for them, the capacity is between 10 to 30 L.

2.2 *Methods of Storing Luggage on Passenger Trains*

There are various different methods of storing luggage onboard of a train. Consider rapid transit and some urban commuter trains, which are designed to serve those people who need to get to work or study fairly quickly in peak hour. Generally, there is no space for the large size luggage, which means baggage that is transported has to go as hand baggage only. However, the demand still exists, typical for those people who need to use commuter trains and rail rapid transit to travel to an airport or a bigger transport junction for example.

Currently, three traditional types of storage methods are available:

1. **Hand luggage rack** (suited to high-speed passenger trains, intercity trains and commuter trains)
2. **Large size luggage rack** (suited to high-speed passenger trains, intercity trains and part of some commuter trains)
3. **Under the passenger's seat**s (suited to high-speed trains, intercity trains, commuter trains and rail rapid transit services) However, the problem here is that space under passenger seat is often too limited and luggage must be turned or tilted for stowage there. In many situations passengers want to avoid this due to lack of comfort.

Different interior structures of high-speed passenger trains show that the luggage racks are normally located near the entrance, making it more convenient for passengers to store their luggage onboard. However, luggage racks close to the passengers wagon doors limit the train boarding area and as a result are not popular with passengers. Also there is no or only limited visual contact from the seat to the luggage stored on these racks. From an operational point of view, the luggage racks in the boarding area are inconvenient, as they cause a backlog after a few passengers boarding and the passenger changeover time is noticeably longer. However, in order for railway companies to reach their expected revenue, the volume of seats are kept at maximum. This means luggage space is relatively small compared to the whole passenger wagon. The railway companies try to restrict the number and size of luggage allowed onboard to control the "get-on get-off the train" situation. There are no thorough evidences suggesting this strategy is efficient, hence further research might be required.

Generally, passengers are allowed to take two large size suitcases and one backpack onboard the train. However, some railway companies like Virgin and Crossroad only allow passengers to bring one large suitcase.

None traditional methods for storing luggage on passenger trains and metro vehicles have been proposed in (LOK PAN LO et al. 2020/21) and (Toal and Marinov 2020). The possibility of promoting foldable seats has been explored to create more space for luggage on metro vehicles. Designing of a new service for passenger travelling to an airport to check their luggage in on a metro vehicle has also been studied. Alternative designs of separate baggage handling systems have also been studied, the results are promising and the interested reader is encouraged to consult Yeung and Marinov (2019), Alliance (2017), Korsun (2016), Express et al. (n.d,), METROLINX (2010), SIEMENS AG (2014), Kelly and Marinov (2017), SEAT61 (2016).

3 Solutions for Luggage Storing Onboard of a Passenger Train

The solutions proposed here include new possibilities for large size luggage racks and a solution for transporting pets onboard of a train are also discussed.

3.1 Transformed Luggage Rack

Generally, the large size luggage rack is used for storing large size luggage, and is usually placed near the entrance or end of the train car. However, it must be remembered that many passengers want their luggage close to them and therefore racks in the boarding area are not popular. If luggage racks are used in the boarding area during busy hours, this normally result in a longer passenger changeover time. There are no exact sizes of luggage racks used. Figure 1 shows the size of a more common luggage racks on passenger trains.

A luggage rack could normally contain 3.078 m^3, under the assumption that the carriage is at full capacity and 30 passengers carry one large size suitcase. The

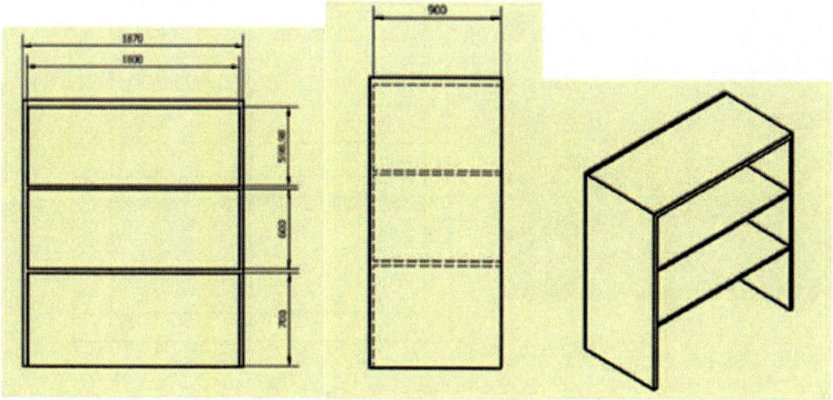

Fig. 1 The size of a more common luggage rack in the boarding area of a passenger carriage

volume for storing 30 passenger's large luggage is 3.7 m^3, and there should be two luggage racks near the start and the end of the carriage respectively. This means there shouldn't be lack of space to store large luggage.

Space could be easily wasted by passengers' luggage that is not placed efficiently on the luggage racks, which is quite common. If improperly placed one large bag can occupy the space for three middle size bags and so on. In order to achieve high space utilization, the luggage racks could be redesigned to be suitable for different luggage types and sizes. Efficiency could be improved if the first deck was fixed and the second deck is designed to be moveable. There are two feasible options for this design:

4. **Transformed luggage rack** (oven)
5. **Transformed luggage rack** (trench)

3.1.1 Transformed Luggage Rack (Oven)

The purpose of this design is to improve the space utilization, heavily inspired form the room you get from an actual "oven", as can be observed in Fig. 2.

The blue line (representing the first deck) is fixed and the size of Part 1 is the same as the old version of luggage rack, meaning the largest size luggage can still be placed into Part 1. Part 2 as shown in Fig. 2.

Consider a trench with decks and levels, Fig. 3. The middle deck could be inserted into three grooves of differing heights. More luggage could then be stored through adjusting the height of second deck, after placing several locks to ensure the stability and safety of the whole rack. This design could achieve high space utilization. As mentioned, the maximum size of suitcases are 85×65×35 (cm), the middle size: 70× 50×30 (cm) and the small size: 48×32×18 (cm), it impractical to place suitcases onto the top floor as many will not be able to safely reach the top lifting a heavy case.

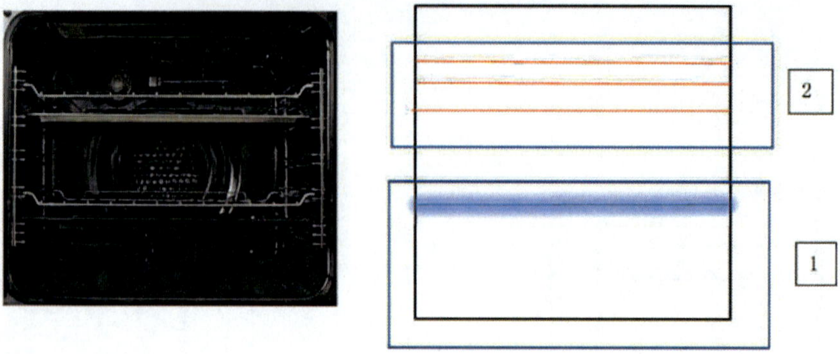

Fig. 2 An oven

Fig. 3 A trench

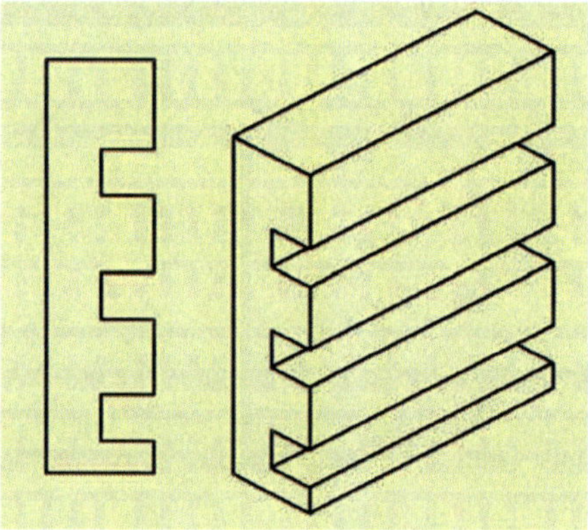

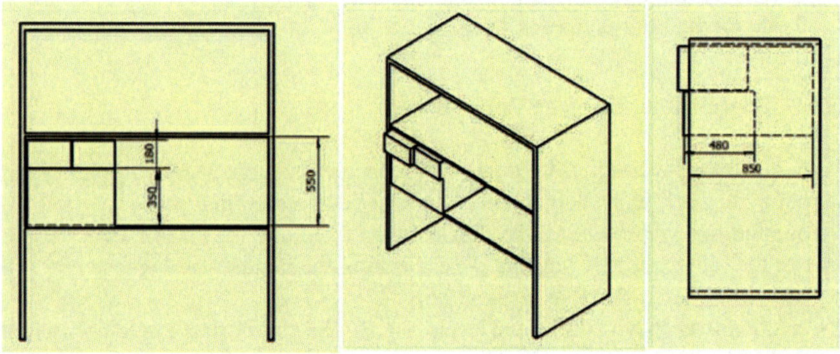

Fig. 4 When moving deck on level one

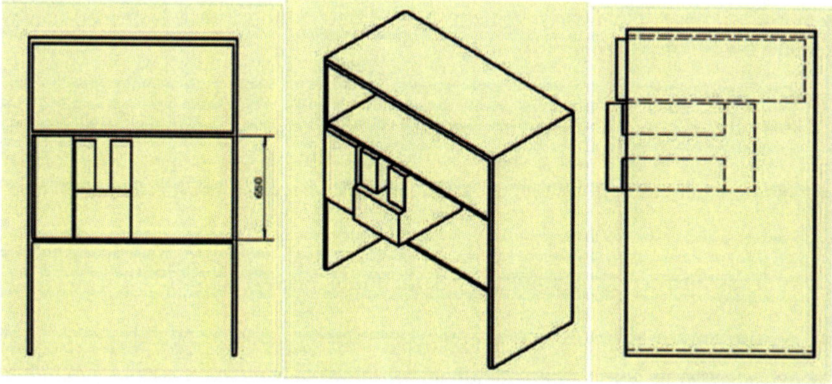

Fig. 5 When moving deck on level two

That is why this design will put heavy luggage on middle levels. Figures 4, 5 and 6 give an accurate representation of how a trench with three levels would work.

Level one (Fig. 4)

Considering the fact that not all large suitcases of passengers could originally be stored, the distance between two decks have been set at 550 mm, see Fig. 4, meaning one large size suitcase and two small size suitcases can fit perfectly into such space. Additionally, passengers also can place small sized suitcases upright here which won't waste much space as an alternative.

Level two (Fig. 5)

The distance between two decks have been improved to 650 mm, see Fig. 5, where one large size suitcase, one middle size suitcase and two small size cases can

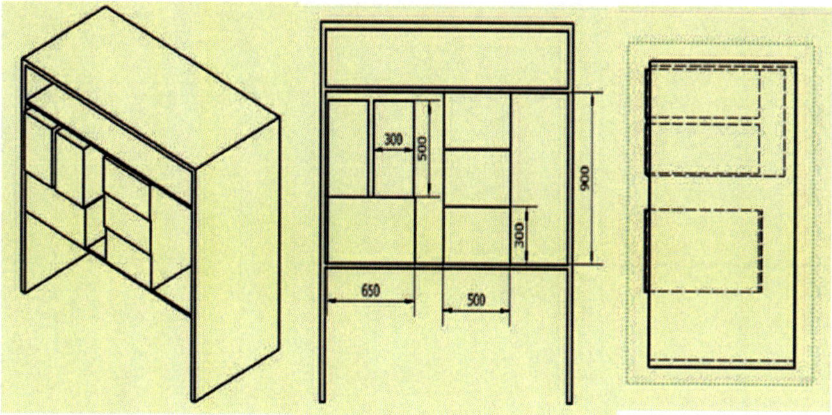

Fig. 6 When moving deck on level three

be stored safely. A clear improvement to level one if more space is needed due to greater volume of passenger luggage.

Level three (Fig. 6)

Generally, not many passengers would not need to use the upper level, Fig. 6. However, when there is an abnormally large volume of passengers with lots of luggage items to store, it may be used. The distance between two decks is 900 mm.

3.1.2 Transformed Luggage Rack (Adjustable Table)

The purpose of this design is similar to the oven version, but the principle is different. This design gets its height from an adjustable table.

As Fig. 7 shows, the green frame acts as a height adjustable table. The blue line is the first deck (fixed) which is similar to the oven version.

The frame of this design is shown in Fig. 8.

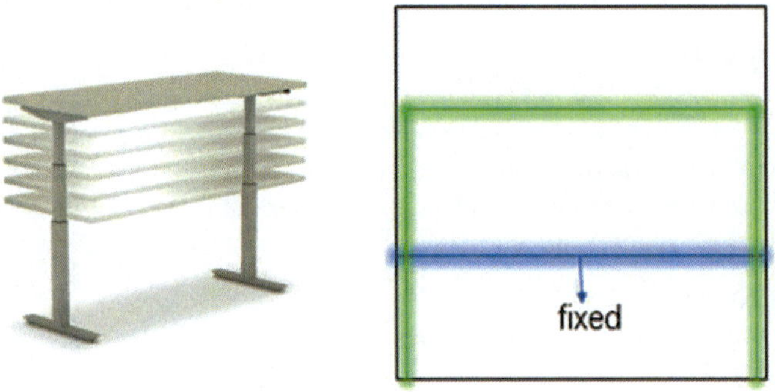

Fig. 7 Height adjustable tables

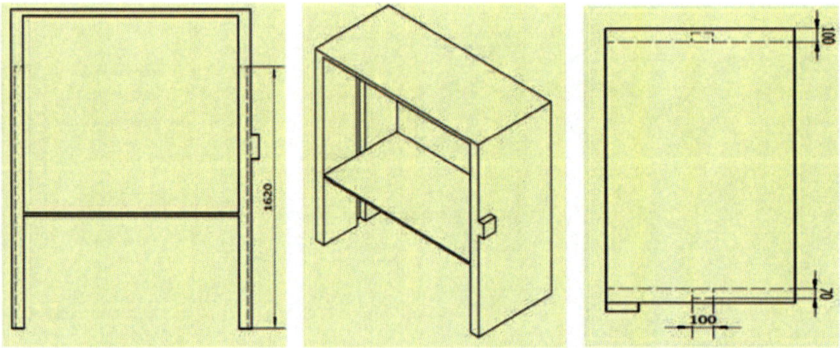

Fig. 8 The frame

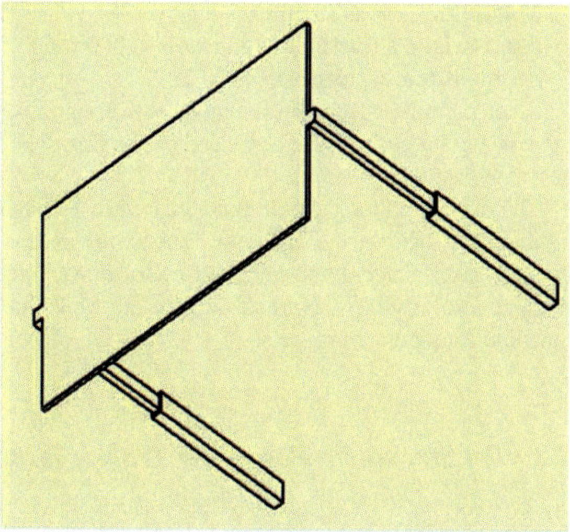

Fig. 9 Height-adjust machine

Compared to the normal luggage rack, there are three main differences that exist. First, in order to meet the height-adjustment the thickness of the board on both sides have been increased to 100 mm. Secondly, two trenches have been set into the board on both sides (as calculated in oven's version), and the largest distance between two decks is 900 mm. This means the second deck should reach 1620 mm. Thirdly, a new controller has been set on the right board to control the second board up and down, which is 1200 mm to the ground, meaning the design is child proof.

Height-adjust machine (Fig. 9)

These parameters are taken from a Chinese company named AOKE.

Input voltage: 230VAC.

Maximum load: 120 kg.

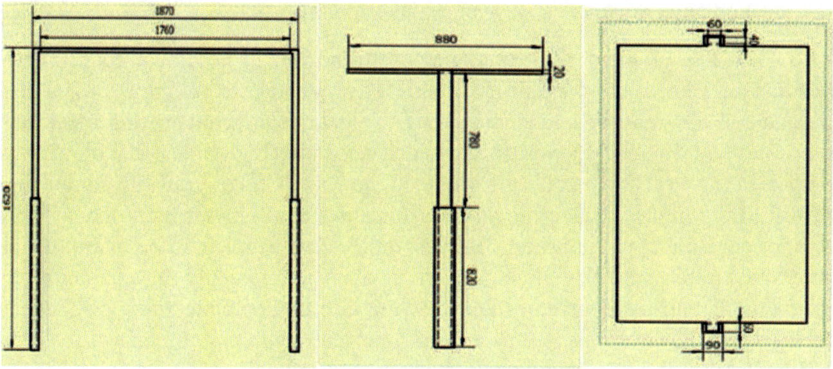

Fig. 10 Accurate size of height-adjust machine

Maximum speed: 38 mm/s.
Shelf height adjustment range: 600-1800 mm.
Ambient temperature: 0–40°C
Some parameters are not perfect for this design, like the maximum load. However most heavy luggage would be placed on the first deck, to ensure maximum safety to passengers.

The maximum load should be increased to 160 kg to allow for more baggage.

Figure 10 shows the accurate dimensions of this height-adjust machine, with enough space reserved so the shelf could move free in its adjustment range. The lifting column will be compatible with the trench so that the whole height-adjust machine can operate properly.

3.2 *Pet Storing on Passenger Trains: Issues and Solutions*

Pets are considered special luggage onboard trains, and more and more people are willing to travel with them. Compared to the technology of transporting pets via aircraft that has become very mature, travelling by train is still a conundrum. The pets are not accepted by Eurostar except guide dogs and there are strict policies on pet friendly trains in UK. There are several reasons why it is difficult to fix this problem:

(1) **Noise**—some pets especially dogs could make a lot of noise, which could disturb other passengers.
(2) **Smell**—the pets and their excretion can be very smelly, and potentially unmanageable onboard the train.
(3) **Stress**—onboard some types of train (especially high-speed train) the stress may increase with the rising speed of train. In such situations, snub-nosed animals such as Pugs, Bulldogs, Boxers, Pekingese, Shih Tzu or Persian cats, are prone to respiratory problems.
(4) **Allergy**—passengers may be allergic to animals' fur, which means they would have to have separate carriages for pets and non-pets.

To solve this problem, the experience of airlines could be drawn upon. There are some differences between airline and railways with pet travelling issues, as the international air transport association (IATA)'s data. The profit margin of airline companies was 4% in 2015 worldwide, which was much higher than that of railway companies (no exact data could be provided, however Chinese railway companies are still losing money), therefore airline companies could invest more money into facilities developing pet transport. Besides, quite a few people choose international airlines when they travel by plane, requiring passengers to apply for a passport for their pets. The airline companies know all the information concerning passengers and their pets, however this is very hard to achieve by railway companies.

What lessons can be drawn from airline companies?

(1) **Medical record**—this is a relatively simple solution which could show if passengers' pets have been vaccinated. This could avoid the possibility of pets transmitting disease to passengers and other pets.
(2) **In the cabin/hold**—there are two methods to transport pets by plane, in the cabin or in the hold. This generally depends on either the pet weighing less than 8 kg/17 lb or more. Railway companies could use a similar method, but part of carriage must be redesigned for pet compartments.
(3) **Booking 'pet seats' online**—although it is used by many railway companies, it is not common in UK. Following the design in (2), two types of pet tickets could be sold to the passengers, in the cabin or in the hold.

3.2.1 The Design of Pet Compartments

This would be the last compartment of the whole train, length: 25000 mm, width: 3380 mm, height: 3650 mm. The purpose of this design is to transport large size pets, and the compartment space could be determined by the requirements of each trainline (this design will use half the space of a single compartment). If there are fewer passengers that need this service, one train trip could have pet compartments for every two or three trips could be arranged.

The design may focus on storage for large dogs or cats (over 8 kg) and the soundproof door could reduce the noise that can be heard from adjacent compartments. Also the tickets for adjacent carriages would be a priority to those passengers who buy tickets for pets. There would also be CCTV set up in the pet compartment, meaning passengers can keep an eye on their pets. The exit/entrance is designed as the final point of the whole train, which means passengers who take their pets onboard can find the pet carriage easily (also would not affect other passengers who want to stay away from pets). The toilet would be used to clear the excrement of the pets.

The design of the pet carrier is based on the IATA (International Air Transport Association) standards (AIRFRANCE 2017) and (IATA 2015):

A = Length of the animal from the nose to the base of the tail.

B = Height from the ground to the top of the leg (elbow joint). The length of the container must be at least equal to A + 1/2 B (½ B = half of the length of the leg).

C = Width of the animal's back. The width of the container must be at least equal to Cx2. **D** = Height of the animal in a natural position, up to the top of the ears or head. The height of the container must be at least equal to D.

Assuming the largest size dog (about 70 kg) is $1000\times600\times400\times800$ ($A\times B\times C\times D$), the size could be calculated as $A + 1/2B = 1300$, $D = 800$, $C\times2 = 800$, so $1300\times800\times800$ could be set as the standard size of the carrier. The size of the toilet is 2200×1500. An accurate graphical representation is shown in Fig. 11.

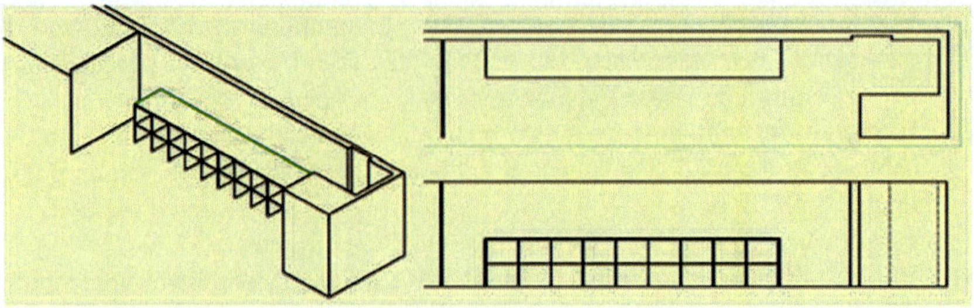

Fig. 11 Graphical representation of pet compartments

4 Evaluation

In order to assess the solutions proposed, we discuss additional aspects with luggage, railway company impacts (cost) and passenger impacts in detail.

4.1 Transformed Luggage Rack (Oven)

Luggage

As designed, level one in an "oven" version could result in large size suitcases and small size suitcases being placed on second floor of the luggage rack. Level two will be more commonly used, as it could accommodate almost any situation (any size luggage) and maintain high space utilization rate. The design of level three, as proposed, mainly considers passengers that are unable to place middle sized and large sized suitcases (which are heavy) on the third floor of the rack. This level, though, can be used to deal with unexpected situations (i.e. more passengers and luggage than anticipated).

However, the safety of this design needs improvement. According to the design, it should have a switch to control the moving deck. When switched on, the deck can be taken out easily to change the storage required. The older trains are more likely to shake as they travel, and if passengers who use the luggage rack forget to turn off the switch the deck may fall out and pose a danger to passengers. To prevent this from happening, an insurance could be set up or companies can arrange train attendants to manage and supervise any operation with the luggage rack.

Passengers should be told to use the second floor first (after the first floor) and only when the height of moving deck have been determined (the arrangement of the first suitcases has been stored securely), then passengers can store their luggage on the third floor. This operation is necessary, and a strict order is needed because it is hard to move the deck when luggage is already stored on it.

Railway company impacts (cost)

Railway companies need to transform their luggage racks which would cost them very little and the design would lead to energy savings, protecting the environment, requiring minimal maintenance. However, a train attendant must make sure the luggage rack's switch is turned on before the train departs. The economic pressure and financial uncertainty would be relatively small for a railway company.

Passenger

There is no doubt that with this new design, passengers would have more space to store their suitcases if they used it properly and followed the rules, but passengers would not be able to keep visual contact on their luggage.

4.2 *Transformed Luggage Rack (Adjustable Table)*

Luggage

Luggage could be safely secured on the luggage rack, adjusting the height of the moving deck within its range meaning passengers could set up their luggage according to bag size/scale. This allows for additional methods of luggage storage, more than what the "oven" version has.

Railway company

The railway company needs to spend time transforming luggage racks to include "adjustable tables", which will cost more money than the "oven" version. The price of the height adjustment machine used in 3.1.2 costs four hundred pounds, and the item is from China meaning shipping costs may be added taking the total much higher. A single compartment should have two electric luggage racks, so the total cost of a compartment is about one thousand pounds (not including labour cost). Additionally, railway companies need to provide power to the machines and perform regular maintenance to keep it working.

Generally, the economic pressure is relatively large for railway companies. However, this design contains many advantages, such as better safety where passengers do not need to move the rack by themselves. Another benefit is space utilization, as luggage takes up minimal space; that protects the interest of railway companies.

Passengers

Compared to the "oven" version, the biggest advantage of "adjustable table" is that passengers do not need to move the rack by themselves. This means the whole luggage rack will be safer and more convenient to use. Passengers will have more options to choose how they arrange their luggage.

4.3 *Pet Compartment*

Luggage (pet)

Pets would have a more comfortable travelling environment with this design. Also, they could be monitored by their owner using the CCTV, which would leave out many passengers traveling with their pets at ease so they could enjoy the journey. Although medical records can prove whether pets are healthy and vaccinated, there is always a risk of other issues such as fleas or viruses spreading.

Railway company

Railway companies need to modify part of the compartment to store pets and install a special toilet.

According to 'Mr. dog's quoted price', a stainless steel large size pet carrier (125 × 80 × 100) costs 35 pounds. This price should be affordable to railway companies, and it would not cost much to modify a toilet suitable for pets. This design needs to remove half of total seats in the compartment which is 44 and add 18 pet carriers. It could be assumed that one in every three train trips has pet compartments, and could be arranged to avoid peak times. Assuming that one train trip has 20 compartments, and every three train trips contains one with pet compartments; the proportion of pet compartments could be calculated as 1 in 120. Besides, railway companies can also charge those seats for pets. This design could even generate additional revenue for railway companies if arranged correctly.

Passengers

The pet compartment would be situated in the last carriage, so the design has no impact on those passengers who want to avoid pets. Only during peak time, will some of them be unable to book tickets due to the removed seats. This design also protects those passengers who are allergic to pets.

The passengers who travel with pets by train will happy to see this, as they do not need to worry about their pets' smell and noise affecting other passengers. This can also mean passengers can keep a visual contact on their pets. Some high-speed rail lines like Eurostar do not allow pets on-board, causing some to protest and campaign to try and change this. If this design could resolve the issue Eurostar has with pets, people in UK would have another alternative if they need to travel with pets. With the upgrades of facilities and services, the price of the pet tickets would be increased, costing passengers more money per journey.

5 Results

Transformed luggage rack (oven)

The cost of this design is not high, however there is a requirement for the track to be flat due to the moving direction of the train. Any shaking could make the luggage rack lose connection causing serious safety concerns. Therefore, the high-speed trains are more suited to this design, as well as some inter-city trains whose tracks are stable and stop off at airports and train stations (where there is a high proportion of large size luggage) are suited to this design too. The distance between stops of rapid transit are too short, which means passengers may not have enough time to use and operate the luggage rack. However, some metro lines to the airport may adopt a similar solution if people using the service would spend more time on the metro vehicle and have enough time to safely secure their luggage.

Transformed luggage rack (adjustable table)

The stability of the "adjustable table" version is more reliable than the "oven" one, which means the luggage rack could be used on more types of rail passenger vehicles. However, the cost of this design is greater than the "oven" version, which may dissuade railway companies from choosing it. High-speed rail is the next generation of the railway development. In general, high-speed rail would be constructed in large passenger traffic areas, which suits the design and would cope with high passenger volumes. The electric version luggage rack would also be suited to those intercity trains that have a large volume of passengers. On the other hand, commuter trains or rapid transit operations would not be suited to this design.

Pet compartment

This solution would be designed for long-distance passenger trains, so rapid transit and commuter rail are not suitable for it. Many factors need to be assessed, in order to see if it is necessary to adopt this design such as passenger opinions, the number of pets on-board, the public spending on the railway sector etc. Creating an experimental unit to observe the benefits/drawbacks of this design is recommended.

6 Conclusions

Issues surrounding luggage storage on-board of passenger trains have been studied. Solutions for luggage and pet storage have been introduced and evaluated against critical aspects related to luggage (pet), railway company and passenger. Suggested is that the following solutions can be applied to different types of passenger train:

- **Transformed luggage rack (oven):** high-speed train, inter-city train (conditional), commuter train (conditional)
- **Transformed luggage rack (adjustable table):** high-speed rail, inter-city train
- **Pet compartment:** high-speed train (conditional), inter-city train (conditional)

Implementation of the solutions proposed can result in:

(1) The space of luggage storage becomes larger for each passenger;
(2) Improving of the rate of space utilization, making interior design of passenger trains more efficient;
(3) Maximizing of the profit of railway companies;
(4) Solving the problem around transporting pets on-board of a passenger train.

References

AIRFRANCE (2017) Conditions for transporting a dog or cat in the hold. [Accessed on 29 Sep 2020] Available at: https://www.airfrance.us/US/en/common/guidevoyageur/pratique/animal_cabine_airfrance.htm

BayRail Alliance (2017) Types of train. [Accessed on 29 Sep 2020] Available at: http://www.bayrailalliance.org/types_trains

Brice D, Marinov M, Rüger B (2015) (2015) A newly designed baggage transfer system implemented using event-based simulations. Urban Rail Transit 1(4):194–214

Bicol Express (2012) Philippine national railways. Pnr.gov.ph. Retrieved 04 Oct 2012

Crosscountry (2017) Where can I store my luggage. [Accessed on 29 Sep 2020] Available at: https://www.crosscountrytrains.co.uk/on-board-with-crosscountry/luggage

Duddu P (2013) Top ten fastest trains in the world. [Accessed on 29 Sep 2020] Available at: http://www.railway-technology.com/features/feature-top-ten-fastest-trains-in-the-world

IATA (2015) Global aviation net profit hit a record high this year. [Accessed on 29 Sep 2020] Available at: http://news.gaotie.cn/hangkong/2015-06-11/247446.html

ICE High Speed Train (2017) ICE high speed train. [Accessed on 29 Sep 2020] Available at: https://www.eurail.com/en/europe-by-train/high-speed-trains/ice

Kelly J, Marinov M (2017) (**2017**) Innovative interior designs for urban freight distribution using light rail systems. Urban Rail Transit 3:238–254

Pavel Korsun 2016 Kiev-Kherson intercity train operate daily. [Accessed on 29 Sep, 2020] Available at: https://ukraine-kiev-tour.com/2016/kiev-kherson-intercity-train.html

Ligao N (2015) The coin did not fall for 9 minutes. [online] Available at: http://www.guancha.cn/Project/2015_07_15_326913.shtml

Lo LP, Marinov M, Rüger B (2020/21) A qualitative study on providing alternative solutions for handling the HSR passenger's luggage. Urban Rail Transit (in press)

METROLINX (2010) Rolling stock technology assessment. [Accessed on 29 Sep 2020], Available at: http://www.gotransit.com/electrification/en/project_history/Appendix%20Files/Appendix%204.pdf

Rüger B (2008) Efficient high-speed train interiors. [Accessed on 29 Sep 2020], Available at: http://www.uic.org/apps/presentation/ruger.pdf

Rüger B (N.D.) Efficient railway interiors concepts—theory versus practice [Accessed on 29 Sep 2020], Available at: http://www.ukintpress-conferences.com/uploads/SPRIX15/d2_s0_p6_bernhard_rueger.pdf

Rüger B (N.D.) Consideration of passengers' basic needs for designing efficient railway stations. [Accessed on 29 Sep 2020], Available at: http://publik.tuwien.ac.at/files/PubDat_165536.pdf

SEAT61 (2016) How to travel in mainland Europe. [online] Available at: https://www.seat61.com/dogs-by-train.htm

SIEMENS (2017) High speed and intercity train. [Accessed on 29 Sep 2020] Available at: https://www.mobility.siemens.com/mobility/global/en/interurban-mobility/rail-solutions/high-speed-and-intercity-trains/Pages/high-speed-and-intercity-trains.aspx

SIEMENS AG (2014) www.siemens.com/mobility. [Accessed on 29 Sep 2020] Available at: https://www.mobility.siemens.com/mobility/global/SiteCollectionDocuments/en/rail-solutions/high-speed-and-intercity-trains/velaro/velaro-e320-en.pdf

Toal J, Marinov M (2020) A check-in and bag drop service on-board light rail vehicles for passengers travelling to the airport. In: Marinov M, Piip J (eds) Sustainable rail transport. Lecture Notes in Mobility. Springer, Cham. https://doi.org/10.1007/978-3-030-19519-9_1

Tripadvisor (2017) Carrying luggage on trains, is it a problem? [Accessed on 29 Sep 2020] Available at: https://www.tripadvisor.co.uk/ShowTopic-g298184-i861-k10188107-Carrying_luggage_on_trains_is_it_a_problem-Tokyo_Tokyo_Prefecture_Kanto.html

Virgin Rail Company (2015) Nobody needs the burden of extra baggage. [Accessed on 29 Sep 2020], Available at: https://www.virgintrainseastcoast.com/rail-travel/your-journey/luggage/

Yeung HK, Marinov M (2019) A systems design study introducing a collection point for baggage transfer services at a railway station in the UK. Urban Rail Transit 5:80–103 (2019). https://doi.org/10.1007/s40864-019-0101-4

The Potential of Kazakh Rail Transport Mode as a Competitive Transit Route Between Europe and China

Dewan Md. Zahurul Islam, Alima Bissenbayeva, and Thomas H. Zunder

Abstract The purpose of this research paper is to explore the potential of the railway network of the Republic of Kazakhstan as an effective and competitive transit transport route between China and Europe, as well as ensuring the accrued economic benefit to the Kazakh economy from the transit flows potentially generated. Kazakh territory was a major part of the old Silk Road, especially the southern regions of the current Republic. This was disrupted during the Soviet period, with borders closed and any transit forbidden. From her independence, Kazakhstan has intended to be an open and transparent state in terms of world trade and business. This desire has led the country to make use of her transit potential between China and Europe. To explore this, the research work applied multiple research methods: Multiple Regression analysis, Correlation analysis, PEST and SWOT analysis. The findings of the research allow us to suggest that Republic of Kazakhstan is on the way to reaping the economic benefits of being a transit state. To achieve the full potential of transit services, the research recommends a number of actions including strengthening the scientific support for the industry; formation of intersectoral coordination with focus on marketing of transport services and logistics; construction and organization of an international transit corridor "Western Europe—Western China"; formation of a flexible tariff policy; the elimination of physical barriers to the organization and operations of domestic transit traffic; and the streamlining of customs procedures; all to facilitate smooth border crossings.

Keywords Kazakhstan · Europe · China · Rail freight transport and logistics · Potential transit transport · Economic benefits · Problems · Development needs

D. Md. Z. Islam (✉)
Bang College of Business, KIMEP University, Office 426, Dostyk Building, 2 Abay Avenue, Almaty 050010, Kazakhstan

A. Bissenbayeva
Logistic Coordinator, LGEAK O/S Part, Baykonyr Tower, 42 Abay avenue, Almaty, Kazakhstan

T. H. Zunder
Newcastle University, Newcastle upon Tyne NE1 7RU, UK

M. Marinov and J. Piip (eds.), *Sustainable Rail Transport 4*, Lecture Notes in Mobility,
https://doi.org/10.1007/978-3-030-82095-4_8

1 Introduction

In the modern globalized economy, the importance of freight transport is increasing for each state, region and continent since both domestic and international freight transport services directly affect the efficient running and competitiveness of the economy (Raimbekov et al. 2018). There is a strong relationship between economic activity and transport demand, it has been suggested that importance of logistics in international trade has made it one of the key elements in the development of a country (Rezaei et al. 2018, Martí et al. 2014). As a derived demand (Rühl and Boltze 2017) freight transport demand in Kazakhstan has grown as has economic development. Logistics transport is particularly important due to the vast territory with a dispersed low-density population. Due to a variety of reasons (e.g. geographic, historical, etc.) Kazakhstan has lower integration compared to other major global markets. With this background, the freight transport plays a crucial role in the development and implementation of economic ties with neighbouring countries. National and regional economic growth requires good transport infrastructure, efficient logistics services, and a business-friendly regulatory environment.

The transport sector of the Republic of Kazakhstan consists of vast railway and road network, ports including dry ports, air transports and limited sea and inland waterways transport mode options. Rail transport is the most important mode of transport in Kazakhstan (TSRK 2015) rail transport is dominant and about 70% of the cargoes are currently transported by railways. The share of road transport is gradually increasing and currently, about 20% cargo is transported by road transport (Bazarbekova et al. 2018; UNECE 2018).

Rail transport is particularly important for the following factors: (1) Kazakhstan is a big country with low population density (9th world's largest country with an area of 2,727,300 km with a population of 18,776,707 in 2020); (2) Kazakhstan is at the center of central Asia; (3) with no or little waterways transport (both inland and maritime); (4) with huge distances from Europe to the west, and the East Chinese; (5) the Kazakh economy is largely raw materials oriented; and (6) has a strong focus on large scale natural resource extraction well suited to rail (BURKITBAYEV 2000; Bazarbekova 2018).

Transit costs, as part of the total transport cost, significantly increase the cost of international trade. According to a report of the World Bank (2009), transit, economics and politics are closely intertwined. Various factors tend to increase such costs: geographical location, costly transport systems, the volume of transit traffic, and the duration and complexity of customs clearance procedures. Many countries have experienced a hindering of economic development due to the expense and complexity of accessing world markets (Kuzmina 2013), and such the transport strategy of the Republic of Kazakhstan (TSRK, 2015) recognizes this as a key issue for the country.

China (the second biggest economy in the world) to the East and European countries to the West have become important factors for the Kazakhstani economy. The purpose of this research paper is to explore the potential of the railway network of the Republic of Kazakhstan as an effective and competitive transit transport route between China and Europe, as well as ensuring the accrual economic benefit to the Kazakh economy from the flows moving through the country.

Section 2 of this paper discusses the literature review followed by Research Methodology in Sect. 3. Section 4 describes Data Analysis that includes: Multiple Regression in subsection 4.1; PEST analysis in subsection 4.2 and SWOT analysis in subsection 4.3. Section 5 discusses the findings of the research followed by conclusion and recommendation in Sect. 6.

2 Literature Reviews

The transport and communication sector has a share in the country's GDP of 11% and about 7% of the economically active population is employed in the transport industry (Republic of Kazakhstan 2010). We conducted a state-of-the-art literature review on the rail freight transport system of the Republic of Kazakhstan focused on four main aspects displayed in Fig. 1.

2.1 Historical Perspective

The priority of the newly independent Kazakhstan in 1990 was to re-align and restructure the transport sector including railways, which had been aligned as per the needs of the Union of Soviet Socialist Republics (USSR). With the changing trade patterns of the new geopolitics the origins and destinations of the rail routes have proven quite different to contemporary needs for national and international trade. Kazakhstan's rail system was designed ignoring inter-soviet-republic borders and to cater to the needs of Moscow led USSR planning and development. This means that the

Fig. 1 The main aspects of literature review

previously designed some of routes are not suitable for modern-day Kazakhstan's economic needs and needs re-alignment or new connections (Bazarbekova et al. 2018). This has caused border problems in the post-Soviet world, for example, the route from Uralsk to Aktobe passes briefly through Russian territory, introducing a potential border delay.

The transit and trade of Kazakhstan was dependent on Russia and other ex-USSR countries. After 1990, this dependency started decreasing slowly but steadily and the international trade with other countries including European countries and China started increasing. There has been and continues to be significant economic growth in Central Asian (comprised of five former Soviet republics: Kazakhstan, Kyrgyzstan, Tajikistan, Turkmenistan, and Uzbekistan) countries and the CIS (with memberships of 12 States—Armenia, Azerbaijan, Belarus, Georgia, Kazakhstan, Kyrgyzstan, Moldova, Russia, Tajikistan, Turkmenistan, Ukraine, and Uzbekistan.) as a whole. Particularly, Kazakhstan and its Central Asian neighbours, as well as Russia, began to rebuild trade ties with China. The rail network was developed taking into account of major and heavy resources such as agriculture, energy and minerals (TSRK, 2015). As with other economies modern transport needs to be able to cater to different needs: for example, transport of high value manufactured goods in containers.

The rail sector was organized for both infrastructure and operations as part of a centralized command economy, and as the Kazakh economy transitions to a market economy, these organizational, structural, monetary and operational frameworks have been re-assessed with a view to financial, social and environmental sustainability and meeting modern day customer requirements with attractive market offerings.

This new geopolitical, economic and trade environment has created a need for both construction and re-alignment of transport networks in Kazakhstan.

2.2 *Current Circumstances*

The latest Logistics Performance Index 2018 (LPI) (World Bank 2018a) suggests that the overall logistics system in Kazakhstan has improved (from a global ranking of 85 in 2012–71 in 2018) as has the big neighbour Russia (from a global ranking of 95 in 2012–75 in 2018) (World Bank 2018a). The global LPI 2018 is prepared based on six key performance indicators such as customs clearance; infrastructure; international shipments; logistics competence; tracking and tracing; and timeliness. A critical review regarding this performance would reveal that the performance in all aspects are in the 'average' position, particular attention is needed in customs clearance and tracking and tracing aspects to improve is performance.

Global international trade created opportunities for transit between European countries and Asia where millions of containers are being transported mainly via maritime transport every year. If the transport system of Kazakhstan could be improved and competitive with global maritime transport option, a significant portion of these movements of containers could pass through the territory of Kazakhstan (Erniyazova et al. 2014). The main problem towards this includes the low level and quality of transport infrastructure, particularly in rural territories where the population is not so big. The most significant problem of the Kazakh transport system is the high cost of transport services, causing lower competitiveness compared neighbouring countries. For example, the cost of exporting one container of cargo in Belarus is estimated at $1.800, in Russia—$1.850, in Kazakhstan—$ 3000 (World Bank 2019). That is, shipping one container from Kazakhstan is 50% more expensive than from Russia or Belarus.

The share of transport sector in GDP constitutes about 9% (1160 billion tenges) (Bazarbekova et al. 2018). However, the level of development of transport and logistics systems as a whole is assessed as insufficient. Particularly, Kazakhstan's transport infrastructure is characterized by a significant lag in the use of modern transport technologies including information systems. The technical and economic characteristics of the majority of vehicles including new ones are significantly lower than the parameters adopted in the developed countries. Low technical characteristics (e.g. lack of container carrying wagons), outdated design manufactured domestically as well as imported from neighbouring countries, leads to a preference for the purchase of vehicles produced in developed countries. Overall, the current stage of development of the market economy of Kazakhstan in transport has not yet led to the effective interaction of various modes of transport based on the development of logistics (Mazhorova 2011). Towards this, the structural adjustment and the development of the transport infrastructure is taking place (Nurly Zhol 2015).

2.3 *Transit Route and Potential*

Transit can be a powerful tool for realizing a country's geographic location and capabilities to provide and maintain international transport flows between different regions. In a number of countries such as the Netherlands, Singapore, etc., transit or trans-shipment is a significant source of income through value adding activities and employment and thus plays an important role in the development of the country's economy. Kazakhstan has a favourable geographical position at the junction of two economically significant parts of the world—Europe and Asia. The main advantage possessed by transit corridors passing through the territory of Kazakhstan is a significant reduction in distances and hence supply lead-times. When transporting between Europe and China through Kazakhstan, the distance of transportation is reduced by half compared with the sea route and by a thousand kilometres compared with transit through the territory of Russia.

The United Transport and Logistics Company—Eurasian Railway Alliance (UTLC ERA), operator of transit container services between China and Europe, was established by the national railways of the transit countries—Russia, Kazakhstan and Belarus, in which the co-founders own 33.33% of the shares (UTLC ERA 2019). Each member has its own competitive advantages, the synergy of which allows them to develop a joint project. Kazakhstan is witnessing a gradual increase in international rail transport in transit from China to Europe and vice versa. In addition, through integration with existing transport corridors, including the Chinese One Belt—One Way, the country gained access to the sea. So, in order to develop unhindered international trade, the formation of the Khorgos-Vostochnye Vorota free economic zone has been completed on the border with China (Sukhoverkova 2019). The infrastructure includes a dry port along with the industrial and logistics zones.

Offering efficient and effective transit to the neighbouring countries is a valuable strategic resource that can give the country the opportunity to fully realize its economic potential, ensure effective integration of the country into the world economic system and thereby raise the country's economy to a new level. Transit time on rail routes from China to Europe and back through Kazakhstan continues to decline, while container traffic is growing. Container transit traffic through Kazakhstan maintained the growth trend and increased only over the 5 months of 2019 by one and a half times, compared to the same period last year, amounting to 240 thousand containers (Grom 2019; Kamaliev 2019).

The projects envisaged by Kazakh Government's programs are aimed at developing and improving infrastructure, the construction of new transport facilities and the repair of existing transport networks and the establishment of public and private partnership (PPP) investments. For example, it was planned to invest 2.8 trillion tenge (about US 6,468,312,760 at current exchange rate) in the implementation of programs for the development of the country's transport system. Realization of the transit potential is one of the priority areas of the economic policy of Kazakhstan. The state programmes, aimed at creating effective ways to use the country's transit potential, include the Strategic Plan for the Development of the Republic of Kazakhstan until 2020, the Development and Integration of the Infrastructure of the Transport System of the Republic of Kazakhstan until 2020, and the Transport Strategy of the Republic of Kazakhstan until 2020.

Overall, there is a noticeable improvement in the socio-economic situation in the Republic of Kazakhstan, a general increase in the solvency of the population, an increase in industrial production, and an increase in goods turnover in foreign trade, wholesale and retail trade. This stimulates, in particular, the growth of investments in the development of the transport and logistics system and the creation of a modern logistics warehouse infrastructure in the country. The corridors will have to contribute to economic development of adjacent midland areas, as well as to economic and political integration of extensive Eurasian space (Omarbekovich et al. 2017).

2.4 *Factors for Future Growth and Actions Needed*

Sukhoverkova (2019) discusses the importance of Chinese initiative of the Silk Road Economic Belt. We already discussed that the volume of container traffic on land routes China—EAEU (Eurasian Economic Union)—Europe is growing annually, albeit from a very low base. Responding to changes in the pricing environment (e.g. competition with maritime route), shippers began to more actively use land transport routes. However, with the growth of containerized cargo flows, we fear that the efficiency of logistics will decrease as there is a lack of extra transit capacity which may affect the preferences of shippers in near future. It is a fact that Kazakhstan lacks modern warehouse and terminals (Tityukhin and Ovcharenko, 2011). To solve this problem, the construction of modern container terminals with value added logistics services (so called logistics hubs) is necessary. Their appearance, along railway routes will increase the total transit capacity. Considering the population density in a few cities, it is estimated that by building three or four modern container hubs connected with railway route in Kazakhstan will be able to achieve more than doubled throughput in transit and reduce the cost (Abdullayev et al. 2016).

But, the presence of a high share of single-track sections in the railway network significantly limits the speed of trains, especially with a high degree of traffic intensity. Semak et al. (2017) argue that due to the single-track section, the carrying capacity of rail routes, its operational indicators (especially the speed of movement) sharply (more than 50%) worsened due to the increase in stops at the crossing or railway sidings. Therefore, there is an urgent need to build and establish double-tracks (on the part of rail route with single track) so that it can be of world standards with for international transport corridors.

Another important issue is that the depreciation of railways has now reached 70% due to the inefficiency of the existing system for their restoration and modernization. This negatively affects the contribution of the rail freight transport component in the cost of goods and services and the competitiveness of Kazakhstan transit routes. In addition, such systems cannot guarantee the necessary reliable transit time and traffic safety. Also, they require significant operational and maintenance costs to maintain them in working condition.

Another important issue is the training and educating management and staff with knowledge and skills who understand the importance of today's national and international trade requirements and the needs of improved coordination among different actors in transport chains and supply chains. Keeping this fact in mind, the Kazakh Academy of Transport and Communications was opened a decade ago with a specialty "organization of transportation and logistics in transport" to train and educate personnel required in transport and logistics sector. But more institutions are needed to cater the needs of the sector.

Warehouse real estate differs in its functions: for example, the requirements for the storage of pharmaceutical products are completely different than for the storage of metal products. The latter can be located in open areas, while for different groups of drugs you need to constantly maintain the required temperature and humidity or other factors, which can affect during the storage period. The same applies to fruits and vegetables and other food products: experts indicate a shortage of warehouses with suitable temperature conditions depend on the level of investments, to each classes of warehouses. Moreover, the division of warehouse real estate in Kazakhstan approximately corresponds to international requirements (Tityukhin and Ovcharenko 2011). There is a significant need of improved warehouses (for example, for storage of apple fruits) in Kazakhstan.

Today, in large cities (such as Almaty) there are no transport and logistics centres (TLCs) that could fulfil the task of distribution centres or wholesale centres (wholesale sites), while acting as a major link between manufacturers, wholesalers and retailers along the supply chain distribution networks. The shortage of TLCs spaces and, consequently, high rental rates, high logistics and storage costs determine the overall inefficiency and non-competitiveness of the entire trading system. Infrastructure issues (space, engineering, logistics and warehousing) are currently the least studied part of the trade industry, although there is significant potential for improving the efficiency of the entire system. It is necessary to create infrastructure and facilities with modern transport and logistics system including TLC network in Kazakhstan (Raimbekov et al. 2016).

There are a number of shortcomings in the national legislation, which negatively affects the use of the transit potential of the Republic of Kazakhstan. In particular, there are no clear legal norms for ensuring multimodal transportation throughout the republic. The implementation of these shipments is regulated by the Civil Code of Kazakhstan and the Law of the Republic of Kazakhstan on transport. But the Civil Code (CC) does not specify the legal conditions for the implementation of these shipments, i.e. in what form the contract of carriage in mixed/combined traffic can be concluded. The provisions of the Civil Code only established that the relations of transport organizations during the carriage of goods, passengers and baggage by different modes of transport are carried out according to a single transport document, and the procedure for organizing these transportations is determined by agreements between organizations of the respective modes of transport concluded in accordance with legislative acts on direct multimodal transport. However, how the unity of transportation is achieved, the Civil Code does not indicate (KazNU bulletin 2015).

The republic has not yet acceded to a number of major international conventions. For example, the Barcelona Convention and the Statute of Free Transit are still among the international documents that are priority for the accession of the Republic of Kazakhstan. Under the provisions of this convention, the transit passage should not be subject to any special charges, with the exception of covering administrative expenses associated with such transit (Kazinform 2016). The implementation of this convention by Kazakhstan would certainly contribute to the development and efficient use of the transit potential of the country.

Kazakhstan has not yet acceded to the ATA Carnet Customs Convention (2004), which is aimed at the free movement of goods across borders and their temporary import into customs territory with exemption from duties and taxes. The successful integration of the country into the world community, of course, involves joining these international documents, which will become an important factor in increasing transit traffic in Kazakhstan and enhancing trade and economic cooperation with external partners.

3 Research Methodology

This research paper consists of literature review (conducted in Sect. 2) on transit transportation (including rail and road) of Kazakhstan to assess and identify current status and issues, problems and prospects. Data was collected from the statistical department documents published in the statistics committee of the Republic. Data was also derived from the annual reports publication in the official publication of JSC Kazakhstan Temir Zholy (KTZ) and from other sources such as World Bank, (see appendix I and II). The current research applies multiple analytical techniques. Multiple regression analysis is complemented with the PEST (Political, Economic, Social and Technological) and SWOT (Strength, Weakness, Opportunity and Threat) analyses. With these methods this research explores the prospects of rail and road transit transportation development in Kazakh railway and road with future transit integration with Europe, Russia and China. In brief, the following research methods have been used:

- Theoretical analysis;
- Documentary analysis;
- Observation;
- Comparison;
- Induction and deduction;
- Quantitative methods;

With the collected data, we conducted calculations in order to analyse Kazakhstan's income from railways and road logistics for 18 years[1] and then applied appropriate methods of analysis.

4 Data Analysis

4.1 Multiple Regression Analysis (in Excel)

Multiple Regressions are way of forecasting the meaning of a dependent variable, leaned on the meaning of two or more independent variables. The multiple regression equation is as follows:

$$\widehat{Y} = b_0 + b_1X_1 + b_2X_2 + \cdots + b_pX_p$$

Quantitative methods such as multiple regression start by identifying dependent & independent variables, which we did and list in Table 1.

We managed and fitted the formula above according of the quotation of Gunst and Mason (1980), where $\hat{Y}$ is forecasted and anticipated value of the dependent

[1] relevant data indicated in Appendix 1 and 2.

Table 1 Dependent variable and independent variable or factors

Dependent variable	Independent variable or factors
Kazakhstan income from railways and road for 18 years (million USD)	Transportation of goods by railways and roads (million tons) for 18 years
	Railway and road freight turnover (billion ton-km) for 18 years
	Export (million USD) for 18 years
	Import (million USD) for 18 years
	Railway and road transport lines length (km) for 18 years
	GDP of KZ (billion.USD) for 18 years
	Population of KZ million for 18 years

variables such as Kazakhstan Income from Railways and Kazakhstan Income from Road transit, $\mathbf{X_1}$ through $\mathbf{Xp}$ are P specific independent variables—Transportation of goods by railways/road, Railway/Road freight turnover, Export turnover, Import turnover, Railway/Road transport lines length, GDP of KZ, Population of KZ, b_0 (intercept) is the value of Y when all of the independent variables (X_1 through Xp) are equal to zero, and b_1 through bp are the estimated regression coefficients. Each regression coefficient represents the change in Y relative to a one-unit change in the respective independent variable. In the multiple regression situation b_1, for example, is the change in Y relative to a one-unit change in X_1, holding all other independent variables constant (i.e., when the remaining independent variables are held at the same value or are fixed). We are estimated output of regression analysis for Railways and Road Income by using data from official source of Kazakhstan Temir Zoly and Department of Statistical Committee and receive output.

The Railway and Road output description of Regression analysis:

SUMMARY OUTPUT (RAILWAYS)

Regression Statistics	
Multiple R	0.997658732
R Square	0.995322946
Adjusted R Square	0.991581302
Standard Error	48116309.04
Observations	19

ANOVA

	df	*SS*	*MS*	*F*	*Significance F*
Regression	8	4.92693E+18	6.15866E+17	266.012234	1.239E-10
Residual	10	2.31518E+16	2.31518E+15		
Total	18	4.95008E+18			

	Coefficients	*Standard Error*	*t Stat*	*P-value*	*Lower 95%*	*Upper 95%*	*Lower 95.0%*	*Upper 95.0%*
Intercept	32.87832546	1104414270	-2.9468674	0.01461278	-5.72E+09	-793774070	-5.715E+09	-7.94E+08
Transportation of goods by railways (mln to	425.8261062	1.61773889	-0.0263223	0.97951815	-3.647129	3.56196425	-3.6471295	3.561964
Railway freight turnover (bln ton-km)	775.1896508	3.321285105	0.233400514	0.82015876	-6.625095	8.175474	-6.6250947	8.175474
Passenger turnover of railway transportatio	46053.13987	38641.59034	1.191802394	0.26085472	-40045.69	132151.968	-40045.688	132152
Export (mln US dollars)	11747.66187	49133.76252	-0.23909551	0.81586045	-121224.5	97729.1829	-121224.51	97729.18
Import (mln US dollars)	6978.332379	20084.53342	0.34744807	0.73545647	-37772.8	51729.4614	-37772.797	51729.46
Railway transport lines length (km)	60421.31338	71723.78719	0.842416662	0.41924645	-99389.24	220231.87	-99389.243	220231.9
GDP of KZ(bln.USD)	2186540.75	1313990.218	1.664046444	0.12707791	-741211.9	5114293.39	-741211.89	5114293
Population of KZ mln.	139.2015007	97.08819156	1.433763452	0.18215875	-77.12447	355.527471	-77.12447	355.5275

Source The data from annual reports of KTZ and Statistical Agency (2018)

Summary Output (Road)

Regression Statistics	
Multiple R	0.973356976
R Square	0.947423802
Adjusted R Square	0.905362844
Standard Error	5.066332644
Observations	19

ANOVA

	df	*SS*	*MS*	*F*	*Significance F*
Regression	8	4625.327799	578.1659748	22.525017	1.96716E-05
Residual	10	256.6772646	25.66772646		
Total	18	4882.005063			

	Coefficients	*Standard Error*	*t Stat*	*P-value*	*Lower 95%*	*Upper 95%*	*Lower 95.0%*	*Upper 95.0%*
Intercept	62.74928435	149.1893677	-4.206015838	0.00181133	-959.9074685	-295.0782184	-959.9074685	-295.0782184
Transportation of goods by road (mln tons)	303.3667156	0.028873647	2.089679649	0.06316302	-0.003997822	0.124671165	-0.003997822	0.124671165
Road freight turnover (bln ton-km)	531.0851968	0.369734855	1.436394728	0.18142767	-0.292735396	1.354905789	-0.292735396	1.354905789
Passenger turnover of road transportation (n	1366.051348	0.000423909	-3.22251375	0.00913613	-0.002310579	-0.000421524	-0.002310579	-0.000421524
Export (mln US dollars)	1303.877534	0.004521519	0.002883716	0.99775585	-0.010061533	0.01008761	-0.010061533	0.01008761
Import (mln US dollars)	1076.796127	0.001383952	-0.007780589	0.99394508	-0.003094405	0.003072869	-0.003094405	0.003072869
Road transport lines length (km)	328.9256563	0.001175074	2.799189978	0.01882068	0.000671028	0.005907486	0.000671028	0.005907486
GDP of KZ(bln.USD)	227.6721984	0.096494965	-2.359420493	0.03999278	-0.442676379	-0.012668018	-0.442676379	-0.012668018
Population of KZ mln.	25.02671904	1.25617E-05	1.992296362	0.07433585	-2.96259E-06	5.3016E-05	-2.96259E-06	5.3016E-05

R Square

R Squares for Kazakhstan Income from Railways and Road equals 0.997 and 0.947, which are a very good fit. 99% and 94% of the variation for Kazakhstan Income from Railways and Road are explained by the independent variables Transportation of goods by railways/road, Railway/Road freight turnover, Export turnover, Import turnover, Railway/Road transport lines length, GDP of KZ, Population of KZ. The closer to 1, the better the regression line fits the data.

Coefficients

Kazakhstan income from Railways and Road = 425.826/303.366 × Transportation of goods by Railways/Road + 775.189/531.085 × Railway/Road freight turnover + 11,747.66/1303.877 × Export turnover + 6978.33/1076.796 × Import turnover + 60,421.31/328.925 × Railways/Road transport lines length + 21,865/227.67 × GDP of KZ + 139.201/25.026 × Population of KZ + 32.878/62.429 × Intercept.

The coefficients indicate that Kazakhstan's income from railways and road logistics is dependent on these factors, which means that if one of these indicators increase to positive side then Kazakhstan's Income from Railways and Road logistics also rise up or vice versa. From this we conclude that the factors are strongly interconnected thereby changing or innovating for one or more factors can significantly affect the positive dynamics of income growth for railways and roads.

Correlation analysis (in Excel)

A Correlation analysis was conducted for Kazakhstan's income from railways and roads.

For the analysis we propose the hypothesis:

- $H_0 = 0 \rightarrow$ income from Railway and Road and factors have a very weak correlation and do not depend on each other;
- $H_1 = 0 \rightarrow$ Income from Railway and Road and factors have a strong correlation, it should be greater than 50%/0.5 and factors depend on each other; (Table 2).

The above findings show that the coefficient for Kazakhstan's Income from Railways and Road transportation has a strong positive correlation between these indicators. Correlation of regression composes more than 50% of the relationship between factors and the results of calculations. From the quantitative analysis we can find that Kazakhstan's income from Railways and Road totally depend on economic and political factors such as the variables transportation of goods by railways/road, Railway/Road freight turnover, Export turnover, Import turnover, Railway/Road transport lines length, GDP of KZ, Population of KZ which indicate and influence to the overall picture of railways and road development.

4.2 PEST Analysis

We then conducted a political, economic, social and technological factor analysis (PEST) of the rail and road transport sector (Table 3).

PEST Analysis Details

Political factors

Table 2 Summary output for Kazakhstan's income from Railway and Road below:

	Kazakhstan Income from Railway and Road for 18 years (million USD)
Transportation of goods by Railways and Roads (million tons) for 18 years	Rail: {1; 0.926715}
	Road: {1; 0.8365232}
Export (million USD) for 18 years	Rail: {1; 0.978868}
	Road: {1; 0.8621599}
Railway and Road transport lines length (km) for 18 years	Rail: {1; 0.753071}
	Road: {1; 0.8441256}
Population of KZ million for 18 years	Rail: { 1; 0.9445374}
	Road: { 1; 0.8813582}
Railway and Road freight turnover (billion ton-km) for 18 years	Rail: {1; 0.9612084}
	Road: {1; 0.82673758}
Import (million USD) for 18 years	Rail: {1; 0.8889783}
	Road: {1; 0.70996941}
GDP of KZ (billion USD) for 18 years	Rail: {1; 0.9533459}
	Road: {1; 0.7192286}

Table 3 PEST analysis of the rail and road industry

Political factors 1. Legislative amendments and acts in the rail and road industry of Kazakhstan 2. State support of major structural projects 3. Sustainability of political power and the existing government	Economic factors 1. To increase/adjust in tariffs and decrease in monopolies 2. Fluctuations in the exchange rate 3. Growth rates of the economy 4. Degree of globalization and openness of the economy
Social factors 1. Training and education in transport and logistics sector 2. Preferences of end-users of products 3. Attitude to work, career, leisure, and retirement	Technological factors 1. Use of new equipment and technologies 2. Research and development, in particular applied researches 3. Access to the IT technologies 4. The degree of use, introduction, and transfer of technologies

(1) Legislative amendments and acts in the rail and road industry of Kazakhstan—This sector is regulated by the following amendment "The Law of the Republic of Kazakhstan dated September 21, 1994 No. 156-XIII On Transport in the Republic of Kazakhstan".

(2) State support of major structural projects—Government is supporting transport field by some investment projects, such as Nurlu Zhol strategy, Development programs of transport until 2020 and Strategy 2050.

(3) Sustainability of political power and the existing government. Actually, Kazakhstan is ruled by the same political power and party for the last almost thirty

years, and because of this stability of the country has opportunities to invite foreign investors without any problems of escalation.

Economic factors:

(1) To increase or adjust in tariffs of natural monopolies—considering cost involved in offering services to customers in China, Russia and Europe, the operators need an increase in tariffs on natural monopolies and steps to decrease the monopolies.
(2) Fluctuations in the exchange rate—until 2015 Kazakhstan had state-controlled exchange policy. Because of this we had several devaluations. Now KZT rate goes according to the external factors such as political stability and oil prices, and volatility rate.
(3) Growth rates of the economy—Kazakh economy is growing at about 4% annually.
(4) Degree of globalization and openness of the economy—Kazakhstan is open country for the FDI and tourism. The country is being part of the globalization and it is stated that there are no limits for the collaborations with the global community.

Social factors:

(1) There is a lack of trained and educated personnel in transport and logistics sector. The Kazakh Academy of Transport and Communications has a specialty "organization of transportation and logistics in transport" (Timofeeva, 2019).
(2) Preferences of end-users of products—the main advantage of the transit through Kazakhstan is that end-user users Asia-Europe will get their products in shorter time than before, facilitating backup routes as well as backbones, and increasing supply chain resiliency to global shocks such as COVID-19.
(3) Attitude to work, career, leisure, and retirement—particularly in freight transport sector, normally freight transport sector male dominated.

Technological factors:

(1) Use of new equipment and technologies—currently Kazakhstan does not use new technologies in the logistics and border transition (for example, for customs clearance). It needs to be updated for the new opportunities.
(2) Research and development—this sector nowadays is not developed in term of applied researches; There are not so many local experts and scientists to enrich the sector for the new approaches. Currently all the new things and ideas come from the international experiences and investments.
(3) Access to the IT—tracking and tracing technologies are there but not so effective, for example, we can only track trades only by reaching destinations, not in real time.
(4) The degree of use, introduction, and transfer of technologies. For example, currently there are no technological updates in the warehouses and transportation hubs, due to human (lack of skilled personnel) factors playing a huge role on the quality of the delivering products, and companies are responsible for it.

4.3 SWOT Analysis

Furthermore, we present a strength, weakness, opportunity and threat (SWOT) analysis for the rail and road transport in the context economy of Kazakhstan (Table 4).

Table 4 SWOT analysis of the rail and road industry

Strengths: Constructive and dynamic domestic and international trade and investment policy of the Republic of Kazakhstan with providing political stability, conducive to the normal operation of rail and road logistics The favourable investment climate, the readiness of foreign investors to invest in rail and road logistics High level of investment activities from investors from China, Europe, etc Potential for the development of new rail and road transit capacities Developed a legislative base in the field of rail and road logistics State support for investors	Weaknesses: Complex and poor rail and road logistics conditions due to its past USRR legacy Remoteness from the main world markets Dependence on transit countries to enter markets such as Europe and third countries Deficiency of investment resources due to the high capital intensity of projects and the financial crisis (oil price) High wear and tear of rail and road logistics due to weather conditions Poor quality of construction materials usage and lack of actual quality assurance system in place in building warehouses, rail, and road compared to, for example, European standards Lack of staff, technical staff, logistics specialists at middle and senior engineers and management
Opportunities: Big economies such as China at the East, Russia at the North and European continent at the West. There is a huge flow of goods and it is likely to grow If the Kazakh freight transit system can offer efficient and effective services, it can attract high goods turnover by using transit transportation, allowing companies to export more Keeping the above opportunities in mind, there is investment opportunities in the construction of new warehouses and reconstruction of rail and road ways that can significantly increase the amount of revenue and the share of cash receipts Achieving higher employment of the population of Kazakhstan	Threats: To exploit the great opportunity of offering transit services to economies such as China and Europe, Russia (i.e. transit route through Russian) is the biggest threat for Kazakh freight transit system Possible changes in the policy and tariffs for the transit of logistics from transit-seeking countries (in Russian, China and European countries) Development of transit transportation projects in Russia bypassing the territory of the republic Extensive experience and technical capacities in neighbouring countries (particularly Russia), which enhances their competitive advantages

5 Discussion

The growth of transit traffic increases the usage of the transport capabilities and assets, thus stimulating their reproduction and improvement. A consequence of the growth in transit traffic may be an increase in national revenues that includes the revenues of transport organizations and, accordingly, their effective development. The transport transit potential has two main components—the presence of demand for transit in the neighbouring countries' markets and the efficiency of the transport system.

The transit tariff policy of Kazakhstan routes, like any business entity, should aim at increasing net income. But we must understand that the development of transit traffic takes place in the conditions of fierce competition with alternative routes (e.g. through Russia) and modes of transport. From the analysis of coefficients, we find that the coefficient for Kazakhstan's income from railways and road transportation has strong positively correlation between these indicators making that income almost totally dependant on these factors. Despite its advantageous territorial location, the winner in this competition for large-scale profit is the one who ensures the optimal level of tariffs and quality of deliveries for transport companies and ultimately shippers and consignees i.e. cargo owners. The size of the possible income is estimated by experts at US $2 billion per year in World Trade Report (WTR, 2018). So far, in Kazakhstan, the cost of road and rail transport remains high due to, amongst other factors, the unsatisfactory condition of infrastructure. The development of Kazakhstan's transit will contribute to the growth of production, employment and other social development in the regions where transport corridors will run.

The active participation of the Republic of Kazakhstan in transport services can be said to have a number of positive consequences for socio-economic development. Transit transportation potentially brings profits to the entire transport sector. This argues for the comprehensive development of the transport sector (e.g. railway line upgrading to double line;) and further modernization or reform of the economy. It is necessary to ensure the accelerated development of infrastructure, logistics, renewal of fixed assets and rolling stock, and implementation of competitive tariff policy (keeping the competitors in mind) as well as the stabilization of transit freight flows passing through the country.

The prospects for the advancement of railway transport in Kazakhstan incorporate the arrangement of a branched transport framework and the development of the modern express railways. The formation of a unified transport system in Russia and Kazakhstan is favourably influenced by the possibility of expanding the 1,520 mm gauge space and the proposal to extend the wide gauge from Kosice, Slovakia, to Vienna by building a €6.5bn 400 km electrified freight-only line.[2]

It is necessary to coordinate the interaction of the state and private owners of railway transport in the name of public private partnership. As part of the transport strategy, railway reform has gone in two directions: on the one hand, the state is gradually creating a full-fledged competitive environment, on the other; it is directing

[2] https://www.railjournal.com/freight/austria-backs-broad-gauge-extension-to-vienna/.

large-scale investments in updating and modernizing the entire infrastructure. Enterprises of repair, service, and social and cultural life were transferred to the competitive environment. The transport sector has created conditions for the development of private carriers. According to Nazarbayev (2015) by the end of 2006, private wagons in Kazakhstan already accounted for about 35% of their total fleet, and by 2010 they had grown to 60%. The "first signs" of private business appeared on the passenger transportation market, where traditionally work was considered unprofitable. In 2018, "JSC NC Kazakhstan Temir Zholy" acquired 152 imported passenger cars. As part of the program to upgrade the locomotive fleet until 2015, it was planned to modernize 75 trunk diesel locomotives, 190 units of trunk electric locomotives. It planned to purchase more than 300 new trunk diesel locomotives, 40 passenger electric locomotives, and about 500 shunting diesel locomotives (KTZ, 2018).

According to the data of "JSC NC Kazakhstan Temir Zholy"(2018), 1,700 km of new railway lines built in recent years, as well as a number of logistics infrastructure facilities, allowed the industry to form a new architecture of transcontinental transport corridors, which gives significant dynamics to transit traffic from China to Europe and vice versa. The high growth dynamics of container transportation on Kazakhstan's transcontinental routes allowed the company to increase the share of transit traffic revenues. In Kazakhstani practice, this type of transport service has been used for more than one year, and during this time it has formed as a whole direction in the sphere of transit cargo transportation. In the future, as representatives of KTZ Express note, the route-speed principle of transporting goods over long distances can take a more significant place in the overall system of Kazakhstan's transit.

According to WTR (2018) currently, about 100 container trains pass through the territory of Kazakhstan in 11 directions. The volume of transit compared to 2011 increased by almost 100 times. In 2018, 105 thousand containers were handled in twenty-foot equivalent, which is twice as much as a year earlier. In the near future, another container train will be launched along the route China—Kazakhstan—Turkmenistan—Iran. This decision was made at the end of May following a meeting in Almaty of representatives of the railway administrations of these four countries. As noted by NC "KTZ" (2018) these agreements will contribute to an increase in transit traffic through the territory of the Republic of Kazakhstan.

A significant annual investment is being directed to the development of the road network of Kazakhstan. However, the pace and quality of reconstruction and rehabilitation of road infrastructure do not keep pace with the pace of destruction of the pavement, which negatively affects the efficiency of transport work, the technical condition of vehicles (Nazarbayev, 2014). The unsatisfactory condition of the road surface leads to a decrease in speed, an increase in transit time and operational transport costs, and an increase in the accident rate. The axle load of modern vehicles exceeds the capabilities of existing roads. The establishment of standards such as limits on the volume of freight and size of trucks should have an impact on maintaining the quality of roads. However, a more realistic way to bring the road infrastructure in line with needs is to search and introduce additional mechanisms for private funding for the development of road, which will allow developing road construction and reconstruction of the road network in accordance with the need.

As a result of the above-mentioned measures, Kazakhstan is steadily improving its performance in transport sector. The International LPI (2018) produced by the World Bank, Kazakhstan has improved (compared to that of 2016) it's by 11 positions, taking 77th place among 160 countries. This indicator of Kazakhstan is better among the CIS countries (Ukraine—80th, Russia—99th place, Uzbekistan—118th, Kyrgyzstan—146th), which once again proves the leading and important role of Kazakhstan in transport the logistics system of Eurasia.

In summary, we find that transport is a complex industry, carrying out its activities in cooperation with all sectors and economic sectors of the republic. Different types of transport are managed within the framework of several ministries and departments, which requires close coordination and ensuring methodological unity in approaches to development strategies. The main constraints in the development of the transport complex are:

- Poor state of transport infrastructure and in some cases poorly aligned infrastructure; and outdated transport fleet that need update slowly but surely;
- Insufficient investment in the sector as a whole and especially in road and rail transport;
- Insufficient use of innovation, technology and learning foreign experience.
- The lack of qualified personnel for transport and logistics sector; it lacks international standards of training and retraining.

To ensure successful long-term development of freight transport sector, we need to increase the quality of transport infrastructure and services. For this, it is necessary to apply an integrated approach to the analysis and determination of a promising strategy for the development of the industry, to develop and implement a system of measures to reform the economic model, improve the institutional structure, and optimize regulation of transport development by the state. It is also necessary to formulate an industry-wide system of training, retraining and continuing education for workers in transport and logistics, taking into account international experience.

6 Conclusions and Recommendations

World trade has developed rapidly over the past two decades. Due to the advantageous geographical location of Kazakhstan for the passage of cargo transit flows between Europe and Asia, the country's different sectors including transport can increase activities and thereby revenues. To achieve such a goal, Kazakhstan needs to create a modern transport infrastructure that ensures the transit of goods between the East and the West, which corresponds to the level of developed countries involved in transport integration. The ability to take advantage of its natural geographical advantages requires bringing the transport system of Kazakhstan in line with international requirements and standards, which implies the need to modernize the industry. Moreover, this work is to be carried out as soon as possible so that competitors do not seize the initiative in their own hands, providing alternative Kazakhstani transport routes and a level of service for freight and passenger transportation that meets international requirements.

Prospects for the development of railway transport in Kazakhstan include the formation of branched transport infrastructure and the construction of new express routes, door-to-door service development (Islam 2014), as well as improving the condition of existing routes in order to increase their actual speed and service level. To increase the level of corporate governance, it is advisable to attract knowledgeable and experts in the field, if needed foreign, managers to the management of JSC "NC KTZ", which would make it possible to raise the company's management to the international level by introducing the best world business and operational practices. For subjects of the railway industry operations, it is advisable to develop a personnel policy. Together with the Ministry of Education and Science, it is necessary to improve personnel training for the railway operations through vocational education institutions and to form a system of continuous training for employed workers in the railway industry.

It is advisable to develop and implementation of economic reform that is suitable for service sector (so that the freight and logistics service can be offered in a free and fair competition basis) and improve the institutional governance structure. As part of this activity, it is necessary to develop the marketing of transport services, which will allow for the formation of a more informed industry strategy. Also, it is necessary to improve corporate governance in the transport sector, in order to bring the activities of transport companies and individual entrepreneurs in line with increased requirements for the transport industry. It is advisable to unite individual entrepreneurs working in transport into unions and associations.

It is necessary to create an industry-wide training system based on modern international training standards adapted for Kazakhstani conditions, as well as to create an industry-wide system of continuous (lifelong) education, retraining, qualifications for transport workers and the logistics sector, taking into account international experience. The transport industry needs to strengthen scientific support for improvement of service offerings. It is also advisable to organize interaction and cooperation between industrial science and educational transport educational institutions, which will increase the quality of training and will contribute to the relative saving of costs for the maintenance of industrial science.

It is necessary to continue work on bringing reforms in legislative and regulatory documentation and technical regulations in line with international requirements and standards, as well as conclude tariff agreements with all partner countries. With the participation of international experts and Kazakhstani scientists, within the framework of pilot projects, it is advisable to work out measures to eliminate the negative factors of modern practice of insufficient efficiency and misuse of investments in transport, as well as improve existing tendering practices that lead to lower road repair quality, and create an independent/state control system in this area. It is advisable to develop a state program of comprehensive measures to improve transport safety in cooperation with experts in the field of social and educational policies. Under this program, provide measures to increase the transparency of the industry, regulatory authorities, as well as measures to improve transport statistics, taking into account international experience.

Kazakhstan is less-integrated with the major global markets. Transport plays a crucial role in the implementation of interstate ties. Their intensive growth presents new, increased requirements for the development of the transport infrastructure and regulatory policy, on the state of which the results of the development of the domestic economy of the republic and international cooperation depend to a large extent. The solutions to the problems of Kazakhstan's transport sector require:

- Strengthening the scientific support for the industry, conducting design studies and developing industry-specific recommendations for future development, taking into account of international experience.
- Formation of intersectoral coordination with other sectors based on marketing of transport and logistics services.
- Construction and organization of transit traffic within the international transit corridor "Western Europe—Western China".
- For the development of transport sector, it is necessary to introduce modern methods of logistics and marketing with the feedback and wishes of the population.
- Introduction of environmentally friendly and energy-saving technologies, increase the economic efficiency of motor vehicles (based on the introduction of modern methods of organizing transportation, use marketing and logistics) and develop technical standards based on international requirements.
- Formation of a flexible tariff policy in the field of transit traffic and the elimination of physical barriers to the organization of transit traffic, in particular, customs posts and border crossings.

Appendix 1

Data for Regression and Correlation Analysis. Retrieved from Statistical Agency (2018) and World Bank (2019)

Year	KZ income from railways mln. USD	Transportation of goods by railways (mln tons)	Railway freight turnover (bln ton-km)	Export (mln US dollars)	Import (mln US dollars)	Railway transport lines length (km)	GDP of KZ (bln.USD)	Population of KZ mln
2000	**275,197,889.2**	**171,896,512**	**125,258,963**	**939.25**	**1909.42**	**14,530**	**18.3**	**15,057,363**
2001	**336,411,609.5**	**183,811,454**	**135,778,825**	**1122.97**	**2803.43**	**14,588**	**22.2**	**15,039,971**
2002	**387,335,092.3**	**178,733,698**	**133,548,963**	**1406.81**	**3737.21**	**14,648**	**24.6**	**15,105,645**
2003	**465,171,504**	**202,741,236**	**147,758,936**	**1579.71**	**3845.38**	**14,648**	**30.8**	**15,232,325**
2004	**548,021,108.2**	**215,632,145**	**163,553,021**	**1874.88**	**5120.73**	**15,081**	**43.2**	**15,385,916**
2005	**650,659,630.6**	**222,789,654**	**171,990,025**	**2087.32**	**7521.29**	**15,021**	**57.1**	**15,541,457**
2006	**855,145,118.7**	**246,947,258**	**191,233,654**	**2676.89**	**8811.08**	**15,082**	**81**	**15,690,861**
2007	**881,794,195.3**	**260,663,214**	**200,814,578**	**3424.83**	**11868.08**	**15,082**	**104.9**	**15,841,355**
2008	**1,086,279,683**	**268,963,388**	**214,988,749**	**4292.41**	**11218.93**	**15082.4**	**133.4**	**16,001,176**
2009	**1,000,791,557**	**248,357,000**	**197,486,015**	**4103.67**	**10081.74**	**15079.1**	**115.3**	**16,184,163**
2010	**1,186,807,388**	**267,789,362**	**213,235,987**	**4118.96**	**11368.54**	**15016.1**	**148.05**	**16,398,976**
2011	**1,392,612,137**	**279,073,858**	**223,633,987**	**4337.73**	**10972.94**	**14892.4**	**200.38**	**16,647,380**
2012	**1,608,707,124**	**294,889,974**	**235,918,325**	**5430.91**	**14,344.55**	**15332.9**	**215.9**	**16,921,179**
2013	**1,708,443,272**	**293,719,654**	**231,336,524**	**5970.58**	**14083.54**	**15341.1**	**243.78**	**17,207,257**
2014	**1,694,195,251**	**390,741,222**	**280,653,322**	**7002.48**	**13845.94**	**15341**	**227.44**	**17,487,779**
2015	**1,483,113,456**	**341,371,852**	**267,362,889**	**6177.43**	**10897.74**	**15341**	**184.36**	**17,749,648**
2016	**1,576,517,150**	**338,928,000**	**238,972,269**	**6084.52**	**9846.95**	**16103.8**	**137.28**	**17,987,736**
2017	**1,661,213,720**	**387,236,788**	**266,611,993**	**6504.88**	**10082.66**	**16614.3**	**162.89**	**18,204,499**
2018	**1,728,443,271**	**397,884,841**	**283,345,254**	**7304.34**	**11978.27**	**16634.8**	**181.2**	**18,354,845**

Appendix 2

Data for Regression and Correlation Analysis. Retrieved from Statistical Agency (2018) and World Bank (2019)

Year	KZ income from roads mln. USD	Transportation of goods by road (mln tons)	Road freight turnover (bln ton-km)		Export (mln US dollars)	Import (mln US dollars)	Road transport lines length (km)	GDP of KZ(bln.USD)	Population of KZ mln
2000	**14.53**	**982.0**	**31**	**939.3**		**1909.4**	**85,867**	**18.3**	**15,057,363**
2001	**21.37**	**1076.9**	**33**	**1123.0**		**2803.4**	**88,045**	**22.2**	**15,039,971**
2002	**24.66**	**1219.3**	**37.6**	**1406.8**		**3737.2**	**88,388**	**24.6**	**15,105,645**
2003	**23.99**	**1318.2**	**40.2**	**1579.7**		**3845.4**	**88,992**	**30.8**	**15,232,325**
2004	**36.36**	**1444.8**	**43.9**	**1874.9**		**5120.7**	**90,018**	**43.2**	**15,385,916**
2005	**39.54**	**1511.1**	**47.1**	**2087.3**		**7521.3**	**90,845**	**57.1**	**15,541,457**
2006	**36.59**	**1582.6**	**53.8**	**2676.9**		**8811.1**	**91,563**	**81**	**15,690,861**
2007	**39.15**	**1667.4**	**61.5**	**3424.8**		**11868.1**	**93,140**	**104.9**	**15,841,355**
2008	**45.12**	**1721.0**	**63.48**	**4292.4**		**11218.9**	**93,612**	**133.4**	**16,001,176**
2009	**55.14**	**1687.5**	**66.25**	**4103.7**		**10081.7**	**96,846**	**115.3**	**16,184,163**
2010	**52.18**	**1971.8**	**80.3**	**4119.0**		**11368.5**	**96,018**	**148.05**	**16,398,976**
2011	**53.31**	**2475.5**	**121.1**	**4337.7**		**10972.9**	**97,155**	**200.38**	**16,647,380**
2012	**44.66**	**2718.4**	**132.3**	**5430.9**		**14344.5**	**97,418**	**215.9**	**16,921,179**
2013	**46.53**	**2,983.4**	**145.3**	**5,970.6**		**14,083.5**	**96,873**	**243.78**	**17,207,257**
2014	**44.11**	**3,129.1**	**155.66**	**7,002.5**		**13,845.9**	**96,421**	**227.44**	**17,487,779**
2015	**66.78**	**3,174.0**	**161.824**	**6,177.4**		**10,897.7**	**96,529**	**184.36**	**17,749,648**
2016	**57.61**	**3,180.7**	**163.262**	**6,084.5**		**9,846.9**	**96,353**	**137.28**	**17,987,736**
2017	**65.08**	**3,322.3**	**166.1461**	**6,504.9**		**10,082.6**	**95,409.6**	**162.89**	**18,204,499**
2018	**77.76**	**3,432.3**	**185.1973**	**7,304.3**		**11,978.3**	**96,245.72**	**181.2**	**18,272,430**

References

Abdullayev S, Kiseleva SO, Adilova N, Bakyt G, Vakhitova L (2016) Key development factors of the transit and transport potential of Kazakhstan. Transport Problems 11(2)

ATA Carnet Customs Convention (2004) Customs convention on the A.T.A. Carnet for the temporary admission of goods. Retrieved from: http://www.wcoomd.org/-/media/wco/public/global/pdf/about-us/legal-instruments/conventions-and-agreements/ata/pf_ata_conv_text.pdf?la=en on 25 Oct 2019

Bazarbekova M, Assipova Z, Molgazhdarov A, Yessenov M (2018) Review of transportation modes in Kazakhstan region, Central Asia. Cogent Eng 5:1. https://doi.org/10.1080/23311916.2018.1450799

Burkitbayev S (2000) Rail transport in Kazakhstan. Transport research board, online accessed on 6 Oct 2020 https://trid.trb.org/view/657050

Erniyazova ZN, Kazbekova LA, Mukhanova AE, Utegenova KA, Smagulova ZB (2014) Particularities of formation of transport-transit cargo-traffic in the Republic of Kazakhstan. Life Sci J 11(2s)

Grom A (2019) Il Mondo dei Trasporti, № 284/Novembre. Interview with Alexey Grom, Chief executive officer UTLC ERA. China a huge growth potential. Retrieved on 20 Nov 2019 from https://www.utlc.com/en/news/il-mondo-dei-trasporti-284-novembre-interview-with-alexey-grom-general-manager-of-utlc-era-china-a-h/

Islam DMZ (2014) Barriers to and enablers for European rail freight transport for integrated door-to-door logistics service. Part 2: enablers to multimodal rail freight transport. Transport Problems 9(4):5–13

Kazinform (2016) The role and importance of the transport and transit potential of Kazakhstan in Eurasia. Retrieved on 5 Nov 2019 from https://www.inform.kz/ru/rol-i-znachenie-transportno-tranzitnogo-potenciala-kazahstana-v-evrazii_a2972127

KazNU Bulletin (2015) International relations and international law series. №3 (71)

Kamaliev B (2019) Kazakhstan plans to introduce fees for transit vehicles. Retrieved from: https://inbusiness.kz/ru/last/kazahstan-planiruet-vvesti-platnost-dlya-tranzitnyh-avtomashin on 20 Nov 2019

Kuzmina EM (2013) Transport development in Kazakhstan. Transport and communications in the newly independent states: features and development factors. - M.: IE RAS

KTZ (2018) "NC" Kazakhstan Temir Zholy. Retrieved from: https://www.railways.kz/ru/infrastructure/2018-zhyl-orytyndysymen-2019-zhyl-a-arnal-an-negizgi-tapsyrmalar on 17 Oct 2019

Martí L, Puertas R, García L (2014) The importance of the logistics performance index in international trade. Appl Econ 46(24):2982–2992

Mazhorova V (2011) Transport in Kazakhstan: modern situation. Retrieved on 25 Oct 2019 from http://www.analitika.kz/images/5773.pdf

Nazarbayev NA (2014) Strategy 2050. Retrieved on 25 Oct 2019 from https://strategy2050.kz/ru/news/53507/

Nazarbayev, NA (2015) The transport strategy of the Republic of Kazakhstan. Retrieved on 30 Oct 2019 from http://www.akorda.kz/en/official_documents/strategies_and_programs

Nurly Zhol (2015) Nurly Zhol—The Path to the future is a large-scale development program of the country. Retrieved on 20 Oct 2019 from https://www.zakon.kz/4692032-poslanie-nurly-zhol-put-v-budushhee.html

Omarbekovich M, Droździel P, Muratovna A, Rybicka I (2017) Integration of Kazakhstan's rail road in the world transport and logistics space, journal of Autobusy : technika, eksploatacja, systemy transportowe, issue 6. pp 1737–1739

Raimbekov Z, Syzdykbayeva B, Zhenskhan D, Bayneeva P, Amirbekuly Y (2016) Study of the state of logistics in Kazakhstan: prospects for development and deployment of transport and centres. Trans Prob 11(4)

Raimbekov Z, Syzdykbayeva B, Sharipbekova K (2018) Economic aspects of freight transportation along the East-West routes through the transport and logistics system of Kazakhstan. In: Sładkowski A (ed) Transport systems and delivery of Cargo on East–West Routes. Studies in Systems, Decision and Control, vol 155. Springer, Cham. https://doi.org/10.1007/978-3-319-78295-9_5

Republic of Kazakhstan (2010) The strategic plan for the development of the Republic of Kazakhstan until 2020. Available at: https://www.akorda.kz/upload/content_files/doc/Gos_programi/Strategy_2020.doc

Rezaei J, van Roekel WS, Tavasszy L (2018) Measuring the relative importance of the logistics performance index indicators using Best Worst Method. Transp Policy 68(March):158–169. https://doi.org/10.1016/j.tranpol.2018.05.007

Rühl F, Boltze M (2017) Freight transport demand management: influencing the freight transport demand within traffic management BT—dynamic and seamless integration of production, logistics and traffic: fundamentals of interdisciplinary decision support. In: Abele E, Boltze M, Pfohl H-C (eds) Springer International Publishing, Cham, pp 163–184. https://doi.org/10.1007/978-3-319-41097-5_9

Statistics Agency (2018). The main indicators of foreign trade of the Republic of Kazakhstan: Retrieved from: http://www.stat.gov.kz/faces/wcnav_externalId/homeNumbersCrossTrade?_afrLoop=3932798357642291#%40%3F_a on 10 Oct 2019

Strategy (2050) Kazakhstan strategy 2050. Retrieved on 25 Oct 2019 from https://strategy2050.kz/en/

Semak YA, Raimbekov BKh, Assanova MK, Kozhabergenova AY (2017) Transport and logistics potential of the Republic of Kazakhstan. Retrieved on 15 Oct 2019 from https://articlekz.com/en/article/17832

Sukhoverkova A (2019) Kazakhstan is too slow for China: what are the problems of transit? Retrieved on 25 Sept 2019 from https://regnum.ru/news/economy/2653580.html

Timofeeva E (2019) Russian railways will create the newest testing center. Transport Magazine, June 2019. Retrieved from: http://www.rostransport.com/transportrf/#a40 on 5 Oct 2019

Tityukhin N, Ovcharenko N (2011) Terminal and warehouse infrastructure of the Republic of Kazakhstan. Retrieved on 10 Oct 2019 from : http://www.ctcs.kz/ru/informaciya/stati/ 229.aspx-2011

TSRK (2015). Transport strategy of the Republic of Kazakhstan until 2015. Astana, 2006, pp14. Retrieved on 14 Sept 2019 from www.tranco.kz/pr/doc/transportstrat.doc

TCITR (2016) Trans-Caspian international route. Retrieved from: https://middlecorridor.com/en on 10 Nov 2019

UNECE (2018) Transport statistics infocard KAZAKHSTAN. Available at: https://www.unece.org/fileadmin/DAM/trans/doc/2018/wp6/Kazakhstan.pdf

UTLC ERA (2019). United transport and logistics company—Eurasian rail alliance. Railway companies of Belarus, Russia and Kazakhstan sign agreement, Belarus Retrieved from: https://uic.org/com/uic-e-news/651/article/belarus-railway-companies-of-belarus-russia-kazakhstan-sign-agreement?page=modal_enews on 15 Nov 2019

World Bank (2009) World Bank supports improved transport efficiency and safety in Kazakhstan. Retrieved from: https://www.worldbank.org/en/news/press-release/2009/04/27/world-bank-supports-improved-transport-efficiency-safety-kazakhstan on 1 Oct 2019

World Bank (2019) World bank in Kazakhstan. Retrieved from: https://www.worldbank.org/en/country/kazakhstan on 1 Oct 2019

World Trade Report (2018b) WTR. Trade and development: recent trends and the role of the WTO. WTO Publications, Geneva

World Bank (2018a) Global ranking 2018 LPI—logistics performance index, online accessed on 6 Oct 2020 at https://lpi.worldbank.org/international/global/2018

Evaluating Russia's Railway Infrastructure—Geography, Demography and Kremlinology

Emily Ferris

Abstract This paper unpacks the sustainability of Russia's railway infrastructure, whose development faces innumerable challenges including its geography, demographic density and specific political considerations. The monopoly of companies such as Russian Railways (RZhD) over railway infrastructure, many of its freight carriers, and construction tenders have reduced business competition in this sector, and have made the railways a restrictive environment for foreign investors. Russia's demographic challenges, which include a much denser population distribution in the western part of the country, have meant that many regions in the central and eastern parts of the country remain remote, under-populated and inaccessible, despite the wealth of natural resources located there. Government plans to improve rail access to these disconnected areas are often unfeasible and unsustainable, and few foreign investors are willing to offer their financial support to projects that likely have low yields and high risks. Poor diplomatic relations with the West have restricted Russia access to western capital markets and alternative sources of loan financing, particularly for large infrastructure projects such as rail, and many western funders have withdrawn from Russia's rail market. Without a significant injection of foreign investment from Asian partners such as China, Japan or South Korea, improvements to Russia's railway network are likely to remain relatively piecemeal. For Russia's railway system to be sustainable in the long term, RZhD's monopoly must be broken up, international partnerships with countries that have similar geographic considerations could be brokered, and Russia must re-evaluate its demographic priorities. Climate change and new geographical conditions are likely to pose an additional challenge for Russia in the coming years, but without accumulating proper engineering expertise to mitigate these issues, the country's rail transport network is likely to lag behind those of its international counterparts.

Keywords Railway infrastructure · Russia · Natural resources · Demography · Geography · Connectivity

E. Ferris (✉)
Research Fellow, Royal United Services Institute (RUSI), London, UK
e-mail: EmilyF@rusi.org

M. Marinov and J. Piip (eds.), *Sustainable Rail Transport 4*, Lecture Notes in Mobility,
https://doi.org/10.1007/978-3-030-82095-4_9

1 Introduction

With a territorial expanse 2.5 times larger than the whole of Europe, Russia has been reliant on its railway networks for decades to convey cargo and passengers across the country, and link up its individual territories.[1] Russia's rail system is one of the largest in the world, spanning from Europe to Central Asia, and from Russia's exclave of Kaliningrad in the west to Vladivostok, on the Sea of Japan in the very east of the country. In total, the track spans more than 85,000 km.[2]

As a natural resource producing country, rail transport is vital for Russia's economy, as a means of ensuring social cohesion and often as an important means of transporting Russia's military forces around the country.

As a major global producer and exporter of natural resources such as oil, gas and coal, Russia is highly reliant on its rail network to transport products across the country, for export abroad—comparatively little of its raw material production is absorbed by domestic consumption. This is why the railway network, and the ports that they are linked to, are vital in guaranteeing the future of Russia's economic security, and are a part of Russia's critical national infrastructure, serving economic, military and social functions.

But Russia's rail systems have three main vulnerabilities—geographic challenges; demography, and the political environment, which has created one company's monopoly over the rail system and reduced business competition.[3]

This paper unpacks the key question as to whether Russia's railway systems are sustainable in the long term. It identifies the way in which the Russian government's strategic thinking about its transport infrastructure has developed, and some of the political frameworks that govern this decision-making. The paper then outlines some of the inadequacies of Russia's existing rail systems, including problems of geography, demography, and the monopoly of powerful companies over the rail sector. Finally, the paper evaluates some of the key ways in which the rail sector could improve, including prospects for international partnerships and greater access to foreign expertise and investment.

---

[1] Rossiskiy Investitsionnoye Agentstvo, (2020), *'Maschtabi Rossii v Tsifrakh'*, https://www.investment-in-russia.com/site/page?view=ROSSIYA-V-CIFRAH, Accessed 2 June 2020.

[2] Railway Technology (2013), *'Russian railways: connecting a growing economy'*, 30 April 2013, https://www.railway-technology.com/features/featurerussia-railways-connecting-growing-economy/, Accessed 3 June 2020.

[3] Grubesic, T. and Matisziw, T. (2013), *'A typological framework for categorizing infrastructure vulnerability'*, GeoJournal, 7(2), Springer, https://www.jstor.org/stable/42006320, Accessed 3 June 2020.

2 Why Railways Matter to Russia

Russia's railway system is one of the largest in the world, spanning eight time zones and linking up parts of Europe, Central Asia and the Arctic, as well as stretching across the country to the Sea of Japan. The network accepts around 85% of Russia's freight transport and is a major employer for the country.

During the early 1990s, railways were responsible for the transportation of 70% of Russia's freight cargo. But the dissolution of the Soviet Union and Russia's subsequent economic decline prompted its rail industry to fall into disrepair, with declining freight volumes, loss-making passenger traffic and low operational productivity.[4]

One of the most significant and well-known infrastructure projects in Russia was the Trans-Siberian Railway (TSR). Construction on the railway began in 1891, and once completed became one of Russia's greatest engineering feats. It was built in part to offset the potential for Chinese territorial expansion into Siberian and Far Eastern regions, as well as serving as a practical link, designed to bridge the growing political and economic gap between Moscow and Russia's provinces.[5]

The route is 10,555 km long and was finally completed in 1903, linking up Siberia and the Far East to the western parts of Russia, with the aim of reducing transport times with minimal freight costs. Engineers battled adverse weather conditions including permafrost and marshy tundra, and were obliged to repave many of the roads they constructed, which had already sunk into the marsh.[6] The single-track railway was built for light loads and engines, with a standard Russian 5ft gage, and at the time was compared favorably with many of the US's trans-continental systems. The TSR was fully electrified by 2002. And modernization efforts have since the 1990s focused on increasing capacity, with new rail bridges eliminating the last single-track sections of the line.[7]Today, the TSR plays an important role in Russia's freight distribution process, and has significantly reduced travel time for cargo– it takes just ten days for containers to be shipped from China to Finland, versus 28 days by sea.[8]

The TSR also had a military use. It was a valuable asset in Russia's war with Japan (1904–1905) as it supplied troops in the Far Eastern region with food and ammunition,

---

[4] World Bank (2017), *'Case Study—Russian Railways'*, 27 December 2017, https://ppiaf.org/sites/ppiaf.org/files/documents/toolkits/railways_toolkit/PDFs/RR%20Toolkit%20EN%20New%202017%2012%2027%20CASE13%20RUSSIA.pdf, Accessed 5 June 2020.

[5] Zoellner, T (2017), *'The Struggle Behind the Trans-Siberian Railway'*, Reconnecting Asia, 5 July 2017, https://reconnectingasia.csis.org/analysis/entries/struggle-behind-trans-siberian/, Accessed 4 June 2020.

[6] Scientific American (1900), *'The Trans-Siberian Railway'*, Scientific American, 82(4), 27 January 1900, https://www.jstor.org/stable/10.2307/26121401, Accessed 3 June 2020.

[7] Istoriya.ru (undated), *'Transsib'*, Russian-Military Historical Society, https://histrf.ru/lenta-vremeni/event/view/transsib, Accessed 4 June 2020.

[8] RZhD (undated), *'Brief description of the Trans-Siberian Transport Corridor'*, https://eng.rzd.ru/en/9627/page/103290?id=11923#main-header, Accessed 13 July 2020.

and was able to place 1,200 troops per day at the front.[9] While Russia's military and economic thinking has shifted somewhat since then, the TSR and Russia's railways still serve an important military and economic purpose, both in connecting up the East and Western parts of the country, and in transporting troops and supplies around during Russia's frequent military exercises. For example, during Russia's Vostok ('*East*' in Russian) military exercises held in 2018 in the Far Eastern regions of the country, the railways were used to demonstrate how Russia could assemble troops quickly in a region where transport infrastructure is more poorly developed.[10]

The TSR still functions as Russia's main railway network, with various offshoots including the Baikal-Amur line (running north from the TSR and linking up to the Pacific Ocean). The BAM line is almost 3,220 km long, and passes through several mountain ranges and rivers, with around half the route running through permafrost. Construction of the BAM line, which was completed in 1989, required hundreds of new bridges, rail stations and tunnels, as well as the establishment of more than 60 new towns along the route.[11]

The railways also fulfill a social and urban function. Russia's centralized form of governance has historically demanded the establishment of regional administrative centers all over the country, which ultimately serve Moscow's political and economic needs. But Russia's urban population did not experience significant growth until the nineteenth century, during the industrial revolution, when there became a more pressing need for larger cities and towns to address the demographic demand.

As a result, many cities during the Soviet Union were specifically established to meet the needs of certain industrial projects, such as Magnitogorsk which was established in 1929 to produce metals, as its name suggests.[12] There are now hundreds of these 'monotowns' all of the country, many of whose economic growth has stagnated and which have fallen into disrepair, with the working-age population moving away to larger cities. The monotowns now present a serious economic burden for the Russian government, as the Kremlin cannot agree on whether to increase rail links and connectivity between these ailing and isolated areas, or whether to abandon them altogether.

[9] Scientific American (1904), 'Military aspects of the Trans-Siberian Railway', Scientific American 90(16), 16 April 1904, https://www.jstor.org/stable/10.2307/24988384, Accessed 4 May 2020.

[10] Norberg. J. (2018), 'Vostok 2018: about the Russian military's brain, not its muscles', MEMO 6470, Swedish Defence Research Agency (FOI), https://www.foi.se/report-summary?reportNo=FOI%20MEMO%206470, Accessed 10 June 2020.

[11] RZhD (undated) '*History*', https://eng.rzd.ru/en/9627, Accessed 12 July 2020.

[12] Markevich, A. and Mikhailova, T. (2012), '*Economic Geography of Russia*', New Economic School, Moscow, February 2012, https://pdfs.semanticscholar.org/16ed/4b58dad340b475d86cffe77bdd91a53921e2.pdf, Accessed 3 May 2020.

3 Political Frameworks for Russia's Transport Infrastructure

There are several state-initiated strategies that govern the development of Russia's transport network, but the successful implementation of many of these plans are subject to changes in Russia's own economic circumstances, as well as global trends.

As part of President Vladimir Putin's so-called 'May Decrees'—action plans that he announced upon his resumption of the presidency in May 2018—he has pressed for improvements to Russia's economic cohesion, by increasing the quality of Russia's roads, railways, waterway and air infrastructure.[13] But as with many of his decrees, his pronouncements in 2018 were vague and imprecise, and did not offer concrete plans to his regional administration about how these upgrades ought to be carried out.

Putin appears to have recognized the importance of addressing the stagnating mono-towns and linking up urban centers through rail improvements. In his annual address to the Federal Assembly in 2018 he noted that urban development would likely be the driving force behind the country's future economic improvement.[14] Following Putin's resumption of the presidency for another term of office in 2018, the Russian government produced a Strategy for Spatial Development up to 2024, which categorized Russian cities based on their economic potential, which could then be linked together through road or rail to form investment and education hubs.[15]

All of this runs in parallel with the Ministry of Transport's strategy up to 2030, the main strategy that governs the development of rail transport. Its main priority appears to be constructing high-speed rail links to connect up urban centers. As part of this strategy, Moscow allocated RUB1, 253bn to develop railway infrastructure, larger than any other amount apportioned for infrastructure upgrades.[16] The construction of thousands of kilometers of new routes, upgrading freight lines for heavy axel loads and acquiring 23,300 new locomotives and almost one million freight cars, are all part of these planned upgrades.[17]

These strategies and the funds apportioned to implement them do indicate the importance with which the Russian government appears to accord these improvements. But as with many major projects in Russia, the outcomes do not always match the stated intentions.

Development plans

---

[13] Putin, V.V. (2018), '*Prezident podpisal Ukaz o natisionalniy tselyakh i strategicheskiy zadachakh razvitiya Rossiiskoi Federatsii na period do 2024 goda,*' 4 May 2018, http://kremlin.ru/events/president/news/57425, Accessed 15 May 2020.

[14] RBK (2018), '*Putin predlozhil razrabotat programme prostranstvennovo razvitiya Rossii*', 1 March 2018, https://realty.rbc.ru/news/5a97ca8a9a79475d3e2a6447, Accessed 4 June 2020.

[15] Pravitelstvo Rossii (2018), '*Ob utverzhdenii Strategii prostranstvennovo razvitiya do 2025,*' 14 February 2019, http://government.ru/docs/35733/, Accessed 6 June 2020.

[16] Ministry of Transport (2018), '*Kompleksniy plan modernizatsiy i rashireniye magistralnoy infrastrukturi na period 2018–2024*', http://government.ru/docs/34297/, Accessed 4 June 2020.

[17] Ibid.

There are several major plans already in the pipeline. Putin signed a decree in May 2020 to accelerate the modernization of Russia's railways, including the upgrading of the TSR and BAM lines, scheduled to be complete by 2023. There are plans to increase the capacity of these lines by 1.5 times, to transfer around 180 m tonnes of cargo annually.[18] Much of this work has been completed and some cargo bottlenecks have been reduced, which restricted the movement of transport in Far Eastern regions such as Zabaikal. These improvements have been supported by the commissioning of five large railway bridges, including across the Lena and Zeya rivers, which the government maintains will help improve connectivity within some of the more isolated Far Eastern regions.[19]

Most recently in 2019, long-awaited work was completed on an offshoot of the BAM line, running north to the remote territory of Yakutsk, for both passenger and freight rail. The concept had been originally proposed in 2011 when former prime minister Dmitry Medvedev visited the far-flung region and had pledged to increase its links with the rest of the country. The Yakutsk region saw a construction boom in the 1970s which had fizzled out by the mid 1980s, and while various Soviet-era committees discussed the importance of connecting up the region and laying new rail track, talks on construction dragged on for decades, with some projects such as road links delayed for more than 20 years.[20] Although it appears that this project is increasingly making headlines in Russia, the long delays to planned construction work epitomizes some of the government's many inconsistencies between its stated goals to improve infrastructure, and the end results that fall short of expectations.

Another major priority for Russia is to increase cargo traffic along transport routes from Siberia to the north of Russia, with the aim of improving access to the Baltic, Bely, Barents and Karsky seaports.[21] Unwilling to rely on and pay customs duties at sea ports owned by the Baltic states—who are NATO members -, Russia is increasingly moving towards self-reliance on its own port infrastructure to handle cargo in the north, particularly on the Baltic and Barents seas.[22] But Russia only has 6 ports in the North-West of the country, which handle a significant chunk of Russia's cargo turnover. In 2019, of Russia's total 309 m tonnes of exports, around a third—131 m tonnes—was funnelled through those six north western ports.[23]Increasing turnover at these ports would likely necessitate an upgrade in Russia's rail capacities in the north-west of the country, particularly if cargo volumes continue to grow steadily in the coming years.

Complications

[18] Zhuravlev, K. (2020), '*Na god ranshe: RZhD ukoryayet modernizatsiu*' 6 April 2020, https://www.gazeta.ru/business/2020/04/06/13038151.shtml, Accessed 4 June 2020.

[19] Ibid.

[20] Solodukhin, O. (2019), Sakha News, 26 July 2019, http://www.1sn.ru/231626.html, Accessed 5 June 2020.

[21] Ibid.

[22] Ferris, E. (2020), '*Unplugging the Baltic states: Why Russia's economic approach may be shifting*', Russia Matters 1 July 2020, https://russiamatters.org/analysis/unplugging-baltic-states-why-russias-economic-approach-may-be-shifting, Accessed 17 July 2020.

[23] Ibid.

However, one major complicating factor is that government departments appear to disagree about how best to allocate funding for infrastructure improvements, and there is little interest from private investors to support these projects. Government departments are divided over which rail lines should be given priority for funding, with economic arguments against upgrading more remote areas where there is little passenger or freight traffic.

Although the government has claimed to be prioritizing the development of Russia's eastern Siberian and Far Eastern urban networks, efforts to do this have been piecemeal, with funding usually unevenly distributed among regions. According to many of the government's key development plans, the priority for most projects has been upgrading high-speed rail to link up Moscow with cities such as Kazan, and the Siberian capital of Ekaterinburg.[24] State-controlled companies are expected to fund just under half of the scheduled upgrades to the TSR, with some contributions from the state, private investors and regional governments.[25] But few wealthy Russian private companies are prepared to invest in long-term rail projects which are high risk and potentially low yield, particularly as it is not clear whether links such as the BAM line are capable of accepting more cargo loads.

An additional issue is that in Russia, the line between business and politics can often be blurred, and large-scale infrastructure projects can occasionally take on a political dimension. This was epitomized by the construction of a rail and road bridge to the territory of Crimea, which was completed in December 2019—Russia had annexed Crimea in 2014 from Ukraine, a move which was internationally condemned at the time. Although the Russian government maintained that the road and rail bridges were a practical project designed to improve economic access to Crimea, they effectively cemented Russia's political and economic control over the peninsula, facilitating an increase in freight and passenger traffic.[26]

The US and EU have introduced sanctions against specific rail companies for their involvement in Crimea—most recently in January 2020 both the EU and US sanctioned a Moscow-based private rail company that launched a passenger service between Russia and Crimea, maintaining that this was a breach of international law.[27] This has increased the risk of falling under sanctions for some private companies working with the Russian government on particularly politicized projects, and so identifying willing co-funders for large projects can be problematic.

Economic challenges

---

[24] RZhD (2020), '*Perspektivi razvitiya vysokoskorostnovo dvizheniya v Rossii*', http://www.rzd-expo.ru/innovation/high_speed_traffic_and_infrastructure/prospects_for_the_development_of_high_speed_movement_in_russia.php, Accessed 4 June 2020.

[25] Ibid.

[26] RZhD Partner International (2018), '*Game of Trains*', 1–2(51–52), https://www.rzd-partner.ru/upload/iblock/9d5/International_1_2018_site_min.pdf, Accessed 6 June 2020.

[27] US Department of the Treasury (2020), '*Treasury sanctions illegitimate Russian-backed Crimean officials and railroad company linking Crimea to Russia*', 29 January 2020, https://home.treasury.gov/news/press-releases/sm889, Accessed 17 July 2020.

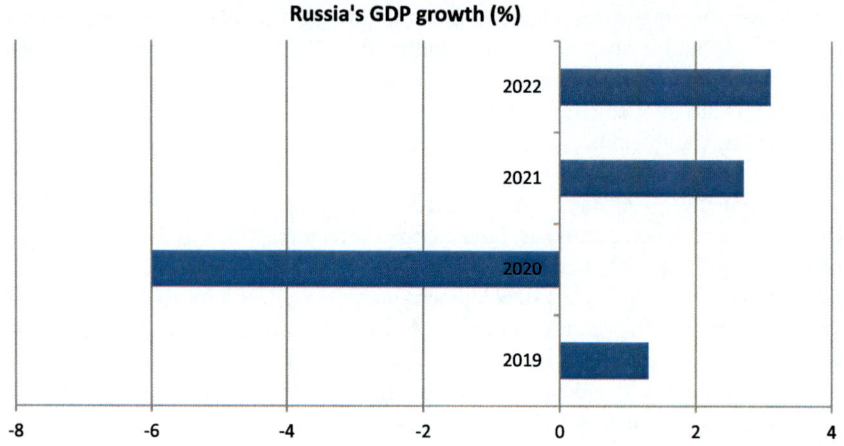

Fig. 1 Russia's projected GDP growth 2019–2022 (%). *Source* World Bank (projected figures for 2020 onwards)

Russia's rail sector naturally depends on the performance of Russia's own economy, as well as the international economies that it exports to.[28] As can be seen from Fig. 1, the outlook for Russia's economic growth in the coming years is poor, owing to a combination of depressed global oil prices, reduced trade activity in Europe—which is still a key trade partner for Russia—and the ongoing impact of the COVID-19 pandemic. In particular, growth in 2020 is likely to remain severely depressed with a contraction of 6%, which is highly likely to have a concomitant impact on the government's ability to finance rail infrastructure programmes.

To mitigate this, the Russian government in June 2020 released a draft action plan to shore up economic growth, at a total cost of RUB7triillion. Of this amount, it appears that a significant sum of RUB2.2trillion has been designated for infrastructure projects, although this includes a range of other important sectors such as the construction of roads, bridges and ports.[29] But Russia's construction sector has also been negatively impacted by COVID-19, and many large projects have been halted as a result. Russia's construction industry is thought to contract by 3.8% in 2020, even though the government has pledged additional financial support.[30]

---

[28] Murray, B. (2014), '*Russian Railway Reform Progra*mme,' European Bank for Reconstruction and Development, Working Paper June 2014, Accessed 3 May 2020.

[29] World Bank (2020), '*Russia: Recession and Growth under the shadow of a Pandemic'*, Russia Economic Report #43, 6 July 2020, http://pubdocs.worldbank.org/en/879631593984780605/July5-ENG-RER43.pdf, Accessed 18 July 2020.

[30] Thomas, E (2020), '*GlobalData: Russia's construction industry is expected to shrink by 3.8% in 2020'*, World Cement, 4 May 2020, https://www.worldcement.com/europe-cis/04052020/global data-russias-construction-industry-is-expected-to-shrink-by-38-in-2020/, Accessed 18 July 2020.

The serious economic downturn in 2020, and the contraction of markets in Asia owing to COVID-19 are likely to have an impact on many of Russia's export capabilities. As the sixth largest coal producer in the world, Russia's coal industry depends on its export, as domestic demand is relatively low.[31] Even as other countries in Europe are reducing their demand for coal, the Russian authorities have insisted that coal production is expected to grow from 550 to 670 million tonnes per year by 2035, an increase chiefly driven by exports. The current modernization programme for Russia's rail network also has a significant cost—the authorities have estimated that around RUB11trillion will need to be invested in the network by 2030 if any of the promised goals are to be achieved, and even with the backing of the regional and federal government, it is not clear how these funds will be raised.[32]

Other projects have also stalled owing to a lack of state or private funding. Putin in 2017 ordered the construction of a rail bridge in the Far Eastern region of Sakhalin, to link up remote territories in this region. But feasibility studies suggested that insufficient cargo volumes are transported through this region, and that the project is not economically viable. Even RZhD's own investment in the project, which totaled around RUB3.5bn, was not part of a state-supported investment project, and there has been no interest from coal companies or other private investors to help fund it.[33]

Investments are not the only economic obstacle to developing Russia's transport network. The international coal market as a whole has declined, prices for the product are low, and costs for the rail transport of coal within Russia are relatively high.[34] Although demand for coal has increased in China, large-scale exports to China require a reorientation of Russia's rail infrastructure towards the east, which is challenging to construct at pace.

As the below sections highlight, remote regions that are difficult to access require a significant amount of investment to construct railways through challenging and often mountainous terrain, and many infrastructure projects tend to be shelved following feasibility studies that suggest the project is unviable.

4 Fundamental Inadequacies

Russia's railway network today suffers from several fundamental inadequacies, which will continue to restrict the system's sustainability if they are not addressed. While there are other factors that affect the rail system, which are beyond the scope of

[31] Putin, V. (2019), '*Meeting of the Commission on the Strategy for the Development of the Fuel and Energy industry and Environmental safety*', 27 August 2018, Kemerovo, http://en.kremlin.ru/events/president/news/58382, Accessed 17 July 2020.

[32] Ibid.

[33] Adamchuk, O. (2020), '*Most na Sakhalin prevrashaetsya v proekt dlya geopoliticheskikh zadach*', 12 March 2020, https://www.vedomosti.ru/economics/articles/2020/03/11/824977-most-sahalin, Accessed 14 July 2020.

[34] Kommersant (2019), '*Vostochny vector rossiiskovo uglya*', 5 September 2019, https://www.kommersant.ru/doc/4080739, accessed 12 July 2020.

this analysis, this section unpacks four key- limitations that are restricting the sustainability of the rail sector, including Russia's challenging geography and terrain; its demographic density; geopolitical relationships that impact on trade relations, and the monopoly of the railways by a single state-controlled company.

Geography

The first issue is geographic and climate-related. Much of Russia's rail network is concentrated in the western part of the country, meeting the demand of a larger population there (see **Demography** below). This means that large swathes of eastern regions are relatively isolated, and occasionally cut off from the western regions seasonally, with few rail links other than the TSR or parts of the BAM line connecting them.

Terrain in the eastern region is particularly challenging—permafrost in parts of the Far East and Northern Arctic makes construction difficult and more expensive, and the humid summer seasons exacerbate the marshy conditions. The terrain also complicates the laying of new tracks. Russia has around 2.8 million rivers, but has comparatively few bridges– just 42,000—and few have been built since Putin came to power in 2000.[35] Moreover, many bridges in Russia are made of wood, a cheaper construction material, but which warps during the wet season. More remote and mountainous regions are also prone to landslides and falling rocks, which makes it channelling to construct rail tunnels, and there is often high groundwater surrounding areas near the rivers, creating an additional construction challenge.

But it was not until the 1950s that Russia began to systematically map some of the important geological considerations that needed to be taken into account when constructing railways, particularly in Far Eastern regions such as the Amur. Some southern parts of the Far East, which border China and are relatively more populated, have been quite well-studied, but the northern parts of the Far East and Arctic, and the more mountainous remote regions, are comparatively poorly understood by Russian geologists and engineers. This makes it more difficult to understand the region's specific construction requirements, and feasibility studies must often be conducted from scratch with little guidance.[36]

Another issue is distance. Most of Russia's coal is mined far from its destination markets, in the Kuzbass Basin in the centre of the country. It must then be transferred to seaports, loaded onto vessels and transported overseas. Most of the countries now importing Russia's coal are in the Asia–pacific region—in 2018 Russia exported 25.6 m tonnes of coal to South Korea, up from 19.3 m in 2015. There are several major pipeline projects underway to continue this reliance on coal, such as the Elegest-Kyzyl-Kuragino railway line, which will link the Siberian republic of Tuva (in the south of Russia) to Krasnoyarsk. Tuva is an important coal basin, with deposits

---

[35] Prestupnaya Rossiya (2018), '*Totalniy mostopad. Pochemu v Rossii mosti padayut sotnyami a novikh prakticheski ne stroyat?*', https://crimerussia.com/gromkie-dela/totalnyy-mostopad-pochemu-v-rossii-mosty-padayut-sotnyami-a-novykh-prakticheski-ne-stroyat/. Accessed 5 May 2020.

[36] Kvashuk, S. (2004), '*Geodinamika Priamurya i problem funktsionirovanniya seti zheleznikh dorog'*, Khabarovsk, https://static.freereferats.ru/_avtoreferats/01002637933.pdf, accessed 4 June 2020.

estimated at 20bn tonnes, but a lack of rail infrastructure has thus far restricted the development of these sites, as air and land transport are currently unsuitable. The idea in principle is not new—it was first mooted in 1950 and not reconsidered until 2009, and has been repeatedly postponed owing to a lack of funding. Despite much fanfare, this latest version of the project not scheduled to become operational until at least 2023, and there are few guarantees that it will be completed on time, given the previous delays.[37]

Other similarly geographically configured countries that also rely on the export of raw materials suffer from the same problems of connectivity and the financial viability of constructing new networks. Finland, for example, has a large population in Helsinki but an otherwise scattered rural population, and has invested significant funds in upgrading its freight transport network. Finland has also struggled with the long distances, as most of its railways are single-track, and many of the lines are already at full capacity. Although discussions to link up Finland's railway to the Arctic Ocean were at one time mooted, these plans have been discarded as the cost of a EUR3bn railway line across Lapland is economically inexpedient.[38]

These geographic challenges, coupled with Russia's ageing rail infrastructure and vehicles, exacerbated by a long history of under-investment in the rail sector, have left a challenging legacy. Transport costs are high in Russia, chiefly owing to the challenging terrain, which makes freight transport an expensive endeavour for international freight carriers.[39] This is highly likely to restrict Russia's ability to boost exports of its key commodities such as coal to its major clients.

Demography

The second issue is demographic. As can be seen by Fig. 2, the majority of Russia's population is located in Western regions, with the largest number of people in the Central, Volga and Southern districts. The Siberian and Far Eastern regions, which the TSR spans, are the least populous, which presents a challenge in identifying qualified local experts to construct new railways. There are only three large companies in Russia that produce equipment for constructing rail tracks, and very few institutions that offer engineering courses that are specific to railways.[40] Although there are some military academies that do teach these skills—indicating that there is still an important security element to the railways—they do not operate on the same scale as they did during the Soviet Union.

---

[37] Rail Freight (2020), *'This is Russia's most ambitious railway project: line to Siberia'*, 17 March 2020, https://www.railfreight.com/railfreight/2020/03/17/this-is-russias-most-ambitious-railway-project-line-to-siberia/, Accessed 15 July 2020.

[38] Ruohonen, K. (2015), *'Finland concentrates on rail network development'*, Global Railway Review, 7 April 2015, https://www.globalrailwayreview.com/article/23378/finland-concentrates-on-rail-network-development/, Accessed 13 September 2020.

[39] Railway Technology (2013), *'Russian railways: connecting a growing economy'*, 30 April 2013, https://www.railway-technology.com/features/featurerussia-railways-connecting-growing-economy/, Accessed 10 July 2020.

[40] RZhD Partner International (2018), *'Game of Trains'*, 1–2(51–52), https://www.rzd-partner.ru/upload/iblock/9d5/International_1_2018_site_min.pdf, Accessed 6 June 2020.

Federal District	Population (million)
Central	39.4
North-West	13.9
Southern	16.4
North Caucasus	9.9
Volga	29.2
Ural	12.3
Siberian	17.1
Far Eastern	8.1

Fig. 2 Russia's population density by Federal District (2020). *Source* Statdata.ru

The current Ministry of Transport Strategy and the Strategy for Spatial Development both appear to sideline Russia's many smaller cities and towns that are dispersed over its large territory. In the interests of political stability and prioritization of economic resources, there is little evidence that Moscow is interested in diverging from this course—although there is little risk of any political 'uprising' from these neglected monotowns that would pose a serious threat to the government. Other sparsely populated countries such as Canada suffer from similar demographic issues. There, most of the population lives close to the US border, and most transport networks are orientated east–west, from the Atlantic to Pacific Ocean, with most transport corridor operating the south of the country, leaving the north much more neglected.[41]

Moscow and St Petersburg are responsible for around 25% of Russia's GDP, and this trend looks likely to increase in the coming years as Russia's economy becomes increasingly dependent on the output, and as the population in these cities grows.[42] There is a considerable risk that many of those residing in monotowns and in more remote parts of Russia will continue to be neglected by the federal center, particularly as there seem to be few projects in the pipeline that aim to attract new sources of economic growth to these areas. A government fund was set up to develop these monotowns for the years 2016–2025, but it was largely ineffective and many smaller economically depressed towns that lacked the capacity to attract investors fell through the net, and did not receive any funding. Ultimately the fund was deemed so ineffective that it was terminated ahead of schedule.[43]

---

[41] Wiegmans, B et al. (2018), *'Rail and road freight transport network efficiency of Canada, member states of the EU and the USA',* Research in Transportation Business and Management (28), 16 October 2018, https://reader.elsevier.com/reader/sd/pii/S2210539517300767?token=825BA99B446DE42C98267F2C60A1251088664B0DF9524C523E42DB961B14005D8871614E865300D5DBEFBCC46D921BC9, Accessed 12 September 2020.

[42] Crowley, S. (2020), 'Global cities versus Russian Rustbelt Realities', PONARS Eurasia Policy Memo no.651, May 2020, https://www.ponarseurasia.org/memo/global-cities-versus-russian-rustbelt-realities, Accessed 5 June 2020.

[43] TASS (2019), *'Schotnaya palata: depressivniye monogoroda ne poluchali subsidii ot gosprogrammi s 2016 goda',* 29 July 2019, https://tass.ru/ekonomika/6713799, Accessed 10 June 2020.

The government has attempted to try to encourage its population to move to less well-developed regions, and has introduced programmes such as the Far Eastern Hectare scheme, which was established in 2016, to offer Russian citizens free land in exchange for permanent migration in more challenging areas.[44] The programme was later expanded to include areas of the Russian Arctic, but there has been very little uptake or interest in the scheme. Those who did apply outlined numerous problems with the project, including bureaucratic issues in organizing the paperwork and poor quality land. Most of the applications for land are from residents who already live in the Far East, eager to expand their agricultural holdings, which has done little to stimulate population growth.[45]

Monopoly on power

The third and perhaps most significant issue is that one state-controlled company monopolises the railway industry, which is vertically integrated. Rossiyskiye Zheleznye Dorogi (Russian Railways, RZhD), headed by Oleg Belozerov, has pushed out all other competitors and reduced opportunities for business competition on the rail transport market. RZhD by its own admission accounts for almost half of all the freight handled by Russia's entire transport system, including energy pipelines.[46]

While there are numerous private companies in Russia that offer freight services or manufacture locomotives, most of them are subcontractors of RZhD, which provides almost all railway infrastructure services, spearheads Russia's transport research and development, and is one of Russia's main employers, with just under a million workers.[47] The largest branch of RZhD is called the Central Directorate of Infrastructure, which oversees the management and technical conditions of the entire network, but the repair and maintenance of tracks and repairs of facilities is overseen by third-party companies. Many of the contractors that RZhD has outsourced to do this job lack the experience, equipment and staff necessary to complete these tasks. Allegations have surfaced in the Russian media that many of the labor costs for maintenance projects have been artificially inflated, with project funds stolen.[48]

Belozerov's predecessor, Vladimir Yakunin, also a long-time friend of Putin's, ran RZhD for almost ten years before he was unceremoniously dismissed in 2015, allegedly for corrupt practices. Yakunin was also under sanctions as a government

[44] Zvorykina, Y. (2019), '*On the prospects of the Far Eastern Hectare program extension to the Russian Arctic regions*', 13 May 2019, http://inveb.ru/en/articles-menu-eng/113-public-eng-2018-03-01, Accessed 18 June 2020.

[45] Deutsche Welle (2018), '*Shto ne tak c 'dalnevostochnim gektarom*', 1 February 2018, https://www.dw.com/ru/%D1%87%D1%82%D0%BE-%D0%BD%D0%B5-%D1%82%D0%B0%D0%BA-%D1%81-%D0%B4%D0%B0%D0%BB%D1%8C%D0%BD%D0%B5%D0%B2%D0%BE%D1%81%D1%82%D0%BE%D1%87%D0%BD%D1%8B%D0%BC-%D0%B3%D0%B5%D0%BA%D1%82%D0%B0%D1%80%D0%BE%D0%BC/a-42404994, Accessed 16 June 2020.

[46] RZhD (undated), '*The Company*', https://eng.rzd.ru/en/9498, Accessed 10 July 2020.

[47] RZhD (undated), '*Overview*', https://eng.rzd.ru/en/9635?redirected, Accessed 5 July 2020.

[48] Khazov-Cassia, S. (2016), '*Russian whistle-blower pulls back cover on railways corruption*', Radio Free Europe, 10 October 2016, https://www.rferl.org/a/russia-railways-yakunin-whistle-blower-corruption/28042893.html, Accessed 17 July 2020.

official for his role in Russia's annexation of the Ukrainian territory of Crimea in 2014.[49] But under Yakunin's leadership the company hemorrhaged funds, losing RUB99bn (USD1.6bn) in 2015, Yakunin's final year.[50][51] There is also evidence to suggest that RZhD, while required to allow purchasing deals to be competitive, often flouts this requirement.[52] Although this practice appeared to be particularly widespread under Yakunin's leadership, the detention of businessman Valery Markelov in 2018, who is thought to have earned billions from state contracts from RZhD, indicates that this may be an ongoing issue.[53]

RZhD also appears to have significant reach across the country. Even in Yakutia, whose rail infrastructure is technically governed by Yakutia Railways, it seems that RZhD has a strong presence. RZhD is the majority shareholders in Yakutia Railways' and many of its members sit on the company board, allowing RZhD a significant say in the policies and direction of the company.[54] Other smaller companies have complained about RZhD's dominance of the sector and have pushed (with minimal success) for better and more equal access to infrastructure.[55]

Geopolitical considerations

The fourth issue is Russia's geopolitical relationships with foreign countries, which have been significantly impacted by Russia's actions abroad over the past few years.

Russia's annexation of Crimea in 2014 and subsequent military involvement in the conflict in eastern Ukraine prompted many western countries to introduce wide-reaching economic and individual sanctions on Russia.[56] The sectoral sanctions in particular have prevented Russia from accessing western capital markets, as well as concluding trade deals in economically significant sectors such as oil, gas and defence. While this has not had an explicit impact on the railway systems—none of Russia's main companies involved in the railways are explicitly under international sanctions—the sanctions have restricted Russia's access to international financing

---

[49] US Department of the Treasury (2014), '*Treasury sanctions Russian officials, members of the Russian leadership's inner circle and an entity for involvement in the situation in Ukraine*', 20 March 2014, https://www.treasury.gov/press-center/press-releases/Pages/jl23331.aspx, Accessed 4 July 2020.

[50] Ibid.

[51] RZhD (undated), '*Results and reporting*', https://eng.rzd.ru/en/9637?redirected, Accessed 5 July 2020.

[52] Ibid.

[53] Mironenko, P. (2018), '*Corruption arrest sheds rare light on illicit big business in Russia*', The Bell, 5 October 2018, https://medium.com/@thebell.io/corruption-arrest-sheds-rare-light-on-illicit-big-business-in-russia-5e3f7bac57f5, Accessed 17 July 2020.

[54] Yakutia Railways (undated), '*Board of Directors*', https://rw-y.ru/okompanii/sovdir/, Accessed 6 July 2020.

[55] MaxConference (2018), '*Conference: Railway transportation of mining and metallurgical cargo*', 4–5 October 2018, http://www.maxconf.ru/event/77/, accessed 6 June 2020.

[56] European Council of the EU (2020), '*EU restrictive measures in response to the crisis in Ukraine*', 1 July 2020, https://www.consilium.europa.eu/en/policies/sanctions/ukraine-crisis/, Accessed 5 July 2020.

and bank loans, which have limited its options for co-financing major infrastructure investment projects.

The European Bank for Reconstruction and Development (EBRD) used to be a major investor in Russia's rail network. One of its largest loans it disbursed was in 2012, worth USD78.6 m, to Globaltrans, a private Russian rail freight company, to purchase 10,000 modern rail cars. However, many of the EBRD's investment deals in Russia have soured—in 2016 the EBRD sold off its stake in TransContainer, which it had purchased in 2008 as part of its attempts to reform Russia's rail industry, at a loss of USD105m, according to exchange rates at the time. TransContainer is a company that operates RZhD's rolling stock. The EBRD's Russia operations have significantly declined since then, and given the challenging relations with the West, there are few indications that the EBRD intends to restart these operations in Russia.[57]

Over the past five years, Russia has mooted the idea of allowing foreign investors such as Japan to jointly transport cargo across the TSR—a topic discussed at the Eastern Economic Forum (EEF), a high-profile investment event held in the Far Eastern city of Vladivostok since 2015, designed to attract FDI from predominantly Asian partners.[58] But any deeper coordination has been stymied by Russia and Japan's difficult political relationship, which hinges on a territorial solution to the disputed Kuril Island chain (known in Russia as the Northern Territories). Most economic deals, including investment plans or allowing Japanese businesses and freight carriers to have less restricted operating access to the Trans Siberian Railway, depend on the resolution to this territorial dispute, which is highly unlikely to be forthcoming.[59]

Russia also has a challenging relationship with China, which makes collaboration and joint investments more difficult. Wary of China's growing demographic and economic clout, as well as some of its military ambitions, Russia is keen to avoid being shoehorned into a role as junior exporter, and so has proceeded with deals very cautiously. There has been some evidence of Chinese investment in rail—in April 2018 China Railway Dongfang Group signed an agreement to finance a high-speed railway link between the Chinese city of Suifenhe and Vladivostok, at a cost of around USD7bn, and confirmed the project's feasibility in August of that year.[60]

But since then there has been very little tangible progress on the project, even as China invested some USD117bn in the development of its own railways in 2018, and completed construction on the last section of a high-speed rail link from Beijing to

---

57 Corcoran, J. (2016), '*EBRD loses $100 m as it exits Russia's TransContainer*', 15 April 2016, https://www.intellinews.com/ebrd-loses-100mn-as-it-exits-russia-s-transcontainer-95307/, Accessed 20 July 2020.

58 RZhD Partner International (2018), '*Game of Trains*', 1–2(51–52), https://www.rzd-partner.ru/upload/iblock/9d5/International_1_2018_site_min.pdf, Accessed 6 June 2020.

59 Brown, J. (2020), '*Russia's revised constitution shows Putin is no friend of Japan*', Royal United Services Institute (RUSI), 6 July 2020 https://rusi.org/commentary/russias-revised-constitution-shows-putin-no-friend-japan, Accessed 13 July 2020.

60 Rosscongress (2019), '*Investitsii v infrastrukturu*', Special for the Eastern Economic Forum, 2019, https://infraone.ru/sites/default/files/analitika/2019/investitsii_v_infrastrukturu_dalny_vostok_2019_infraone_research.pdf, Accessed 14 June 2020.

Hong Kong, strongly suggesting that its rail links with Russia are further down its priority list.[61]

Part of Russia's resistance to foreign investment is enshrined in law. The Russian government considers its railways transport networks to be of 'strategic significance'. This refers to a specific law that designates certain sectors of the Russian economy as having 'strategic importance', upon which Russia's long-term economic or existential security depends. The law limits foreign investment or joint ventures with Russian partners in these sectors, which include defence, oil and gas. More recently, railway transport was added to the list, which has stymied opportunities for international partners to invest in the Russian rail network.[62] Given the security status that the Kremlin has conferred upon transport infrastructure, there is highly unlikely to be any easing of the restrictions that govern foreign investment or ownership of the network.

5 What Has to Change?

While there are some insurmountable obstacles to improving Russia's rail systems, such as its geography and climate, there are three key ways in which its transport network could improve its sustainability in the coming years.

Breaking the mold

The first and most important step would be for RZhD's monopoly to be broken up, and tenders for construction and freight services to be made freer and fairer, as well as open to foreign investors. This is a considerable task which is highly likely to face serious resistance from the highest levels of government.

This could include the separation of infrastructure from operations, to create a space for competition—a model that the EU follows. Vertically integrated and state-owned companies such as RZhD are a model that many countries are increasing moving away from, and network utilities could be unbundled, under separate ownership from RZhD.[63] A perhaps more palatable option for RZhD could be a 'partial unbundling', whereby some activities would be separated from infrastructure, but with RZhD technically still in charge. This would allow some third party access without entirely breaking up RZhD's dominance of the sector.[64]

---

[61] DV Ross (2020), '*Skorostnuyu zheleznuyu dorogu v kitaiskom Heihe postroyat k 2024 gody. A Vladivostok-Suifenhe kogda?*' 26 June 2020, http://trud-ost.ru/?p=714189, Accessed 28 June 2020.

[62] Clifford Chance (2020), '*A legal overview of foreign investment in Russia's 'strategic' sectors*', May 2020, https://www.cliffordchance.com/content/dam/cliffordchance/briefings/2020/05/2005_Client%20Briefing%20-%20A%20Legal%20Overview%20of%20Foreign%20Investment%20in%20Russia's%20Strategic%20Sectors.pdf, Accessed 10 June 2020.

[63] Pittman, R. (2005), '*Make, Buy, or Some of Both? The Case of Russian Railways,*' SSRN, http://dx.doi.org/10.2139/ssrn.849504, Accessed 6 June 2020.

[64] Ibid.

But for Russia's railway sector to truly reform, there must be greater business competition to stimulate the private sector and foreign investment. Steps must be taken to improve the competitiveness of Russia's business environment, particularly in construction and freight tenders, to allow better access for foreign companies and smaller Russian companies. Although the 'strategic sector' law only limits—and does not entirely restrict—foreign investment in transport infrastructure, there are numerous other formal and informal means by which foreign operators are prevented from accessing Russia's transport network.

While Russia has improved its position on global ranking indices such as the World Bank's annual *Doing Business* report—which evaluates the ease of conducting business operations across a range of indicators -, conditions for foreign operators in Russia still remain extremely challenging. Large state-controlled enterprises such as RZhD continue to dominate Russia's financial resources, tax benefits and favourable contractual relationships. This picture is complicated by Russia's method of regional governance, with heads of regions choosing to implement and apply federal laws and regulations in a non-uniform way, often colored by personal interests. In particular, applying for and being granted operating and export licenses, as well as construction permits—the kinds of bureaucratic endeavors required for Russian and foreign operators to participate in the rail sector—are well known for being particularly burdensome.[65]

Moreover, any moves to ease access to Russia's rail network for foreign investors—particularly from the West—would naturally require an improvement in diplomatic relations between Russia those governments. There is unlikely to be a significant rapprochement between Russia and the Westin the coming years, or at least as long as President Putin continues in office.

There are certainly many smaller steps that the Russian government could take to improve rail efficiency or the quality of its performance, such as the electrification of larger sections of track, or increasing the number of rail cars to meet cargo demands. But barring an overhaul of political structures in Russia, the Kremlin and RZhD are unlikely to take measures to reduce their control over the transport network, which will unfortunately ensure that the sector remains uncompetitive and challenging to invest in.

Brokering international partnerships

The lack of expertise in Russia's railway construction, geological mapping and engineering could be mitigated by increasing Russia's partnerships with other countries, particularly states who are as similarly reliant on their rail network.

Russia has already attracted interest in rail partnerships from countries such as Finland, who have expressed willingness to invest in a high-speed train link from Moscow-Helsinki route to promote tourism. This particular project is in a

---

65 Kuszmir, J. (2016), '*Doing business in Russia: The main political risks and challenges for international companies'*, in State Capture, Political Risks and International Business Cases from Black Sea Region Countries, Routledge, ed. Leitner, J, Messner, H. November 2016.

nascent stage and is unlikely to be completed within the next three years.[66] But learning from countries such as Finland whose geographies are similarly configured—with one large urban centre (Helsinki) and a geographically dispersed population—would likely assist Russia in improving the sustainability of its own rail services. Unlike other Scandinavian countries, Russia's diplomatic relationship with Finland is reasonably positive, and there could be significant scope for both countries to share expertise in rail transport.

Prior to Russia's annexation of Crimea, Russia had numerous successful joint ventures with European countries such as Germany and France. France's Alstom Transport does have a large (20%) stake in the privately-owned Transmashholding (TMH)—which RZhD also has a stake in—to manufacture freight and passenger locomotives for the Russian market. But western sanctions have put companies such as TMH under pressure, as there is significant uncertainty about the security of Russian investments, which has prompted TMH to attempt to become more self-sufficient, rather than reliant on its international partners.[67]

Russia could also learn from examples of other countries such as Australia, which must also contend with a concentration of natural resources located in remote regions, far from its coastal port infrastructure. With few large lakes, river systems or natural harbors, rail plays an important role in long-distance freight transport. In the 1990s, the Australian rail system also recognized the need for diversification of ownership and underwent a major structural overhaul, with many of the state-controlled railways split up into smaller authorities.[68] Should the Russian leadership decide to unbundle RZhD, it could look to the Australian model to assist its diversification of ownership.

Adjusting demographic expectations

As outlined above, one of Russia's key issues is its demographic distribution, and the prioritization of economic and rail resources accordingly.

Many of Russia's strategies prioritize investments in the western part of the country, but the government could save a considerable amount of its budget by focusing on Russia's more economically promising cities and towns—some of which are located in remote regions, but have the potential to become drivers of economic growth, due to their proximity to large quantities of natural resources. Instead of attempting to jump-start failing mono-towns, or offering incentives for people to repopulate hostile regions, Russia might focus on shifting its workforce from regions in decline to those that are expanding (or have the potential to expand). This could include tax incentives or social security guarantees for people with specific types of engineering or rail expertise who choose to move to these regions.

---

[66] RBK (2020), *'V Khelsinki zainteresovalis skorostnoy zhelezoy dorogi iz Moskvi'*, 20 January 2020, https://www.rbc.ru/politics/20/01/2020/5e20b7499a79477a76433e80, Accessed 10 June 2020.

[67] Smith, K. (2019), *'Transmasholding targets further growth at home and abroad'*, International Railway Journal, 10 January 2019, https://www.railjournal.com/in_depth/transmashholding-targets-further-growth-at-home-and-abroad, Accessed 15 July 2020.

[68] Webb, R. (2000), *'Issues in Rail Reform'*, Parliament of Australia Research Paper 14, 1999–2000, 7 March 2000, https://www.aph.gov.au/About_Parliament/Parliamentary_Departments/Parliamentary_Library/pubs/rp/rp9900/2000RP14, Accessed 5 July 2020.

Adjusting and reevaluating Russia's own demographic expectations would dictate the course of its railway construction, and could help organize the Ministry of Transport's priorities, moving away from investing in constructing rail bridges in regions that have an increasingly smaller population and few industrial prospects. The authorities could then redistribute their workforce to match those regions whose economies are growing, as outlined by Russia's own Strategy for Spatial Development up to 2024.[69] However, this process could be met with some resistance from locals, and would have to occur slowly over time for there to be positive effects.

6 Conclusions

Ultimately, the sustainability of Russia's transport network depends on the government's ability to correctly prioritize and allocate funding to the most feasible and viable projects. Part of the problem is that many of the rail projects that the Kremlin is attempting to promote in Far Eastern and Siberian regions are economically unviable or practically unfeasible, and often serve as purely symbolic gestures—either to assure locals that Moscow is paying attention, or to send a message to foreign partners such as China, to indicate that Russia is interested in boosting trade relations.

Disagreements amongst federal departments, which often vie amongst one another for funding and attention from President Putin, are likely to exacerbate the uneven way in which transport projects are considered and implemented.

Looking further ahead, climate change will continue to have an increasingly dramatic impact on Russia's geography, and its rail network will be obliged to adapt accordingly. The Russian government has had a fairly inconsistent approach to climate change, often appearing unwilling to implement changes that would reduce carbon emissions, even as Russia annually experiences numerous natural disasters such as floods, wildfires and melting Arctic ice that routinely cost the federal budget millions of rubles in repairs.[70]

Thanks to powerful business lobbies within Russia such as the Union of Industrialists and Entrepreneurs (RSPP), which have fought to maintain high production levels of gas, oil and coal, there is unlikely to be a downturn in the transport volumes of natural resources, even as many other countries move to reduce their consumption in the interests of environmental protection.[71] This means that Russia's rail engineers will be obliged to confront new geographical challenges such as melting permafrost, while at the same time attempting to maintain high cargo turnovers.

---

[69] World Bank (2011), *'Russia-reshaping economic geography'*, 1 June 2011, General Economy, Macroeconomics and Growth Study, Vol 1, https://documents.worldbank.org/en/publication/documents-reports/documentdetail/863281468107678371/russia-reshaping-economic-geography, Accessed 20 June 2020.

[70] Kozin, D. (2020), *'Is Russia finally waking up to climate change?'*, 9 March 2020, The Moscow Times, https://www.themoscowtimes.com/2020/03/04/is-russia-finally-waking-up-to-climate-change-a69517, Accessed 23 June 2020.

[71] Ibid.

Without significant investment in engineering expertise, most likely by establishing new technical colleges across different regions of Russia, and allocating more government funding to support student programmes specializing in railway engineering, Russia's ability to improve the sustainability of its rail networks will likely fall behind its international counterparts in the coming years.

References

Adamchuk O (2020) Most na Sakhalin prevrashaetsya v proekt dlya geopoliticheskikh zadach. 12 Mar 2020, https://www.vedomosti.ru/economics/articles/2020/03/11/824977-most-sahalin. Accessed 14 July 2020

Brown J (2020) Russia's revised constitution shows Putin is no friend of Japan. Royal United Services Institute (RUSI), 6 July 2020 https://rusi.org/commentary/russias-revised-constitution-shows-putin-no-friend-japan. Accessed 13 July 2020

Clifford Chance (2020) A legal overview of foreign investment in Russia's 'strategic' sectors. May 2020, https://www.cliffordchance.com/content/dam/cliffordchance/briefings/2020/05/2005_Client%20Briefing%20-%20A%20Legal%20Overview%20of%20Foreign%20Investment%20in%20Russia's%20Strategic%20Sectors.pdf. Accessed 10 June 2020

Corcoran J (2016) EBRD loses $100m as it exits Russia's TransContainer. 15 Apr 2016, https://www.intellinews.com/ebrd-loses-100mn-as-it-exits-russia-s-transcontainer-95307/. Accessed 20 July 2020

Crowley S (2020) Global cities versus Russian Rustbelt Realities. PONARS Eurasia Policy Memo no.651, May 2020, https://www.ponarseurasia.org/memo/global-cities-versus-russian-rustbelt-realities. Accessed 5 June 2020

Welle D (2018) Shto ne tak c 'dalnevostochnim gektarom. 1 Feb 2018, https://www.dw.com/ru/%D1%87%D1%82%D0%BE-%D0%BD%D0%B5-%D1%82%D0%B0%D0%BA-%D1%81-%D0%B4%D0%B0%D0%BB%D1%8C%D0%BD%D0%B5%D0%B2%D0%BE%D1%81%D1%82%D0%BE%D1%87%D0%BD%D1%8B%D0%BC-%D0%B3%D0%B5%D0%BA%D1%82%D0%B0%D1%80%D0%BE%D0%BC/a-42404994. Accessed 16 June 2020

Ross DV (2020) Skorostnuyu zheleznuyu dorogu v kitaiskom Heihe postroyat k 2024 gody. A Vladivostok-Suifenhe kogda? 26 June 2020, http://trud-ost.ru/?p=714189, Accessed 28 June 2020

European Council (2020) EU restrictive measures in response to the crisis in Ukraine. 1 July 2020, https://www.consilium.europa.eu/en/policies/sanctions/ukraine-crisis/. Accessed 5 July 2020

Ferris E (2020) Unplugging the Baltic states: why Russia's economic approach may be shifting. Russia Matters 1 July 2020, https://russiamatters.org/analysis/unplugging-baltic-states-why-russias-economic-approach-may-be-shifting. Accessed 17 July 2020

Grubesic T, Matisziw T (2013) A typological framework for categorizing infrastructure vulnerability. GeoJournal 7(2), Springer, https://www.jstor.org/stable/42006320. Accessed 3 June 2020

Istoriya.ru (undated) Transsib. Russian-Military Historical Society, https://histrf.ru/lenta-vremeni/event/view/transsib. Accessed 4 June 2020

Khazov-Cassia S (2016) Russian whistle-blower pulls back cover on railways corruption. Radio Free Europe, 10 Oct 2016, https://www.rferl.org/a/russia-railways-yakunin-whistle-blower-corruption/28042893.html. Accessed 17 July 2020

Kommersant (2019) Vostochny vector rossiiskovo uglya. 5 Sept 2019, https://www.kommersant.ru/doc/4080739. Accessed 12 July 2020

Kozin D (2020) Is Russia finally waking up to climate change?, 9 Mar 2020. The Moscow Times. https://www.themoscowtimes.com/2020/03/04/is-russia-finally-waking-up-to-climate-change-a69517. Accessed 23 June 2020

Kuszmir J (2016) Doing business in Russia: the main political risks and challenges for international companies. In: Leitner J, Messner H (eds) State capture, political risks and international business cases from black sea region countries, Routledge

Kvashuk S (2004) Geodinamika Priamurya i problem funktsionirovanniya seti zheleznikh dorog. Khabarovsk, https://static.freereferats.ru/_avtoreferats/01002637933.pdf. Accessed 4 June 2020

Markevich A, Mikhailova T (2012) Economic geography of Russia. New Economic School, Moscow, Feb 2012, https://pdfs.semanticscholar.org/16ed/4b58dad340b475d86cffe77bdd91a53921e2.pdf. Accessed 3 May 2020

Max Conference (2018) Conference: Railway transportation of mining and metallurgical cargo. 4–5 Oct 2018, http://www.maxconf.ru/event/77/. Accessed 6 June 2020

Ministry of Transport (2018) Kompleksniy plan modernizatsiy i rashireniye magistralnoy infrastrukturi na period 2018–2024. http://government.ru/docs/34297/. Accessed 4 June 2020

Mironenko P (2018) Corruption arrest sheds rare light on illicit big business in Russia. The Bell, 5 Oct 2018, https://medium.com/@thebell.io/corruption-arrest-sheds-rare-light-on-illicit-big-business-in-russia-5e3f7bac57f5. Accessed 17 July 2020

Murray B (2014) Russian railway reform programme. European bank for reconstruction and development, working paper June 2014. Accessed 3 May 2020

Norberg J (2018) Vostok 2018: about the Russian military's brain, not its muscles. MEMO 6470, Swedish defence research agency (FOI), https://www.foi.se/report-summary?reportNo=FOI%20MEMO%206470. Accessed 10 June 2020

Pittman R (2005) Make, buy, or some of both? The case of Russian railways. SSRN. https://doi.org/10.2139/ssrn.849504. Accessed 6 June 2020

Rossii P (2018) Ob utverzhdenii Strategii prostranstvennovo razvitiya do 2025. 14 Feb 2019, http://government.ru/docs/35733/. Accessed 6 June 2020

Prestupnaya Rossiya (2018) Totalniy mostopad. Pochemu v Rossii mosti padayut sotnyami a novikh prakticheski ne stroyat?, https://crimerussia.com/gromkie-dela/totalnyy-mostopad-pochemu-v-rossii-mosty-padayut-sotnyami-a-novykh-prakticheski-ne-stroyat/. Accessed 5 May 2020

Putin V (2018) Prezident podpisal Ukaz o natisionalniy tselyakh i strategicheskiy zadachakh razvitiya Rossiiskoi Federatsii na period do 2024 goda. 4 May 2018, http://kremlin.ru/events/president/news/57425. Accessed 15 May 2020

Putin V (2019) Meeting of the commission on the strategy for the development of the fuel and energy industry and environmental safety. 27 Aug 2018, Kemerovo, http://en.kremlin.ru/events/president/news/58382. Accessed 17 July 2020

Rail Freight (2020) This is Russia's most ambitious railway project: line to Siberia. 17 Mar 2020, https://www.railfreight.com/railfreight/2020/03/17/this-is-russias-most-ambitious-railway-project-line-to-siberia/. Accessed 15 July 2020

Railway Technology (2013) Russian railways: connecting a growing economy. 30 Apr 2013, https://www.railway-technology.com/features/featurerussia-railways-connecting-growing-economy/. Accessed 10 July 2020

Railway Technology (2013) Russian railways: connecting a growing economy. 30 Apr 2013, https://www.railway-technology.com/features/featurerussia-railways-connecting-growing-economy/. Accessed 3 June 2020

RBK (2018) Putin predlozhil razrabotat programme prostranstvennovo razvitiya Rossii. 1 Mar 2018, https://realty.rbc.ru/news/5a97ca8a9a79475d3e2a6447. Accessed 4 June 2020

RBK (2020) V Khelsinki zainteresovalis skorostnoy zhelezoy dorogi iz Moskvi. 20 Jan 2020, https://www.rbc.ru/politics/20/01/2020/5e20b7499a79477a76433e80. Accessed 10 June 2020

RIA Novosti (2020) Putin poprosil RZhD po minimumu sokrashats investitsii v realniy sector. 25 May 2020, https://ria.ru/20200525/1571951548.html. Accessed 18 July 2020

Rosscongress (2019) Investitsii v infrastrukturu. Special for the Eastern Economic Forum, 2019, https://infraone.ru/sites/default/files/analitika/2019/investitsii_v_infrastrukturu_dalny_vostok_2019_infraone_research.pdf. Accessed 14 June 2020

Rossiskiy Investitsionnoye Agentstvo, (2020) Maschtabi Rossii v Tsifrakh, https://www.investment-in-russia.com/site/page?view=ROSSIYA-V-CIFRAH Accessed 2 June 2020

Ruohonen K (2015) Finland concentrates on rail network development. Global railway review, 7 Apr 2015, https://www.globalrailwayreview.com/article/23378/finland-concentrates-on-rail-network-development/. Accessed 13 Sept 2020

RZhD (2020) Perspektivi razvitiya vysokoskorostnovo dvizheniya v Rossii. http://www.rzd-expo.ru/innovation/high_speed_traffic_and_infrastructure/prospects_for_the_development_of_high_speed_movement_in_russia.php. Accessed 4 June 2020

RZhD (undated) History. https://eng.rzd.ru/en/9627, Accessed 12 July 2020

RZhD (undated) Brief description of the Trans-Siberian transport corridor. https://eng.rzd.ru/en/9627/page/103290?id=11923#main-header. Accessed 13 July 2020

RZhD (undated) Overview, https://eng.rzd.ru/en/9635?redirected. Accessed 5 July 2020

RZhD (undated) 'Results and reporting. https://eng.rzd.ru/en/9637?redirected. Accessed 5 July 2020

RZhD (undated) The company. https://eng.rzd.ru/en/9498. Accessed 10 July 2020

RZhD Partner International (2018) Game of trains. 1–2(51–52), https://www.rzd-partner.ru/upload/iblock/9d5/International_1_2018_site_min.pdf. Accessed 6 June 2020

Scientific American (1900) The Trans-Siberian railway. Scientific American, 82(4), 27 Jan 1900, https://www.jstor.org/stable/10.2307/26121401. Accessed 3 June 2020

Scientific American (1904) Military aspects of the Trans-Siberian railway. Scientific American 90(16), 16 Apr 1904, https://www.jstor.org/stable/10.2307/24988384. Accessed 4 May 2020

Smith K (2019) Transmasholding targets further growth at home and abroad. International Railway Journal, 10 Jan 2019, https://www.railjournal.com/in_depth/transmashholding-targets-further-growth-at-home-and-abroad. Accessed 15 July 2020

Solodukhin O (2019) Sakha News, 26 July 2019, http://www.1sn.ru/231626.html. Accessed 5 June 2020

TASS (2019) Schotnaya palata: depressivniye monogoroda ne poluchali subsidii ot gosprogrammi s 2016 goda. 29 July 2019, https://tass.ru/ekonomika/6713799. Accessed 10 June 2020

Thomas E (2020) GlobalData: Russia's construction industry is expected to shrink by 3.8% in 2020. World Cement, 4 May 2020, https://www.worldcement.com/europe-cis/04052020/globaldata-russias-construction-industry-is-expected-to-shrink-by-38-in-2020/. Accessed 18 July 2020

US Department of the Treasury (2014) Treasury sanctions Russian officials, members of the Russian leadership's inner circle and an entity for involvement in the situation in Ukraine. 20 Mar 2014, https://www.treasury.gov/press-center/press-releases/Pages/jl23331.aspx. Accessed 4 July 2020

US Department of the Treasury (2020) Treasury sanctions illegitimate Russian-backed Crimean officials and railroad company linking Crimea to Russia. 29 Jan 2020, https://home.treasury.gov/news/press-releases/sm889. Accessed 17 July 2020

Webb R (2000) Issues in rail reform. Parliament of Australia research paper 14, 1999–2000, 7 Mar 2000, https://www.aph.gov.au/About_Parliament/Parliamentary_Departments/Parliamentary_Library/pubs/rp/rp9900/2000RP14. Accessed 5 July 2020

Wiegmans B et al (2018) Rail and road freight transport network efficiency of Canada, member states of the EU and the USA. Research in transportation business and management (28), 16 Oct 2018, https://reader.elsevier.com/reader/sd/pii/S2210539517300767?token=825BA99B446DE42C98267F2C60A1251088664B0DF9524C523E42DB961B14005D8871614E865300D5DBEFBCC46D921BC9. Accessed 12 Sept 2020

World Bank (2011) Russia-reshaping economic geography. 1 June 2011, General Economy, Macroeconomics and Growth Study, vol 1, https://documents.worldbank.org/en/publication/documents-reports/documentdetail/863281468107678371/russia-reshaping-economic-geography. Accessed 20 June 2020

World Bank (2017) Case study—Russian railways. 27 Dec 2017, https://ppiaf.org/sites/ppiaf.org/files/documents/toolkits/railways_toolkit/PDFs/RR%20Toolkit%20EN%20New%202017%2012%2027%20CASE13%20RUSSIA.pdf. Accessed 5 June 2020

World Bank (2020) Russia: recession and growth under the shadow of a pandemic. Russia economic report #43, 6 July 2020, http://pubdocs.worldbank.org/en/879631593984780605/July5-ENG-RER43.pdf. Accessed 18 July 2020

Yakutia Railways (undated) Board of Directors. https://rw-y.ru/okompanii/sovdir/. Accessed 6 July 2020

Zhuravlev K (2020) Na god ranshe: RZhD ukoryayet modernizatsiu. 6 Apr 2020, https://www.gazeta.ru/business/2020/04/06/13038151.shtml. Accessed 4 June 2020

Zoellner T (2017) The struggle behind the Trans-Siberian railway. Reconnecting Asia, 5 July 2017, https://reconnectingasia.csis.org/analysis/entries/struggle-behind-trans-siberian/. Accessed 4 June 2020

Zvorykina Y (2019) On the prospects of the far Eastern hectare program extension to the Russian Arctic regions. 13 May 2019, http://inveb.ru/en/articles-menu-eng/113-public-eng-2018-03-01. Accessed 18 June 2020

Development of Sustainable Transport Corridors the Scandria-Case

Lisa Hartmann, Martin Jung, Philip Michalk, and Sorin Sima

Abstract The paper describes the status quo of freight flows in the Scandria Corridor (compare SCANDRIA-Alliance), based on a number of surveys and studies, conducted throughout the Scandria2Act project (2016–2019). Specifically this paper describes the freight flows in detail, by combining surveys conducted within the project, with statistical data from different sources, creating a comprehensive picture of freight flows between continental Europe a Scandinavia. A multifactor-decision model for freight forwarders is implemented, to calculate the boundary condition that need to be met in order to facilitate an actual shift from road to rail. These conditions are described in terms of transport costs and time, which enable the analysis of effects of different political measures, and of different market conditions on the transport flows. In conclusion this paper discusses the political boundary conditions and the administrative and spatial framework, in which such shifts are being fostered.

Keywords Decision modelling · Urban nodes · TEN-T · Freight flows · Transport corridors

1 Introduction

It is an explicit goal of the European Union, to foster a modal shift from road to rail. The study and the accompanying model we present in this paper, shall help to understand the boundary conditions of the transport market and inform political decision makers on 'which measures will lead to which outcomes', when deciding on actions towards the promotion of rail freight transport.

The novelty of our paper lies in, on the one side, in a survey, conducted by our partners (ports and public authorities) in German and Swedish ports. We used the survey data and additional data from road counting stations, shipping companies and timetables and combined them with data-fusion-methods to produce a comprehensive picture of transport flows in the corridor.

L. Hartmann · M. Jung · P. Michalk (✉) · S. Sima
Forschungsgruppe Verkehrslogistik, Technische Hochschule Wildau (TU Wildau), Wildau, Germany
e-mail: philip.michalk@th-wildau.de

M. Marinov and J. Piip (eds.), *Sustainable Rail Transport 4*, Lecture Notes in Mobility,
https://doi.org/10.1007/978-3-030-82095-4_10

Additionally we developed a multi-factorial decision model in order to understand which effects and interactions of effects, as well as measures could have on the transport flows throughout a corridor, with a focus in a modal shift from road to rail.

2 The Scandria®Corridor

The Scandria®Alliance is a transnational communication and cooperation platform which promotes the shortest geographic link between Scandinavia and the Adriatic Sea. It was developed around the Scandinavian-Mediterranean Corridor (one of the nine Core Net Corridors (CNC) of the TEN-T Core Net). This corridor is a crucial axis for the European economy, linking the major urban centres and economic zones in Germany, and Italy to Scandinavia and the Mediterranean.

The corridor is set to experience some changes in its infrastructure. Two projects with major influences on the transport flows of this particular corridor are the Brenner Base Tunnel, which will probably be operational by 2028, and the fixed Fehmarn-Belt link, which will not become operational before 2028. However corridors need to be developed holistically on multiple level. This includes the infrastructure and the regulatory level, but also more or less informal measures, fostering cooperation in numerous areas. The Scandria Alliance, for example, works on solutions for the areas of Clean Fuels, Multimodal Transport, Digitalization and Cross Border Infrastructure. All in order to lower barriers for green transport and also to foster a shift from road to rail transport.

In order to develop and employ effective strategies, a fundamental understanding of freight transport flows and the factors that influence them is necessary.

This was one of the main goals of the Scandria2Act project which ran from 2016 to 2019. Some of the main findings are presented in this paper.

3 Status Quo of Freight Flows in the Scandria® Corridor

The quantification of freight flows in the Scandria® corridor underlines the need for multimodal solutions to generate a sustainable modal shift.

The aim of the freight flow analysis is to gain insight into the existing transport volume in the Scandria® Corridor in order to make recommendations for a modal shift to rail. Research aspects such as these contribute to a long-term relief of road traffic for a CO2-efficient transport system and achieve broad acceptance amongst key decision makers.

In the past, transnational data from both public and private organizations have been collected and consolidated in EU projects to determine the existing freight flows in the Scandria® corridor.

By systematically combining the data, it is possible to obtain a representative overview of freight flows in the [northern] Scandria® corridor.

3.1 Methodology

The combination of survey results, public statistics and traffic counts as well as statistical modelling provided information on cross-border sea and land transport in the [northern] Scandria® corridor.

The modelling focused on the development of a holistic approach for the visualization and quantification of freight flows as well as the European-Scandinavian trade flows. For this purpose, the modes of transport and the analysis components were considered in consideration of the geographical conditions.[1]

This included a comprehensive consideration of the traffic across the Baltic Sea, from which the transport intensities and trade relations in the Scandria® corridor between the countries were derived. Through the involvement of operational participants in the surveys as well as public transport counts and trade balances, the multimodal capacities and flows along the Scandria® Corridor could be evaluated.

At the operational level, the data is based on a Swedish survey, which included almost 2.500 interviews with truck drivers on 14 ferry lines and 166 telephone interviews with haulage companies operating on the Öresund Bridge. Within this survey, more than 3.300 observations were also carried out, giving total results for around 4.000 vehicles. The interviews were carried out both on site in each relevant port in the survey area in connection with outbound transport and via telephone interviews with road freight transport companies. The questions in the interviews focused mainly on the points of departure and destination, the roads used, the type of goods and data on vehicles and drivers.

The dataset served as a basis for freight flows across the Baltic Sea. The study provided a representative sample of 823 thousand vehicles annually passing through the ferry routes analyzed.

These were then allocated to the trade relations of the countries, which were determined from the survey results. Although the data made it possible to determine which respondent loaded how much and to where or from where, the respondents indicated that stopovers were planned along the route and that the loading capacity varied until the destination. Therefore, 'minimum and maximum' scenarios were created using trade balances of the countries,[2] which allowed a range of potential fright transport volumes to be assumed.

For the quantification of land transport, the focus was on the Jutland-Land link. This European transport route via Flensburg generates international transport services by road and rail between Germany and Denmark to Malmö and Helsingborg in Sweden.

[1] In previous studies a more detailed geographical mapping of the Scandria® Corridor (Western Corridor, Scandria® Core Corridor and Eastern Corridor) was carried out to highlight the focus for stakeholders.

[2] Statistical office of the European Union (Eurostat) (2017): *"EU Handel by SITC since 1988"*.

For the quantifications, automatic counting points[3] that monitor cross-border traffic between Germany and Denmark were used as a data basis. The vehicles counted are the tôtal number of vehicles entering Germany and Denmark, with the vehicle type classification of HGV > 3.5t zGG with/without trailer and semi-trailer.

In order to determine the potential amount of freight to be transported by rail in the German Baltic Sea ports, data sets (including data from terminals and timetables) were used, which were based on statements by experts and employees from the operational business of the ports. In this way, a clear picture of the traffic relations and their frequency was created.

The destinations of the considered ports lead to Germany, Czech Republic, Austria or Italy. In the course of the analysis common parameters such as train load and length, wagon type, rail network or load limits were defined.

Even though there is a rail connection via the Jutland-Land link, the traffic relations did not provide any information on cross-border traffic with Scandinavia. In this geographical section only very rough statements can be made. The reason for this is the intransparency of the data base regarding the number of cross-border freight trains crossing the German-Danish border.

A representative report of the Federal Ministry of Transport and Digital Infrastructure published in 2017 stated that the rail network in the German-Danish border region was loaded with 101−10.000 freight trains in 2015.[4] However, this range does not provide any information on the train connections, which meant that the traffic relations of the trips and thus the classification of transit traffic could not be identified.

The elaboration and combination of the correlating data describes the current state of knowledge about the cargo flows and the most important ports and terminals along the European North–South Axis. It also provides an overview of the operation of multimodal services and available information on the potential for multimodal shift.

3.2 Modal Split

Because the surveys at the operational level were the starting point for the modelling, it was possible to segment very practical data for sea and land transport.

[3] These are automatic counting stations placed on motorways and national roads. At these counting points the number of all passing vehicles is permanently recorded and then published by the Federal Highway Research Institute.

[4] Other evaluations of this kind have not become known during the analysis. However, these statistics are considered to be representative, because according to the Federal Ministry of Transport and Digital Infrastructure, freight traffic has not increased in 2016. In the area of "Railways - Passenger, freight and combined transport", the Ministry recorded a decline in traffic of approx. 0.1%. It is therefore assumed that the indicated range can be regarded as characteristic in this geographical section.

The study showed that large shares of regional and international transit transports are managed via Roll on Ro off (RoRo) ports. On average, 70% of the points of departure or destination of the transport relations are outside the district in which the port itself is located. An average of 10% of the traffic relations start or end in the same municipality as the location of the port in question. Three out of four vehicles are at least 75% full. Loading rates depend on the direction of the traffic, through the ports and on the Øresund Bridge.

The limit for the loading capacity of the vehicle varies according to the type of goods, usually weight, volume or loading metres. The survey showed that more than 75% of the vehicles are used to at least 75% capacity (load capacity utilization). About 60% of the transports are carried out with a full load and about 10% as empty trips, whereby a total of about 3% were marked as dangerous goods transports.

Although the participants' vehicles are relatively modern, they are primarily powered by diesel (98–99%). It was revealed that on average only 1–2% of the vehicles are operated with alternative fuels. The study also found that about 40% of the vehicles were produced in 2014 or later.

Furthermore, of the various vehicle types, the articulated truck (up to 18.75 m) was by far the most common (84%). On the Øresund Bridge, however, the truck with trailer (up to 25.25 m) is the most common configuration.

Denmark, Germany and Poland are the countries that used the ferry connections across the Baltic Sea most frequently. It is assumed that this is mainly due to the national transport network (road and rail) to the respective ports.

The analysis has also shown that there is a high proportion of trade with the Netherlands due to the overseas freight flows to and from the ARA ports.[5]

In the modelling, the utilization of the ferries was also considered. This resulted in an average ferry load factor of 60%. The variance around this mean value is significant. It is assumed that factors such as travel frequency and loading meter are strongly correlated.

For road freight transport along the Jutland-Land link, four relevant counting points recorded around 3.618 thousand vehicles crossing the German-Danish border in 2016. If the empty trips from the Swedish study of about 10% are taken into account and a minimum load of 10 tons and a maximum load of 26 tons are assumed, the transport volume amounts to 26.239 thousand tons—85.278 thousand tons (Fig. 1).

In general, it can be concluded from the results of this study that road freight transport within the RoRo segment has shown a strong development from 2005 to 2016 with a higher growth rate than Sweden's total foreign trade. This indicates that road freight transport by truck and trailer is highly competitive for the distances relevant to Swedish exports and imports to and from Europe. At the same time, it can be said that there is great potential for a shift to rail and shipping and for an increase in alternative fuels.

---

[5] It should be mentioned that in a parallel German study the Netherlands were not so strongly represented: The largest proportions of vehicles using the port analysed come from Germany, Italy and Poland. This indicates a certain specialisation of the Scandria® core corridor towards southern and eastern Europe, while the western corridor via the Jutland-Land link and the ports of Schleswig–Holstein rather covers western Europe.

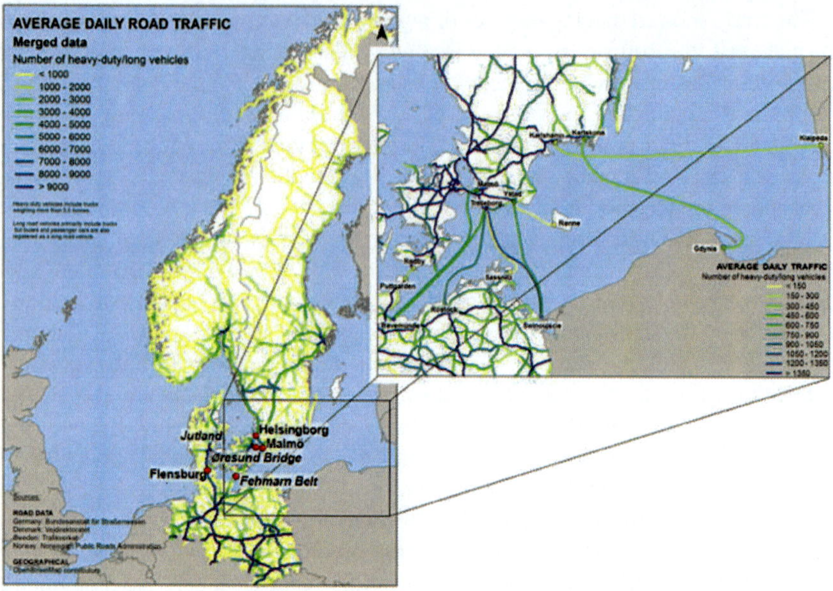

Fig. 1 Average transport intensity by road and ferry in the northern Scandria® corridor

In addition to the route and capacity information, the study also found that at the operational level, the proportion of RoRo units transported to the ferry terminals by rail was very low, although most of the Swedish ferry terminals studied (with the exception of Strömstad and the terminals in Gothenburg, Denmark and Germany) have the infrastructure to support ferry and RoRo operations with land transport by rail.

In summary, even for distances of 300–600 km, rail solutions can hardly compete with road transport, both in terms of time and cost. On the basis of freight volume, there is therefore great potential to increase the share of rail freight transport to RoRo ports.

These findings help create knowledge, to inform regional and national decision-makers about the consequences that can be drawn for strategic decision-making in infrastructure and transport policy, and to provide transport infrastructure planning and logistics administrations with information about the dynamics of freight flows in the Scandria® corridor.

The results of the modelling of freight flows have enabled a differentiation of transport modes in the [northern] Scandria® corridor, assuming a modal split of about 75–81% by road, about 17–18% by sea and about < 1–8% by rail.

4 Possible Influences on a Modal Shift

We formulated three hypotheses on future developments, that could influence a modal shift towards rail transport:

Firstly, an increase of the average speed of freight trains on European rail networks could also make rail transport more attractive for shippers and thereby lead to a shift towards rail transport. The reasoning being, that road transport has a significant transport time advantage over intermodal-rail-transport chains.

Secondly, a longterm increase of diesel prices could lead to an increase in road-transport rates. As intermodal rail-transport chains would be less influenced by a diesel price change, it could be expected, that rail transport becomes comparatively more attractive, at least on certain routes.

Thirdly, the construction of new infrastructure -in the Scandria corridor, this would be the new Fehmarn-Belt-tunnel- could lead to an increase of rail speeds. Furthermore it would lead to an elimination of transhipment processes, currently necessary in order to cross the Baltic Sea.

5 Multifactor-Decision Model for Freight Forwarders

5.1 Model Architecture

In order to test our hypothesis, we used a multifactor decision model, that includes detailed cost- and time-effects.

The model we used, was developed as part of the Scandria-project, running from 2009 to 2013, originally to be used as a planning tool for intermodal transport chain (Michalk 2012). For the study at hand, the model was further refined, in order to reflect changes of state variables, mainly changes of crude-oil-price changes (which also can be representative for changes of fossil fuel-prices as perceived by the transport operator) and of average freight-train speeds on rail networks, as well as different routing, which also shall reflect new infrastructure constructions (e.g. Fehmarn-Belt in the Scandria Corridor).

A number of studies have been concerned with the question, which product features are important to the customer of a transport service.

Bühler (2005) found costs, time of delivery and transport time to be the most important determining factors for the selection of a transport service, whereby transport costs were clearly more important. Beute et al (2003) and Leyn (2010) found costs to be most important, followed by transport time. Geiger (2011a,2011b) found long transport times and costs to be the strongest barriers for using alternatives to road transport, again indicating that these two factors are the two most important to a customer. Ludvigsen (2006) found costs to be as important as reliability, but this study was conducted in a European context, while the aforementioned were conducted within a German context. Most of this studies also identified other decision factors, but the importance of this factors were smaller than that of costs and transport time and the results on the ranking order were not as clear. This can be partly explained by the structure of the studies, which all examined the cost and transport time factor but laid different emphasis on the examination of other factors.

5.2 *Modelling Costs*

The annual costs for operating trains as well as trucks are being calculated through the following function:

$$C_{Ai} = C_{SIai} + C_{Mai} + C_{Di} + C_{Ei} + C_{Ii} + C_{Si} + C_{Moi} + C_{REi} + C_{Oi} + C_{Ti} + C_{Ini}$$

The costs per roundtrip can then be calculated as:

$$C_{Ri} = C_{Ai}/n_{Ri}$$

If the average load of a train or truck is given as l_{Ri}, the costs for the transport of one TEU can be calculated as:

$$C_{CRi} = \frac{1}{2} * C_{Ri}/l_{Ri}$$

As the transport chain of one load unit is described by the costs of truck transport, train transport and transshipment, in this model, the transport costs of one load unit can be calculated as:

$$C_{IT} = \sum C_{tj} + \sum C_{CRi}$$

whereby

C_{Ai}	Annual operating costs
C_{Ri}	Roundtrip costs
C_{CRi}	Transport costs per load unit
C_{SIai}	Annual investment cost
C_{Mai}	Annual maintenance costs
C_{Di}	Personnel costs for driver
C_{Ei}	Energy costs
C_{Ii}	Infrastructure costs
C_{Si}	Shunting costs
C_{Moi}	Train monitoring costs
C_{REi}	Costs of reserve

C_{Ini} Insurance
p_{Ej} Price of energy from source j
s_{Ej} Share of Energy source *j* in Energymix *i*.

With

$$C_{Ei} = \sum s_{Ej} * p_{Ej}$$

5.3 Modelling Transport Time

Utilization is an important input value for the cost calculation. Utilization depends on the schedule of a train or truck. A determining parameter for the utilization of a train is the schedule and the route it is running on. In the simplest case a train would run on only one route between two points throughout the whole year. This is not merely a model assumption, but indeed quite realistic if one thinks of intermodal shuttle trains, running between two terminals on a given schedule. In such a case the utilization during a year would be determined by the rail-distance between the terminals served and the number of trips the train performs. Customers usually demand regular schedules, a fact that finds expression in most intermodal train schedules that usually offers the same departure and arrival times on every served day of the week. This is a restricting factor in the utilization of a train, as the distance a train can cover during a roundtrip is restricted by its speed and the next run it is scheduled for. For example, if a train runs on a regular schedule, with an average speed of 50 km/h between two terminals 400 km (rail distance) apart, it will need about 8 h to run from one terminal to the next. If this train starts its first run on Monday at 8:00 h at terminal A, it can be expected to arrive at the opposing terminal B at about 16:00 h. Given a turnaround time of three hours at terminal B, it could start its way back at 19:00 and would arrive at 3:00 h on Tuesday at Terminal A. As the same departure time shall be offered, the train now has a five hours stopover at terminal A, until it departs again at 8:00 h.

To simplify the calculations, it will be assumed that a terminal is open 24 h a day on seven days per week.

The total time per trip, including the turnaround time for the train and the necessary slack time to anticipate possible delays and other operational irregularities is:

$$t_{TR} = t_T + t_{TT} + t_S$$

with:

$$t_T = \frac{d_{TR}}{v_{TR}}$$

The number of trips that are possible per 24 h period can be calculated as:

$$n_{TR24} = \lfloor \frac{n_h * 24}{t_{TR}} \rfloor \; for n_h = 1, 2, 3, 4, 5, 6, 7$$

As the train would need to start a new cycle after each roundtrip at the same time of the day, n_h needs to be an integer. Also as a non-integer number of trips within a 24 h period would be pointless (under the given parameter of a regular schedule, as described above), n_{TR24} also needs to be rounded down to an integer. Trips that take longer than 24 h would be anticipated with $n_h > 1$.

$$n_w = \lfloor \frac{7}{n_h} \rfloor$$

The total annual utilization of a train can then be calculated as (under the assumption of 52 operational weeks per year):

$$U_a = d_{TR} * n_h * n_{TR24} * n_w * 52$$

The actual travel time of a load unit t_{AC} shall be defined as the time between loading the load unit onto a vehicle and the earliest possible delivery time at the receiver.

$$t_{AC} = \sum_{i=1}^{n} t_{li} + \sum_{j=1}^{n} t_{TSj}$$

With

t_{li}	Time per trip *i* on each leg of the intermodal chain
t_{TS}	Additional time for each transshipment and storage process j along the chain
t_T	Travel time
t_{TT}	Turnaround time per trip
t_S	Slack time per trip
d_{TR}	Distance per trip
v_{TR}	Average traveling speed
t_{TR}	Total time per trip
n_h	Number of 24 h periods
n_w	Number of multiples 24 h periods per week
n_{TR24}	Number of trips per 24 h period
U_a	Annual utilization of one train set (mileage in km)

As industrial companies as well as companies selling goods often have given production or selling timeframes, there is a useful delivery time, for a product. If a company produces or sells products between 6:00 AM and 8:00 PM, it can be argued that a delivery early within this timeframe is of more use, than a delivery late within this timeframe. Often, deliveries outside this timeframe could be of even lesser use, as they might cause additional cost for handling, which within the timeframe would be done by staff anyway on duty. So for this model a "useful delivery timeframe" shall be defined between the beginning of a useful delivery time frame T_{wB} and the end of this timeframe T_{wE}. Deliveries outside this timeframe would always lead to additional waiting time t_w which is added to the actual traveltime t_{AC} and results in the total delivery time t_D:

$$t_D = t_{AC} + t_w$$

The waiting time can be calculated as the time-difference between the actual arrival time T_A and the beginning of the useful delivery timeframe:

$$t_w = T_{wB} - T_A$$

5.4 *Combining Demand Effects*

One indicator shall include all three factors (transport time, transport costs and service frequency) to allow for comparisons of different transport services.

The reaction of customers to alterations of one feature in a given service can be described with an elasticity function (compare for example Bücker 1998):

$$\varepsilon = \frac{\Delta x}{\Delta y} * \frac{y_1}{x_1} \quad (1)$$

With

ε Elasticity
Δx Absolute change of demand
Δy Absolute change of demand-factor
x_1 Demand before change of demand factor
y_1 Demand factor before change of demand factor.

The above equation can be transformed in order to calculate a change of demand based on a given elasticity:

$$\Delta x = \varepsilon * \Delta y * \frac{x_1}{y_1} \quad (2)$$

The demand factor could be a price or transport time, i.e. changing the price would change the demand.

In the study at hand, changes of transport costs, shall be equated with changes of service prices, e.g. an increase of the operational costs by 5% would lead to an increase of the service price of 5%. This assumption is quite realistic, as expert knowledge implies that the transport market is a buyer market. Different intermodal transport operators have explained, that they aim at prices, which allow for a return on sales of about 4%, which means, that they indeed orientate prices on the operational costs.

In the study at hand, the possible transport modes a shipper can choose between shall be pure truck transport or intermodal transport. Transport prices and transport time are assumed to be the most important decision factors for a shipper, when deciding for a transport mode, this assumption is backed up by a number of studies, such as the studies of Bühler (2005), Beute (2003) or Geiger (2011a,2011b).

By inserting cost and time factors into Eq. (2), demand changes can be calculated based on transport-cost, transport-time and transport frequency changes:

$$\Delta x_c = \varepsilon_c * \Delta c * \frac{x_1}{C_1} \tag{3}$$

$$\Delta x_t = \varepsilon_t * \Delta t * \frac{x_1}{t_1} \tag{4}$$

$$\Delta x_f = \varepsilon_f * \Delta f * \frac{x_1}{f_1} \tag{5}$$

With

x_1	Demand before change of demand factor
Δx_c	Demand change based on transport cost change
Δx_t	Demand change based on total delivery time change
Δx_c	Demand change based on transport cost change
Δx_f	Demand change based on service frequency change
Δc	Absolute Change of transport costs
Δt	Absolute Change of total delivery time
Δf	Absolute Change of frequencies
c_1	Operational cost of service 1
f_1	Service frequency for service 1
t_1	Transport time of service 1
ε_c	Transport cost elasticity
ε_t	Transport time elasticity
ε_f	Service frequency elasticity.

As different elasticities can be found for different transport modes, the model also considers different elasticities for different modes in the calculations.

In the study at hand, demand effects of transport cost and transport time changes, which occur through changing from one transport system to another, shall be calculated by adding the demand changes:

$$\Delta x_{Total} = \Delta x_t + \Delta x_c + \Delta x_f \tag{6}$$

An attractivity indicator shall be defined as:

$$A = 1 + \Delta x_t + \Delta x_c + \Delta x_f \quad (7)$$

Furthermore, a Baseline Indicator $A_{Base} = 1$ shall be defined for a given baseline transport service. A transport service superior to the baseline transport service would be indicated by a demand indicator A > 1.

As road transport is not bound to schedules, in cases where road transport services are part of the comparison, the frequency variable is $\Delta x_f = 0$.

6 Determining Effects on a Possible Modal Shift in the Scandria Corridor

In order to examine possible influences on a modal shift, we concentrated on a transport relation between Padua and Malmö. We then determined seven possible routes and transport chains for this relation:

- Padua-Rostock (rail)—Rostock-Trelleborg (ferry)—Trelleborg—Malmö (road)
- Padua-Travemünde (rail)—Travemünde-Malmö (ferry)—Malmö-last mile (road)
- Padua—Jütland—Malmö (rail)
- Padua—Jütland—Öresund—Malmö (road)
- Padua—Rostock (road)—Rostock-Trelleborg (ferry)—Trelleborg—Malmö (road)
- Padua—Swinoujscie (road)—Swinoujscie-Trelleborg (ferry)—Trelleborg—Malmö (road)
- Padua—Puttgarden (road)—Puttgarden-Malmö (ferry)—Malmö—Malmö (road)

We calculated the attractivity indicator A, for each of these transport chains (a larger indicator-value implies a larger attractivity of this route for a shipper):

	A
Padua-Rostock (rail)—Rostock-Trelleborg (ferry)—Trelleborg—Malmö (road)	1.0
Padua-Travemünde (rail)—Travemünde-Malmö (ferry)—Malmö-last mile (road)	1.0
Padua—Jütland—Malmö (rail)	1.1
Padua—Jütland—Öresund—Malmö (road)	0.7
Padua—Rostock (road)—Rostock-Trelleborg (ferry)—Trelleborg—Malmö (road)	1.0
Padua—Swinoujscie (road)—Swinoujscie-Trelleborg (ferry)—Trelleborg—Malmö (road)	0.3
Padua—Puttgarden (road)—Puttgarden-Malmö (ferry)—Malmö—Malmö (road)	0.7

In general the rail-transport chains appear somewhat more attractive. However this is mainly the case for goods, with a low need in transport-time flexibility. In these cases low transport costs are of higher importance. We countered this effect in parts, by using different elasticities for the calculation of road-and rail transport chains, reflecting the different goods-types. As we want to compare changes in the attractivity indicator, these differences are of no significance.

An increase of average rail speeds on the corridor did have a comparatively small influence on the attractivity of rail-transport chains (and of course no influence of train-speeds by 50% only resulted in a maximum increase of attractivity of rail transport of 10% (in case of the rail-transport chain via Travemünde). An increase of the rail speed by 100% only resulted in a maximum increase of 15% of the attractivity.

This is not too surprising, as travel time is only part of the total travel time of a freight train. The rather low influence is of course also due to the necessary time-expenditure for transhipment, shunting and the time penalty incurred by scheduled services.

An increase of diesel-prices had a much stronger effect on the attractivity of road-transport services: An increase of Diesel prices by 50%, led to a decrease of road transport attractivity by 26% to up to 40%. Intermodal transport chains were much less effected and only experienced a decrease of attractivity by 0.5−1.1%.

The largest increase of attractivity for rail transport was achieved by the introduction of the Fehmarn-Belt tunnel into the model. The tunnel resulted in the omission of transshipment processes along the Baltic sea cost. So transport time-effects had very strong influence on the attractivity. This also led to lower transport costs. Transport time was still about 20% higher, than the fastest chain (transport by road via the ports of Rostock and Trelleborg), but faster than all other examined transport chains.

7 Conclusions

The majority of road vehicles are still powered by Diesel. This is not completely surprising, as the road vehicle-market still lacks viable options for long range trucks with alternative drive trains.

Also the modal split is mainly focused on road transport, with up to 81% of the freight volumes being transported by road.

Though a significant share of the freight volumes between the continent and Scandinavia flows via the fixed links over Jütland, a very large portion of the transport chains are "naturally" broken along the sea ports of the Baltic Sea, thus leading to an increase of costs and transport time.

This actually carries the chance for a change in the modal split. As we found, eliminating or reducing factors that increase cost and time-expenditures on rail transport-chains, significantly increases the attractivity of rail-transport relations. This is especially apparent for the case of the Fehmarn-Belt fixed link. While increasing rail-speeds has no significant effect, organisational approaches can have a noticeable impact, if they address effects related to increasing the efficiency of the transhipment points (or eliminate them altogether as in the case of the Fehmarn-Belt link).

Possible recommendations for a modal shift, based purely on political measures, should be focussed on the costs of transport e.g. increasing the costs of road transport or lowering the costs of rail transport.

However, certain infrastructure projects can have a stronger (though locally limited) effect. This underlines the rational behind projects, such as the Fehmarn-Belt and the Brenner-Basis-crossings.

This work could be further expanded on, by conducting additional surveys in the nort- and west-range ports. Currently the model does not include transport flows, that circumvent the corridor by sea.

Bibliography

Beute M, Bouffioux C, Maeyer JD, Santamaria G, Vandresse V, Vandaele E (2003) A multi criteria analysis of stated preferences among freight transport alternatives. Jyväskylä

Bücker R (1998) Mathematik für Wirtschaftswissenschaftler. Oldenbourg

Bühler G (2005) Verkehrsmittelwahl im Güterverkehr. Mannheim

Geiger C (2011) Carbon Footprint—Abbau von Hemmnissen zur Emissionsminimierung

Geiger C (2011) Carbon Footprint Abbau von Hemmnissen zur Emissionsminimierung Not published as of June 2011

Leyn U (2010) Verkehrsmittelwahl im Güterverkehr—Entscheidungsverhalten zugunsten des kombinierten Verkehrs auf Basis der Fallstudiendaten des Verbundprojekts SPIN-ALP. Hochschultagung Verkehrswesen 2010. Berlin

Ludvigsen J, Grue B (2006) Decision factors underlying transport mode choice in European freight transport. Oslo

Michalk P (2012) Planning competitive intermodal services—A multifactorial decision model for service planning. Wildau

The Effects of Sustainability-Driven Policies on Transport CO_2 Production: High-Speed Rail Transportation as an Alternative to Passenger Air Transport

Alessio Tardivo and Armando Carrillo Zanuy

Abstract The purpose of this study is to evaluate to which extent the rail transport mode can improve the environmental situation in Europe if it were to be at the centre of sustainability-driven policies. In particular, it aims at estimating a relative picture of the CO_2 emissions generated by short-distance air passenger transportation in Europe, which could have been transferred to high-speed rail and produce less CO_2. This study follows a three-step methodology. Firstly, it calculates the number of passengers travelling on each route between cities and estimates the total CO_2 emissions. Subsequently, it leverages the current literature on CO_2 consumption from railway passenger transport. Lastly, it estimates the possible scenarios in terms of CO_2 emissions that would have followed adequate sustainability-driven policies. The study found that short-range aviation in EU28 produced 9.2 million tons of CO_2 in 2017, which represents about 5% of total aviation emission, about 1% of total transport emission and about 0.2% of total CO_2emission. Furthermore, the CO_2 production on the 175 routes analysed increased until 2019, while precise policies could have allowed saving 582 MT CO_2. The effects of the COVID-19 outbreak on the European transport sector increases the relevance of this study. To avert the "return to normality" vis a vis Greenhouse Gases (hereinafter "GHG") emissions from the sector, it will be necessary to introduce structural changes. As Austrian Airlines or KLM bailouts show, environmental concerns might finally influence the decision-making process on public transportation. In the context of a green recovery, this study not only lays the foundation for further contributions addressing the CO_2 production from EU-wide sectors but also underlines the role the railway can play in environmentally friendly transportation.

1 Motivation

Traditionally, economic development was thought to be achieved at the expense of the environment and, as a result, the objective of economic growth frequently

A. Tardivo (✉) · A. Carrillo Zanuy
Eurnex, Berlin, Germany
e-mail: alessio@eurnex.eu

M. Marinov and J. Piip (eds.), *Sustainable Rail Transport 4*, Lecture Notes in Mobility,
https://doi.org/10.1007/978-3-030-82095-4_11

got ahead of the social and environmental objectives. Mobility and transport have represented perhaps the best cases in which this trade-off has been evident. Despite the irruption of new technologies, services and approaches in the mobility sector which are radically transforming the concept of transportation, the transport sector is steadily increasing its GHG emissions since 1990.

As carbon dioxide emissions are directly related to fossil fuels consumption, the role of specific transport modes in improving the sector's share of GHG emissions is clear. The aviation industry plays a prominent part in this trend: studies revealed that 3.16 tons of CO_2 are released out of the consumption of one ton of kerosene, a hydrocarbon liquid commonly used as a fuel.[1] As of 2015, Europe is the second-largest region in the world for commercial passenger flights.[2] In this sense, commercial airlines play a relevant role in the European transport sector GHG emissions. Moreover, studies as the one by Alonso et al. (2014) underline how air traffic in Europe is concentrated on short distances below 1000 km, including almost 60% of all flights and 46% of all passengers. These findings are indeed puzzling, since the Union offers a broad range of surface transport alternatives.

The GHG emissions resulting from the transport sector are not the only cause for concern. In the context of heavy rains, rising temperatures and storms arising from global warming with predicted impacts on the air transport sector, the sustainability of the current trend vis a vis air passengers transport is challenged. The railway's higher resilience, on the other hand, might play a more suitable role in the context of climate change unfortunate effects. In the context of intra-European mobility, can railway transport offer a better, less carbon-intensive alternative than air transport for a sufficient number of cases? As the increase in awareness for environmental issues and carbon dioxide production can have impacts not only on end-user choices of mobility and transport, but also on the policymaking behind transport planning, this study aims at exploring novel possibilities other than aeroplanes to reconcile green policies and high levels of mobility.

The relevance of the question this study aims to address is high not only in relation to the EU goal vis a vis GHG emission, but also in the context of the COVID-19 pandemic aftermath. In particular, despite the Union has been characterised by a low production of GHG emission in the spring of 2020, this trend is expected to reverse course once recovery measures are working at full throttle. Transport will have an important role in the predicted rebound effect of GHG emissions; thus, the development of green new mobility and a high involvement of railway is essential (Tardivo et al., 2020).

---

[1] https://www.atmosfair.de/wp-content/uploads/atmosfair-flight-emissions-calculator-englisch-1.pdf.

[2] .https://ec.europa.eu/transport/sites/transport/files/2016_eu_air_transport_industry_analyses_report.pdf.

2 Problem Formulation

The rail transport can improve the environmental situation in Europe if it were to be at the centre of sustainability-driven policies. By calculating the amount of carbon dioxide equivalent produced in the years 2017–2019 by passenger air transport between major cities within the European Union, this study aims at providing a sound evaluation of the amount of carbon dioxide equivalent which could have been saved if better mobility policies would have been in place enhancing the railway's passenger share.

After providing the data for a solid background, this study calls for a more profound cost–benefit analysis not only of traditional air transport in itself but also of low-cost airlines. This analysis is highly relevant for the EU given both its post-COVID-19 pandemic recovery as well as its position in the international system, which as a whole is striving to reach the Sustainable Development Goals (hereinafter "SDGs") by 2030.

This study looks at the results of the lack of policies over last two years with the objective to keep them as a light for the future, since it will be necessary to match the need for governments to avert a deep recession and the needs for safeguarding the environment. Since the transport sector plays a fundamental role in GHG production, it is time to acknowledge this fact and act accordingly.

3 The Position of This Contribution Within Academic Literature

This study is not the first contribution to compare CO_2 emissions from air and rail transport. Prussi and Lonza (2018) calculated in detail the CO_2 emission from rail and air transport on seven routes within the EU. We take advantage of their work and build upon their so-called high-rail scenario, including 175 routes and calculating the precise CO_2 emissions from the air sector. Always related to the topic, Alonso et al (2014) investigated the distribution of air transport traffic and CO_2 emissions within the EU. Despite their focus on the year 2010, their findings must be held into consideration as they calculated total CO_2 emissions from the air sector in 216 million tons, with a large concentration of emissions in a few countries.

On the same line with the previous studies, this contribution aims at providing a solid base for policy measures capable of curb the sector emissions. In this context, Mendes and Santos (2008) provided a forecast on the impacts of incentive-based regulation, suggesting that results were likely to be minimal. The failure of the EU emission trading system (EU ETS) in curbing CO_2 emissions from the aviation sector is evident in the 2020 Commission decision to amend the EU ETS regarding aviation. While academia focused greatly on the effects of aviation's inclusion in the EU ETS (Anger, 2010; Mendes and Santos, 2008; Morrell, 2007), it is clear that the academic literature lacks cross-sectoral analysis vis a vis CO_2 emissions, probably due to the complexity of detailing addressing these trends on a European scale.

Lastly, this contribution aims at developing an understanding of the implications for the CO_2 production by air transport on an international level. Following the footprint of Chèze et al (2013) —which estimated the scenarios for the air transport to reduce emissions to comply with the IPCC scenarios before the Paris Agreement—this contribution stresses the significance of reducing CO_2 emission from the transport sector in regard to the UN Sustainable Development Goals.

4 Objectives

As highlighted above, this study aims at picturing the amount of CO_2 that could have been saved in the 2017–2019 period provided that some of the passenger transport were to be shifted from air travel to railways.

Currently, an increasing number of Europeans are moving towards major urban centres which are interconnected via air, road and rail. The user's choice in transport mode has clear effects on the transport sector CO_2 emissions, thus we aim at understanding the magnitude of the effects of the transport sectors between these urban centres.

There is a number of these conurbations that are less than 800 km away from each other, a distance assumed to be still competitive for rail, and yet have dense air transport connections between them. The threshold of 800 km has been chosen on the blueprint of the Japanese bullet train Shinkansen, capable of having a greater market share than airlines on routes up to 965 km (Albalate and Bel 2012). Since the Shinkansen performance is deeply related to the type of infrastructure present in Japan, this study deemed feasible to take a more conservative approach in relation to the European infrastructure and limit its scope to routes up to 800 km long. The present study calculated the total amount of carbon dioxide equivalent produced between these major cities resulting only from air transportation. The objective is then to compare the level of emissions generated from air and railway transport and understand how much CO_2 could be saved if a modal shift to railway was in place.

Addressing an exact value to CO_2 emission from air transport is not particularly straightforward, as fuel consumption and therefore emission levels are not only related to physical characteristics of the aircraft, as engine types, winglets and number of seating, but also to how the aircraft is operated. Number of passengers carried, cargo loaded, flight distance, airspeed, landing procedure are factors that play a role in how environmentally impacting a flight can be. The flight distance deserves particular attention. At first glance, fuel consumption and carbon emissions are directly proportionate to the total flight distance. However, studies reveal that short-haul flights compared to medium-haul flights consume more fuel per 100 km. This result is based on the fact that the departure and take-off procedures require the use of a large amount of energy, and their implementation is a very energy-intensive step in a flight. Since those flights which are less than three hours long are considered short-haul flights, most domestic and intra-European flights are of this nature. At the same time, long-haul flights consume more fuel than the medium-haul flights per 100 km, as the fuel must be carried for most of the flight.[3]

5 Research

This study aims at identifying the effects of sustainability-driven policies in the transport sector vis a vis its production of carbon dioxide equivalent. After having acknowledged that air transportation plays a crucial role in both Europe's mobility and in the Union's global share of CO_2 emissions, this study undertakes three steps.

Firstly, it calculates the number of passengers travelling between the major European cities and estimates the total CO_2 production from the sector.

Secondly, it takes advantage of the current literature on CO_2 consumption from railway passenger transport and identifies the difference in CO_2 emissions between the two transport modes.

Lastly this study estimates the possible scenarios in terms of CO_2 production that would have followed the application of adequate sustainability-driven policies to the transport sector, analysing the potential carbon dioxide equivalent savings resulting from an ideal transfer of passenger transport from air carriers to the rail network within the European Union.

This study employs quantitative analysis, elaborating data from EU sources as Eurostat, the European Commission, EASA, EEA; from private sources as the Centre for Asia Pacific Aviation Pty. Ltd., Eurocontrol, Ryanair and Lufthansa; and scientific publications as Prussi and Lonza (2018), and Albalate and Bel (2012).

Data, which was produced by different stakeholders prior to this study, has been extracted by open-access websites and included into a new database. This database has been developed and manipulated through Microsoft Excel software.

We recognise that no data set is perfect, thus the present study has limitations in addressing factors such as the number of each type of aeroplanes from any EU country. However, data has been selected amongst public entities' databases with the aim of providing the most possible transparent data. As such, this study, despite focusing on theoretical losses in CO_2 savings which have not been achieved, does employ sound data and it expects to stimulate a new approach to the quantification of GHG emission and its effects on a cross-national EU-wide level.

---

[3] https://www.atmosfair.de/wp-content/uploads/atmosfair-flight-emissions-calculator-englisch-1.pdf.

6 Implementation of the Research

The study takes into consideration the routes between European cities with a population larger than 500.000 inhabitants as well as European urban areas with a metropolitan population of over 1 million inhabitants. These thresholds are arbitrary; however, the researchers consider that this selection is representative enough for the target of the study. Having this study as objective the evaluation of the possible effects of sustainability-driven policies on an EU-wide scale, it focuses on major routes and macro-trends.

The selected cities and urban areas together with their respective countries are the following: (Table 1)

As this study will analyse the potential carbon dioxide equivalent saving resulting from a transfer of the passenger transport from air carriers to the rail network, only some air and rail routes were taken into account, and only the current infrastructure

Table 1 Cities with over 500.000 inhabitants and urban areas with a metropolitan population of over 1 million inhabitants within the EU

Country	City						
Austria	Vienna						
Croatia	Zagreb						
Czech Republic	Prague						
Belgium	Brussels	Antwerp					
Bulgaria	Sofia						
Denmark	Copenhagen						
France	Paris	Marseille	Lyon	Lille	Bordeaux	Toulouse	Nantes
Germany	Berlin Dortmund	Hamburg Essen	Munich Leipzig	Cologne Bremen	Frankfurt Dresden	Stuttgart Hanover	Dusseldorf Nuremberg
Greece	Athens	Thessaloniki					
Hungary	Budapest						
Ireland	Dublin						
Italy	Rome	Milan	Naples	Turin	Palermo	Florence	Genoa
Latvia	Riga						
Lithuania	Vilnius						
Poland	Warsaw	Krakow	Lodz	Wroclaw	Poznan	Katowice	Gdansk
Portugal	Lisbon	Porto					
Romania	Bucharest						
Spain	Madrid	Barcelona	Valencia	Seville	Zaragoza	Malaga	
Sweden	Stockholm	Gothenburg					
The Netherlands	Amsterdam	Rotterdam	The Hague				
United Kingdom	London	Birmingham	Leeds	Glasgow	Sheffield	Manchester	Bradford

has been evaluated. As such, Bucharest, Dublin, Helsinki and Palermo were not included in the study. In fact, although being suitable as far as population is concerned, given the long distances between them and the other major EU cities and/or the lack of infrastructure as bridges to connect them to continental Europe, this study would not have benefited from their inclusion.

The data regarding the number of passengers carried on each air route has been obtained from the European Commission´s statistical office Eurostat. Since the provided data display certain discrepancies regarding the passengers' traffic between airports according to each national database, the study provided a mean between the given passengers' data. In those cases in which the Eurostat database: (https://ec.europa.eu/eurostat/web/transport/data/database) does not present information on one of the two airports involved in a selected route, the only value available has been considered. More information on Eurostat and the transport statistics can be found at the following address: https://ec.europa.eu/eurostat/web/transport.

The data regarding each flight´s carbon dioxide equivalent emission has been found employing the Atmosfair online calculator. Atmosfair, a German non-profit organisation, designed a software tool able to calculate precisely the amount of carbon dioxide and non-carbon emission from each flight. In particular, since aircraft engines emit various pollutants that contribute to rising global temperatures, Atmosfair calculates both CO_2 and other pollutants as well, such as methane, perfluorocarbons, nitrous oxide and others. These pollutants and their effects are summarised by Atmosfair and then converted into CO_2. From the "atmosfair Flight Emissions Calculator" of 2016: "first, the Emissions Calculator calculates the fuel consumption per passenger and based on this result, determines the amount of CO_2 that has a comparable effect to that of all other pollutants emitted by the flight added together (effective CO_2 emissions)". Therefore, the calculator's final output is expressed in carbon dioxide.

At the same time, the production of pollutants by air traffic is three times higher than that of carbon emissions alone, due to the high altitude in which they are released. To be able to compare the effects on the environment resulting from CO_2 production at high altitude with the effects of CO_2 production on the ground (as produced by railways or cars), the calculator multiplies by a factor 3 all carbon emissions produced during a flight at over 9 kms to correctly render the flight's climate impact in CO_2. Carbon emissions emitted at altitudes of less than 9 kms are not submitted to any alterations and are directly included in the flight's carbon footprint. The "atmosfair Flight Emissions Calculator" explains how this is a "conservative, quantitative–qualitative average value based on two metrics (RFI and GWP) and their bandwidths. Both metrics present the same numerical value, whereby the higher-value GWP even has a smaller bandwidth. This actual value of 3 is exactly in the middle of the old IPCC bandwidth of the RFI, which was indicated to be 2–4 by the IPCC in 1999". The validity of this assumption has been confirmed both by Lee et al. (2021) and by the European Commission (2020) and EASA, which state "the CO_2-warming-equivalent emissions based on this method indicate that aviation emissions are currently warming the climate at approximately three times the rate of that associated with aviation CO_2emissions alone".

Table 2 Most common aircraft by airline company

Company	Passengers	Most common aircraft–units
Lufthansa group	130,04	Airbus A320–194 units
Ryanair	128,77	Boeing 737–800–430 units
IAG	104,83	Airbus A320–203 units
Air-France KLM	98,72	Airbus A320–36 units
EasyJet	81,63	Airbus A320–168 units

More information on the atmosfair calculator can be found at the following address: https://www.atmosfair.de/en/offset/flight/.

As the fuel consumption and therefore the carbon dioxide production varies between which aircraft is being analysed, the study took the Airbus A320 as an aircraft model. In fact, as far as the five largest airlines companies in Europe are concerned, the Airbus A320 is the most common aircraft with more than 600 operating units. Furthermore, according to the Centre for aviation's forecast, the Airbus A320 NEO leads the orders for narrowbodied aircrafts in Europe, with 1.058 aircrafts ordered (CAPA 2018).[4] The largest airline companies in Europe, together with the number of passengers carried globally in 2017 and the most common aircraft in their fleet are shown in the following table. (Table 2).

The second most common aircraft is the Boeing 737–800 with 450 units and 611 orders for its variants for the future fleets in Europe. The airline company with more Boeing 737–800 in its fleet is Ryanair.

The overall amount of carbon dioxide production between the two aircrafts does not change dramatically. However, the Airbus is slightly less efficient on distances below 340 km, while its greater efficiency on the Boeing is noticeable starting from 360 km. Furthermore, the difference in carbon dioxide production between the two aircrafts increases steadily from the 420 km threshold onwards. A short overview of the differences in carbon dioxide production between the Airbus A320 and Boeing 737–800 with regard to the route distances is provided below. The data has been extrapolated from the Atmosfair calculator (Table 3).

Furthermore, in order to be able to estimate the feasibility of shifting means of transport from air carriers to the rail network, the distances between city centres have been calculated with the online software "Maps" from Google. Three steps have characterized the approach to distance measurement: first, the distance in a straight line between the two cities has been measured, then the distance between such cities has been measured on the already existing railway network. Lastly, those routes which exceed the threshold of the 800 kms on the rail network have been compared with the same route on the existing road network and have been considered in the study in case the distance on the road network was less than 800 km. Ultimately, those routes which are no longer than 800 km on the existing rail network have been classified as "short-distance route", while those which exceed the threshold of 800 km on the

[4] https://centreforaviation.com/analysis/reports/aircraft-fleets-western-v-easterncentral-europe-airbus-leads-orders-410122.

Table 3 Differences in CO_2 production between airbus A320 and Boeing 737–800 per passenger

Route	Distance in km	Airbus A320 CO_2 production (in Kg, per passenger)	Boeing 737–800 CO_2 production (in Kg, per passenger)
Nuremberg NUE—Munich MUC	150	28	27
Paris CDG—Brussels BRU	260	49	47
Frankfurt FRA— Brussels BRU	315	61	60
Paris CDG—London LHR	340	73	73
London LHR—Amsterdam AMS	360	79	80
Frankfurt FRA—Berlin TXL	420	98	100
Berlin TXL—Cologne CGN	475	106	109
Munich MUC—Berlin TXL	500	110	112
Milan MXP—Frankfurt FRA	520	112	115
Turin TRN—Paris CDG	580	130	135
Budapest BUD – Bucharest OTP	640	146	151
Berlin TXL—Stockholm ARN	810	187	198
London LHR—Naples NAP	1.615	326	357

existing rail network but are no longer than 800 km on the existing road network have been classified as "medium-distance route". These last routes were included in this study in order to evaluate, in terms of CO_2 savings, the results arising from an enhanced rail network. Routes longer than 800 km on the road network have not been selected for the study. A short explanation of the selection process is shown in the following (Table 4).

All the routes, distances, passengers and estimated carbon dioxide production can be found in the Annex section.

Table 4 Selection process of the suitable routes

Route	Straight distance in km	Rail network-based distance in km	Road network-based distance in km	Classification
Paris CDG—Brussels BRU	260	315	307	Short-distance route
Wroclaw—Frankfurt FRA	600	844	725	Medium-distance route
Hamburg HAM—London LHR	720	1000	934	Not selected

Lastly, this study takes into consideration the basis set by Prussi and Lonza (2018),[5] which identify an annual passenger increment of 3.5%. It also applies the so-called high-rail scenario, in which 25% of the expected aviation passenger growth is shifted to High Speed Rail (hereinafter "HSR") service. A shift of 25% of the expected passengers from air to rail transport would allow a 20% greenhouse gas emission saving. Prussi and Lonza (2018) identified a 20% saving only from analysed routes within five European countries. This study employs the same result in CO_2 saving on an EU-wide analysis. The reason behind accepting this value for the routes and countries other than the originally analysed, is based upon the estimation that a shift in 25% of the expected passenger increment is not an ambitious-enough target for the European transport sector at this time.

7 Analysis of Results Through Comparisons

The following table includes the ten routes with higher CO_2 impact in Europe. It is possible to notice, interestingly, the prevalence of national routes amongst the top 10 most environmentally harmful. While only two of the 10 routes are international (London—Amsterdam and London—Frankfurt), the remaining eight routes are within the borders of different European countries. Of these, Germany is the country with the highest number of national routes (three), followed by France with two (one of which is the most polluting route in the EU) and The United Kingdom, Spain, and Italy, with one national route each (Table 5).

Overall, in 2017, the passengers that flew between major European cities on routes which do not exceed the 800 km threshold on the rail network have been 78.3 million. The estimated carbon dioxide equivalent production resulting from these movements was 7.62 MT.

Combining both short and medium distance flights, the total amount of carbon dioxide equivalent produced in 2017 within these routes alone is 9.2 MT. It is an impressive amount of carbon dioxide, especially when considered that it results from short flights within the major cities in the European Union alone. According to the European Environment Agency (EEA 2019),[6] in 2017, the CO_2 equivalent production in the EU amounted to 4.483 MT. Therefore, if we take into consideration the data resulting from this study, it is possible to notice that the 175 analysed routes accounted for more than 0,21% of the total CO_2 equivalent production within the European Union. If only the 144 routes which have been classified as short-distance routes were taken into account, the carbon dioxide equivalent production would still reach a 0,17% of the total European production.

---

5 https://www.hindawi.com/journals/jat/2018/6205714/abs/.

6 https://www.eea.europa.eu/data-and-maps/data/data-viewers/greenhouse-gases-viewer.

Table 5 Ten most environmentally harmful routes in Europe

Route	Distance in a straight line (km)	Distance on rail network (km)	Distance on road network (km)	Passengers 2017 (millions)	Passengers 2016 (millions)	Passengers 2015 (millions)	CO_2 produced Millions of tons in 2017
Paris—Toulouse	586	816	680	3.25	3.26	3.21	0.45
London—Amsterdam	357	597	534	4.49	4.14	3.85	0.34
London—Glasgow	558	650	657	2.52	2.61	2.51	0.31
London—Frankfurt	636	793	786	1.86	1.95	2.00	0.27
Munich—Hamburg	610	825	775	1.74	1.80	1.81	0.27
Madrid—Barcelona	504	602	621	2.34	2.33	2.25	0.26
Paris—Marseille	660	746	775	1.67	1.65	1.59	0.24
Berlin—Munich	500	732	586	2.06	2.03	1.97	0.22
Berlin—Frankfurt	420	512	547	1.95	1.93	1.91	0.19
Milan—Naples	657	777	774	1.18	0.87	1.03	0.17

As highlighted by Eurostat (2019),[7] the transport sector represents 25% of the carbon dioxide equivalent total production within the European Union with 1.120 MT of carbon dioxide, second only to fuel combustion without transport. Followed by agriculture (10%), industrial processes and product use (8%) and waste management (3%). The carbon dioxide equivalent production in the 175 routes analysed in this study, therefore, accounted for 0,82% of the total transport sector emissions.

According to the data reported by the European Member States to the UN Framework Convention on Climate Change (EASA 2019),[8] the total carbon dioxide emissions of all flights departing from EU28 and EFTA in 2017 have been 163 MT. Those routes which are classified as short distance in this study, therefore, amounted to 4.67% of the total production. Including into this evaluation also the medium distance routes, 5.64% of the total CO_2 production to and from European airports has been produced on the routes analysed in this study. These figures might seem irrelevant to consider in the big scheme of things, but it is worth remembering that, as a reference, the EU greenhouse gas emission in 2016 decreased by 0.4% compared with 2015 and later increased again in 2017 by 0.6% compared to 2016, according to preliminary data.[9] So far then the EU target of reducing greenhouse gas emission by 20% compared to the 1990 levels looks within reach, even if by a narrow margin. However, as set in the 2030 climate and energy policy framework, a binding target of at least 40% cut in greenhouse gas emissions compared with 1990 levels have been determined in 2014, and the possibility of achieving a 0.17% cut in emission by enhancing the rail network on routes shorter than 800 km alone should be fully considered (Graph 1).

Building upon the contribution from Prussi and Lonza, this study considers the possible CO_2 emission savings resulting from shifting 25% of the expected aviation passenger growth to HSR service, in a so-called "high-rail" scenario. Although a total transferability scenario would be highly preferable, at the same time it is also extremely unlikely. Therefore, an achievable target for the reduction of carbon dioxide production must be considered, in order to better identify the steps to undertake towards a sustainable future for the sector.

Prussi and Lonza identify a 20% saving in CO_2 emissions by shifting 25% of the expected passenger growth to the rail sector. As they also identify an annual passenger increment of 3.5%, this study builds upon this projection employing the collected data. As a result, the estimated saving of carbon dioxide equivalent on the 175 routes analysed would have been of around 1.84 MT CO_2, lowering the emission production from 9.2 MT to 7.36. If such a trend were to be implemented systematically, the CO_2 saving on these routes by 2019 could have been of an additional 1.47MT. Such a saving would have allowed the analysed routes to decrease their share in the European CO_2 production resulting from the transport sector from 0.82% to a hypothetical 0.53%. Since 31 of these 175 routes need investments in the rail sector

[7] https://ec.europa.eu/eurostat/statistics-explained/pdfscache/1180.pdf.

[8] https://www.easa.europa.eu/eaer/topics/overview-aviation-sector/emissions.

[9] https://www.eea.europa.eu/data-and-maps/indicators/greenhouse-gas-emission-trends-6/assessment-2.

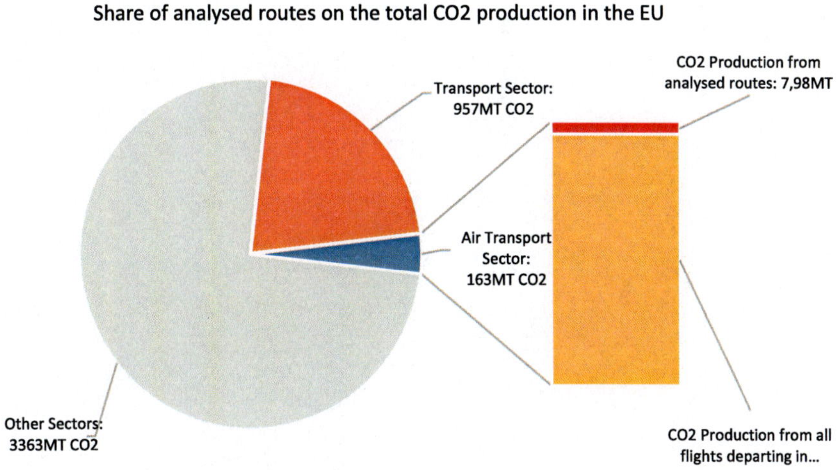

Graph 1 Share of analysed routes on EU CO_2 production

to be competitive with the air transport, enhancement of the network should be considered as an alternative way of cutting CO_2 emission. Furthermore, investments in the current rail network would enlarge the possibility for shifting mean of transport for more end-users even on routes longer than the one analysed in this study. However, it should be noted that the current contribution does not include a detailed calculation of the costs, both in monetary terms and CO_2 production, necessary for the railway infrastructure to accommodate the forecasted modal shift of the passengers.

If 25% of the expected passenger increment on other routes which sits outside the scope of this study were to be substituted with rail transport as well, the saving would be even greater. To give a reference, if the four routes between the five most populous cities which are less than 1500 km long on the rail network were taken into consideration—see in the following table—the production of carbon dioxide equivalent would have been 1.55 MT. In the case of a high-rail scenario, the possible saving on these routes would have been 0.31 MT. These four routes alone would have saved almost one-sixth of all the other 175 routes combined during the year 2018 alone. If both these four high-traffic routes and the previously analysed ones were combined, the total CO_2 production in 2017 would amount to 10.69 MT: 0,95% of the entire greenhouse gas production from the transport sector in the EU. The possible saving in a high-rail scenario from these routes alone could have reached 2.14 MT of CO_2 (Table 6).

On the other hand, however, what did happen as a result of a lack of policies able to alter the CO_2 emission rate? Considering an annual passenger increment of 3.5% on the 175 routes analysed in this study, and not considering the eventual CO_2 production saving from the newest technologies in the air transport field, the results are dangerous. In the first year, the CO_2 emissions increased by 14.24%, with a total carbon dioxide equivalent production of 10.51 MT, from the 9.2 of 2017.

Table 6 CO_2 production on four routes between the five most populous cities in EU

Route	Distance in a straight line (km)	Distance on rail network (km)	Distance on road network (km)	Passengers 2017 (millions)	Passengers 2016 (millions)	Passengers 2015 (millions)	Passengers 2014 (millions)	Passengers 2013 (millions)	Passengers 2012 (millions)	CO_2 produced Millions of tons in 2017
Berlin–Paris	877	1.095	1.053	1.16	1.16	1.17	1.13	1.09	1.07	0.22
Paris–Madrid	1.053	1.431	1.272	2.25	2.24	2.19	1.98	1.93	2.17	0.51
Rome–Paris	1.105	1.438	1.423	1.81	1.86	1.91	1.95	1.89	1.90	0.42
London–Berlin	930	1181	1097	1.99	2.08	1.83	1.61	1.47	1.34	0.34

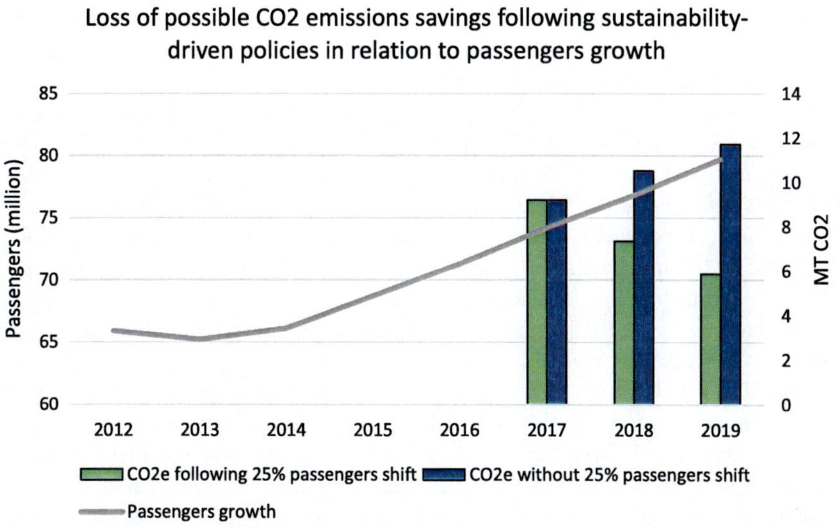

Graph 2 Effects of sustainability-driven policies on CO_2 emissions

By 2019, assuming that a stable passenger increment of + 3,5% remained, the CO_2 emissions reached 11.71 MT. Furthermore, this growth does take into account only the production resulting from the greater number of flights per se, without considering the emission from infrastructure investments (Graph 2).

On the four high-traffic routes alone, in 2019, CO_2 emissions reached 1.66 MT, with an increment of 7.1% over the 1.55 MT resulting from 2017. If this growth is not to be corrected timely, the transport sector contribution to the achieving of the 2030 objectives would not only be limited, but even harmful.

8 Lessons Learnt

In conclusion, this study highlights the influence of airlines on the GHG emissions resulting from the European transport sector. More importantly, it also analyses to what extent it would be possible to lower these emissions by shifting a relatively small percentage of passengers to railways.

We recognise it is vital to safeguard environmental protection and mobility necessity at the same time, thus this study underlined the importance of actively promoting a shift from transport modes which are not environmentally sustainable anymore to transport modes which are ecologically friendly and can play a great role in tomorrow's mobility.

Enhancement of the rail network, a change of mindset in the end-users regarding air transportation and sustainability-driven policies could start a ripple effect in the entire transport sector. Given the fact that since 1990 the emission levels from this sector constantly increased, a change of paradigm together with a re-consideration of different modes of transport is necessary.

This study does not address the question regarding which airline does produce more of the carbon dioxide equivalent on the analysed routes. However, it should be mentioned that in April 2019 Ryanair, according to data from the EU Emissions Trading System statistics,[10] has become the only airline to be included in a list of Europe's top 10 polluters. According to the data, Ryanair's carbon dioxide emissions rose by 6.9% in 2018. The news produced quite a clamour since it has been the first time a company that does not run a coal power plant has entered in the top 10. Indeed, Ryanair has been identified as one of the top polluters, rather than Lufthansa, which has been the largest airline in Europe in number of passengers carried globally in 2017, or IAG which currently is the third airline in Europe. In response to the public opinion, the company stated that Ryanair is "Europe's greenest and cleanest airline". Furthermore, from June 2019 the company became the "first EU airline to release monthly CO_2 emissions statistics, which show an average of 66 g CO_2 per passenger/km in May 2019". However, this "new transparent course" should not be misleading. In fact, a quick look at the data given by the two major companies will allow noticing that albeit Lufthansa carried more people in 2017 than Ryanair (130.04 versus 128.77 million), the difference in number and length of their respective routes is stark. While Lufthansa (2017) connects 288 airports worldwide,[11] Ryanair (2019) flies to 210 which are all located within Europe, North Africa—where it has nine destinations—and the middle east—with four destinations.[12] Thus, Ryanair's statement claiming to have "the lowest CO_2 emissions per kilometre travelled than any other airline", might be correct but certainly is ambiguous. As such, the fact that 14.1 million people flew with the company in May 2019 alone, according to Ryanair figures, should be highlighted. As already noted earlier in the "methodology" section, Ryanair is the leading company in Europe for Boeing 737–800 presence in its fleet, which presents a higher environmental footprint than the Airbus counterpart from routes longer than 420 km. As a Boeing 737–800 can seat up to 189 passengers, even by taking into account an unrealistic full load scenario through an entire month, in May alone 74.603 flights took off, on average more than 2.406 flights per day.

The medium-term sustainability of these numbers must be addressed by further researches.

[10] https://ec.europa.eu/clima/policies/ets/registry_en#tab-0-1.

[11] https://newsroom.lufthansagroup.com/english/newsroom/lufthansa-group-airlines-to-offer-many-new-destinations-worldwide-in-winter-2017-18/s/7ded97ba-a414-4e04-8c8a-a9fa906b63d4.

[12] https://corporate.ryanair.com/network/?market=it.

9 Conclusions

This study showed that the share of the short and medium-distance airline flights in the GHG emissions from the European transport sector is not negligible.

The analysis of the short-distance routes alone shows that 78.3 million passengers travelled between major European cities, while the estimated cost of these movements was 7.62 MT of CO_2. These routes alone accounted for 4.67% of the total production carbon dioxide emissions of all flights departing from EU28 and EFTA in 2017, while if the medium-distance routes are also included, the 175 routes are expected to have produced 5.64% of the total CO_2 production to and from EU28 and EFTA airports.

In 2017, the flights on the 175 analysed short and medium-distance routes carried an estimated 90.56 million passengers and produced 9.2 MT CO_2. As a result, these analysed routes accounted for more than 0.21% of the total CO_2 production within the European Union.

The introduction of sustainability-driven policies would have allowed the European transport sector to reduce the share of the analysed routes from 0.82% of the European transport CO_2 production to a hypothetical 0.53%, while ensuring the service without complication for the end-users. In case these policies were implemented as to shift 25% of the expected passenger growth from the air sector to railways on the 175 routes, it would have been possible for the European transport sector to save 1.84 MT of CO_2, lowering the emission production from 9.2 MT to 7.36.

Keeping in mind that the existing capacity of the European railway system might not be able to accommodate the passenger shift outlined in this contribution, it is nevertheless important to recognise the importance of airlines share in the transport sector CO_2 emissions. While further investments will be necessary to expand the rail network accordingly, the advantage of the sector in terms of GHG emissions over air transportation is undisputable.

10 Reflections Beyond—Policy Recommendations

Given the magnitude of the CO_2 values resulting from air passengers transport, we believe it is time to implement effective policies addressing climate challenge, especially regarding the transport sector. The urgency of these policies is evident when considering the aftermath of the Coronavirus crisis. The pandemic had profound effects on the global air transport sector and on European airlines, which have been forced to accept bailout packages from the respective home governments. Although these bailouts are not expected to focus on the environmental damage airlines pose but rather on economic and societal aspects, the aspect of green mobility still influences key policymakers. The development of significant state bailouts can give governments a rare chance to shift transport away from planes to the greener rail sector. In particular, the decision by Austrian Airlines to replace the air transportation on the Vienna—Salzburg route with train service to meet environmental requirements in the recently accepted government bailout package is noteworthy. This development will enhance railway performance by allowing travellers to choose amongst 31 trains between the two domestic cities instead of the previous three rail connections per day. However, the significance of government bailouts might be lower, as in the case of KLM. In this instance, the bailout package requires the general objective of reducing CO_2 emissions per passenger per kilometre by -50% in 2030 compared to 2005. Such a target not only was already set by KLM itself in 2019 but it also only influences the efficiency levels, thus allowing unlimited passengers growth.

Ultimately, scientific research serves as the compass for environmental policy, and legislation can deliver the forward momentum. A re-evaluation of air mobility services is necessary for the European Union not only to cope with the crisis triggered by the COVID-19 pandemic but also to achieve a long-term vision arising from commitments in international cooperation. In particular, this study shows how the nature of air transportation and its constant increase in passengers share hinder the process of reaching the objectives identified by the UN SDGs 12 "Ensure sustainable consumption and production patterns" and 13 "Take urgent action to combat climate change and its impacts".

SDG 12 highlights the following target: "rationalize inefficient fossil-fuel subsidies that encourage wasteful consumption by removing market distortions, in accordance with national circumstances, including by restructuring taxation and phasing out those harmful subsidies, where they exist, to reflect their environmental impacts, taking fully into account the specific needs and conditions of developing countries and minimizing the possible adverse impacts on their development in a manner that protects the poor and the affected communities". It is clear that, in light of this objective, the pre-COVID-19 situation vis a vis CO_2 emission by airlines cannot be restored. The low level of taxation that the air transportation sector benefited from actively pushed forward the trend of high CO_2 emissions. A fairer, more level playing field amongst transport modes is, therefore, necessary to reduce the emissions from the transport sector to reach this Sustainable Development Goal. Furthermore, SDG 13 underlines the target of "integrate climate change measures into national policies, strategies and planning". In this context, the continued growth of air transport, and thus emissions, indicates how national policies have failed in containing the growth of GHG emissions through selected mobility services. Despite its impact on the environment, air transportation has been at least condoned—if not actively promoted—by national measures sacrificing the environment on behalf of economic growth. Sustained steps towards the achievement of these two SDGs by European leaders would not only be further evidence that railways can accommodate the needs for environmental protection and high levels of mobility better than air transportation but would also firmly establish the EU as global leader in transportation and environmental policies.

In line with these objectives, policymakers should consider the development of ad-hoc financial measures for big polluters such as airlines. In particular, measures which might be introduced to curb CO_2 emissions from the air sector could be either taxes on jet fuel, or rather distance-based air passenger taxes or again an increased ticket prices with the aim to reduce demand for air travel and thus reduce emissions. However, the effectiveness of any carbon tax on airlines depends on the reinvestment of income within the transport sector. As the purpose is not taxation per se but rather development of green mobility, policymakers should consider curbing financial support from environmentally harmful travel modes to more eco-friendly options.

Other options, which however should be investigated regarding their possible outcome, could include a quota obligation for biofuels or mandatory electric vehicles for the airport infrastructure. Further studies must examine the results, costs and opportunities of alternative sustainability-oriented policies.

Following the European Union aim to reduce carbon dioxide emissions substantially, the EU climate action strategy requires a change of paradigm in approaching transportation and mobility: the 2030 targets require efforts beyond the currently implemented measures. The transport sector, in particular, cannot continue its path of steadily increasing emissions, since 1990. The need to meet sustainability targets has to lead to a reconsideration of different modes of transport where the railways might play a key role in the mobility's 'promising' future. The EU cannot afford to maintain the current course of action in respect to air transport regulations and taxations within its member states if it wants to lead the global change towards an ecologically sustainable future.

11 Avenues for Future Work

In the context of the existing literature, this study aimed at starting a debate over the need for research focusing on CO_2 emissions from for EU-wide sectors. As this contribution does not focus on infrastructure development but rather on operations, further research might consider the costs in terms of monetary expenses and CO_2 emissions resulting from enhancement of the infrastructure network for the European transport sector on a continental scale. At the same time, further studies will need to provide a detailed evaluation of the feasibility for the railway system to adapt to the passenger shift outlined in this contribution, together with a detailed evaluation for such a shift.

Lastly, this study wanted to fill in the current lack of cross-sectoral analysis by providing a novel contribution. Further researches might focus on the comparison and analysis of other transport solutions and their CO_2 share within the Union. Greater emphasis, we argue, must be placed on the analysis of transport and economic policies in light of CO_2 emission and the 2030 target.

Annex

France: national short-distance routes

Route	Distance in a straight line (km)	Distance on rail network (km)	Passengers 2017	Passengers 2016	Passengers 2015	Passengers 2014	Passengers 2013	Passengers 2012	CO_2 produced in 2017 (Kg)
Paris–Marseille	660	746	1.675.607	1.647.533	1.587.239	1.536.832	1.625.083	1.653.360	242.329.343
Paris–Lyon	392	427	696.547	666.803	646.480	627.489	601.033	595.637	63.158.506
Paris–Bordeaux	498	558	1.520.990	1.641.466	1.598.309	1.544.990	1.603.251	1.534.959	184.747.443
Toulouse–Lyon	359	561	406.982	379.790	372.006	360.098	372.010	393.205	34.593.470
Toulouse–Nice	468	647	151.511	156.932	149.677	140.040	149.227	142.111	17.120.799
Bordeaux–Marseille	505	671	296.067	258.398	214.120	244.975	216.938	277.605	35.231.973
Bordeaux–Lille	700	784	150.241	151.191	139.417	136.357	151.532	83.549	24.940.006
Lille–Lyon	557	639	76.463	71.431	69.437	70.172	80.255	84.146	10.169.579
Lille–Nantes	507	633	75.874	49.205	48.073	48.501	55.278	43.220	9.256.689

Germany: national short-distance routes

Route	Distance in a straight line (km)	Distance on rail network (km)	Passengers 2017	Passengers 2016	Passengers 2015	Passengers 2014	Passengers 2013	Passengers 2012	CO_2 produced in 2017 (Kg)
Berlin—Frankfurt	420	512	1.956.290	1.935.596	1.907.300	1.792.971	1.850.787	1.815.438	191.716.371
Berlin—Cologne	475	580	1.661.630	1.871.736	1.458.740	1.322.289	1.288.102	1.381.906	174.845.621
Berlin—Dusseldorf	475	540	1.144.435	1.146.999	1.115.068	1.115.126	1.083.065	1.012.837	122.454.545
Berlin—Munich	500	732	2.061.817	2.032.997	1.975.195	1.871.623	1.833.360	1.923.488	226.175.162
Berlin—Nuremberg	375	534	212.288	229.846	228.075	248.152	248.073	302.320	17.195.288
Berlin—Stuttgart	510	699	1.037.259	986.828	961.166	1.000.407	933.613	808.545	122.396.562
Hamburg—Frankfurt	392	513	1.395.166	1.371.938	1.360.826	1.345.147	1.367.068	1.394.538	128.355.226
Hamburg—Cologne	356	439	486.021	438.330	392.791	400.809	382.832	419.119	37.909.638
Hamburg—Dusseldorf	338	422	607.220	607.249	560.614	606.191	598.510	621.865	43.112.620
Hamburg—Nuremberg	461	623	175.525	231.428	226.248	226.621	244.731	277.231	18.781.122
Hamburg—Stuttgart	533	705	689.911	720.864	718.690	740.429	707.934	710.551	86.928.786
Frankfurt—Dresden	370	493	344.396	328.775	317.276	329.897	321.610	341.149	29.273.660
Frankfurt—Munich	304	426	1.178.390	1.129.027	1.148.611	1.107.868	1.113.259	1.084.311	70.703.400
Frankfurt—Leipzig	293	375	306.124	271.832	263.657	247.182	211.703	207.099	18.367.440
Frankfurt—Hannover	262	337	375.187	349.458	346.818	361.589	350.180	349.527	20.260.098
Frankfurt—Bremen	330	459	341.196	330.319	323.550	333.720	309.052	319.651	23.883.685
Cologne—Dresden	474	679	144.042	126.722	128.823	136.992	157.723	163.760	14.404.200
Cologne—Munich	456	641	989.158	955.343	971.682	1.005.352	973.743	1.084.043	97.926.593

(continued)

(continued)

Route	Distance in a straight line (km)	Distance on rail network (km)	Passengers 2017	Passengers 2016	Passengers 2015	Passengers 2014	Passengers 2013	Passengers 2012	CO_2 produced in 2017 (Kg)
Cologne—Leipzig	380	560	92.856	83.098	83.687	84.605	95.039	102.790	7.057.056
Munich—Dresden	360	687	234.106	224.564	238.994	241.234	247.704	227.409	16.855.596
Munich—Duesseldorf	486	649	1.554.584	1.566.252	1.557.412	1.530.577	1.560.861	1.557.746	172.558.824
Munich—Leipzig	360	567	166.597	161.116	158.689	164.548	168.233	164.787	11.994.948
Munich—Hannover	488	647	545.365	519.145	517.341	508.449	523.545	600.159	59.990.095
Munich—Bremen	583	769	375.354	348.709	337.134	338.563	362.171	358.065	48.045.248
Munich—Dortmund	477	722	183.150	169.861	166.346	168.462	174.842	166.493	18.498.150
Dusseldorf—Dresden	486	716	200.761	233.936	224.068	214.963	210.108	249.954	22.284.471
Dusseldorf—Nuremberg	363	451	233.419	223.557	202.167	219.736	264.927	294.059	18.440.101
Dusseldorf—Leipzig	389	597	113.915	96.585	95.907	90.439	103.218	107.076	9.454.945
Dusseldorf—Stuttgart	322	408	121.437	158.450	154.985	149.681	189.576	214.141	8.622.027
Stuttgart—Hannover	400	522	182.972	203.222	201.134	202.169	219.685	236.601	17.199.321
Stuttgart—Leipzig	365	532	88.944	78.931	75.351	81.104	91.564	83.692	7.204.424
Stuttgart—Bremen	478	616	164.174	156.535	159.092	150.556	155.076	118.353	18.223.314
Stuttgart—Dresden	412	678	102.073	98.670	97.136	97.221	108.725	118.553	9.288.643
Frankfurt—Nuremberg	185	228	273.862	264.188	265.900	264.630	210.922	206.340	10.132.894
Frankfurt—Dusseldorf	185	223	425.068	376.701	386.257	396.664	374.904	383.904	15.727.516
Frankfurt—Stuttgart	152	185	375.818	349.253	342.845	316.972	256.464	238.582	11.650.343
Munich—Nuremberg	150	198	136.984	134.984	134.810	121.407	121.921	122.063	3.835.538

(continued)

(continued)

Route	Distance in a straight line (km)	Distance on rail network (km)	Passengers 2017	Passengers 2016	Passengers 2015	Passengers 2014	Passengers 2013	Passengers 2012	CO_2 produced in 2017 (Kg)
Munich—Stuttgart	190	239	184.724	179.809	171.522	154.579	183.009	182.294	6.834.788

Greece: national short-distance routes

Route	Distance in a straight line (km)	Distance on rail network (km)	Passengers 2017	Passengers 2016	Passengers 2015	Passengers 2014	Passengers 2013	Passengers 2012	CO_2 produced in 2017 (Kg)
Athens—Thessaloniki	302	501	1.607.930	1.810.271	1.642.959	1.286.429	890.823	976.667	101.299.590

Italy: national short-distance routes

Route	Distance in a straight line (km)	Distance on rail network (km)	Passengers 2017	Passengers 2016	Passengers 2015	Passengers 2014	Passengers 2013	Passengers 2012	CO_2 produced in 2017 (Kg)
Rome—Milan	475	563	1.298.769	1.519.296	1.728.454	1.867.949	2.176.579	2.328.471	140.116.030
Milan—Naples	657	777	1.181.186	877.902	1.036.253	1.080.744	1.139.585	1.333.868	177.219.784
Rome—Turin	523	710	552.145	638.287	669.837	402.491	663.540	881.145	66.809.485
Rome—Naples	188	214	297.844	328.805	303.773	292.241	280.083	283.027	11.318.053
Rome—Genoa	400	709	346.399	377.302	438.227	268.578	402.430	451.464	30.829.467
Rome—Florence	230	257	246.872	227.474	214.799	196.722	223.567	231.382	11.849.856

Poland: national short-distance routes

Route	Distance in a straight line (km)	Distance on rail network (km)	Passengers 2017	Passengers 2016	Passengers 2015	Passengers 2014	Passengers 2013	Passengers 2012	CO_2 produced in 2017 (Kg)
Warsaw—Krakow	251	295	351.701	321.018	300.984	275.411	246.385	278.692	17.233.349
WarsawvWroclaw	300	404	527.684	321.395	232.249	256.077	249.317	359.468	33.771.776
Warsaw—Poznan	275	325	165.556	122.490	92.392	100.427	77.999	110.988	9.436.692
Katowice—Warsaw	260	294	101.861	83.008	76.692	83.646	53.875	79.373	4.889.352
Gdansk—Krakow	485	617	102.437	89.501	40.917	64.153	53.605	76.892	11.063.196
Gdansk—Warsaw	283	322	523.500	312.062	242.069	277.237	276.259	407.019	34.027.532

Portugal: national short-distance routes

Route	Distance in a straight line (km)	Distance on rail network (km)	Passengers 2017	Passengers 2016	Passengers 2015	Passengers 2014	Passengers 2013	Passengers 2012	CO_2 produced in 2017 (Kg)
Lisbon—Porto	273	337	1.084.907	1.031.420	669.207	451.418	401.071	389.137	60.754.792

Spain: national short-distance routes

Route	Distance in a straight line (km)	Distance on rail network (km)	Passengers 2017	Passengers 2016	Passengers 2015	Passengers 2014	Passengers 2013	Passengers 2012	CO_2 produced in 2017 (Kg)
Madrid—Barcelona	504	602	2.341.255	2.328.056	2.251.699	2.204.737	2.213.182	2.550.462	257.538.050
Madrid—Valencia	302	390	342.007	311.192	300.372	262.715	254.557	264.416	19.152.392
Madrid—Seville	387	472	301.028	284.390	264.203	244.866	238.928	351.566	26.189.436
Madrid—Malaga	415	539	319.069	318.579	318.343	330.809	278.051	401.834	31.268.762

Sweden: national short-distance routes

Route	Distance in a straight line (km)	Distance on rail network (km)	Passengers 2017	Passengers 2016	Passengers 2015	Passengers 2014	Passengers 2013	Passengers 2012	CO_2 produced in 2017 (Kg)
Goteborg—Stockholm	396	457	1.342.789	1.332.425	1.356.789	1.360.005	1.309.350	1.270.131	115.117.133

United Kingdom: national short-distance routes

Route	Distance in a straight line (km)	Distance on rail network (km)	Passengers 2017	Passengers 2016	Passengers 2015	Passengers 2014	Passengers 2013	Passengers 2012	CO_2 produced in 2017 (Kg)
London—Leeds	271	313	161.553	159.972	148.106	131.738	118.368	:	8.239.203
London—Glasgow	550	650	2.523.565	2.607.611	2.510.159	2.270.005	2.245.095	2.214.407	314.890.145
London—Manchester	260	320	663.022	692.631	774.266	876.694	797.233	1.014.816	31.162.011
Birmingham—Glasgow	405	466	221.689	226.364	227.171	230.354	204.348	209.183	19.508.588

European Union: international short-distance routes

Route	Distance in a straight line (km)	Distance on rail network (km)	Passengers 2017	Passengers 2016	Passengers 2015	Passengers 2014	Passengers 2013	Passengers 2012	CO_2 produced in 2017 (Kg)
Berlin—Vienna	522	793	861.549	799.171	779.209	786.821	737.622	733.861	105.648.491
Berlin—Brussels	650	803	621.887	702.091	509.558	438.072	420.897	411.814	87.734.121
Berlin—Amsterdam	577	639	898.328	871.104	788.082	722.859	664.718	597.733	117.048.584
Berlin—Krakow	530	687	107.356	112.611	125.071	127.230	131.155	60.023	13.097.432
Berlin—Warsaw	515	582	108.086	111.309	119.051	125.214	113.879	:	13.078.346
Berlin—Rotterdam	615	672	50.162	60.006	45.424	:	:	:	6.671.546
Berlin–Prague	280	393	87.517	:	:	:	:	:	87.517
Hamburg—Amsterdam	366	487	456.955	499.678	466.508	392.010	370.496	383.474	37.927.224
Hamburg—Brussels	487	662	281.535	177.984	152.531	152.063	140.189	142.414	30.968.795
Hamburg—Prague	490	679	110.445	:	:	:	:	:	12.480.285
Hamburg—Copenhagen	288	509	179.783	182.586	203.819	224.341	164.867	146.944	10.247.631
Frankfurt—Vienna	597	773	1.180.077	1.164.447	1.177.776	1.349.957	1.310.658	1.193.168	166.390.787
Frankfurt—Brussels	316	409	548.283	465.754	517.106	472.273	458.586	449.965	33.445.233
Frankfurt—Prague	410	706	519.591	513.934	551.300	506.908	498.017	484.873	47.282.781
Frankfurt—Amsterdam	363	442	841.591	816.077	799.336	752.964	725.609	716.479	66.485.689
Frankfurt—Lyon	560	717	321.089	295.168	293.225	296.434	269.111	267.986	39.814.974
Cologne—London	497	640	594.321	630.692	662.696	483.048	472.746	482.740	65.859.258
Frankfurt—Paris	477	583	944.070	906.179	964.443	1.053.915	1.041.456	1.114.330	97.239.159

(continued)

(continued)

Route	Distance in a straight line (km)	Distance on rail network (km)	Passengers 2017	Passengers 2016	Passengers 2015	Passengers 2014	Passengers 2013	Passengers 2012	CO_2 produced in 2017 (Kg)
Frankfurt—Milan	517	668	824.949	766.169	766.448	745.606	721.675	692.098	94.832.521
Frankfurt—Poznan	627	770	90.610	83.795	76.820	70.880	37.312	52.576	13.138.378
Frankfurt—London	636	793	1.864.255	1.948.862	2.004.808	1.970.538	1.929.011	1.903.271	270.390.653
Munich—Vienna	355	460	547.029	487.231	555.875	542.825	562.934	563.954	41.574.166
Munich—Prague	300	435	208.823	206.119	210.289	198.640	231.447	208.319	10.858.796
Munich—Budapest	561	717	343.143	330.084	323.050	321.154	321.670	316.537	44.951.733
Munich—Milan	348	587	490.865	422.178	449.664	328.091	335.460	382.039	40.699.460
Munich—Turin	453	734	180.986	165.822	167.375	150.007	158.812	129.568	19.546.488
Munich—Florence	486	645	189.128	170.895	176.012	183.717	159.752	148.268	23.073.616
Munich—Genoa	460	733	92.441	90.247	94.869	57.061	94.390	86.788	10.445.777
Munich—Zagreb	423	579	191.842	181.833	181.593	180.355	181.528	174.846	19.184.150
Duesseldorf—Paris	410	576	466.309	445.904	436.439	479.986	475.470	491.932	40.568.883
Duesseldorf—Amsterdam	180	219	234.348	226.624	205.169	184.800	171.676	182.285	8.202.180
Duesseldorf—London	478	630	895.321	754.168	691.468	655.562	645.832	621.847	101.719.803
Nuremberg—Vienna	411	508	77.401	67.332	66.770	66.010	76.761	96.040	7.585.298
Nuremberg—Budapest	625	772	119.181	:	:	:	:	:	16.446.909
Nuremberg—Milan	465	785	102.908	:	:	:	:	:	9.673.305
Nuremberg—Amsterdam	541	670	188.110	191.908	186.317	168.381	160.861	159.534	23.325.640
Hannover—Amsterdam	327	382	178.804	177.510	170.371	168.236	166.610	162.048	12.516.280

(continued)

(continued)

Route	Distance in a straight line (km)	Distance on rail network (km)	Passengers 2017	Passengers 2016	Passengers 2015	Passengers 2014	Passengers 2013	Passengers 2012	CO_2 produced in 2017 (Kg)
Stuttgart—Vienna	533	699	427.883	367.782	344.447	338.863	338.549	342.851	53.485.375
Stuttgart—Brussels	416	597	99.357	76.923	77.337	:	:	:	9.339.511
Stuttgart—Milan	368	619	155.424	139.455	130.075	81.343	88.323	89.912	11.345.952
Stuttgart—Paris	500	602	182.358	178.710	186.075	184.014	199.562	228.251	20.424.040
Stuttgart—Amsterdam	500	627	325.281	312.612	284.093	228.398	210.300	206.128	38.057.819
Dortmund—London	525	720	213.763	207.802	217.228	236.209	160.585	162.044	23.844.358
Dortmund—Wroclaw	666	788	31.843	31.704	30.516	29.754	29.316	27.658	4.489.863
Bremen—Amsterdam	274	372	164.296	155.210	152.309	141.149	143.921	138.104	9.036.280
London—Paris	343	460	2.181.247	2.245.164	2.194.613	2.042.799	1.831.887	1.751.282	157.571.551
London—Amsterdam	357	597	4.488.263	4.141.696	3.847.757	3.442.148	3.229.507	3.117.265	336.498.375
London—Rotterdam	320	529	178.052	219.274	219.991	272.696	235.109	112.983	10.327.016
Birmingham—Paris	502	640	398.933	380.981	388.357	388.862	365.691	358.896	44.281.563
Birmingham—Amsterdam	460	697	676.693	633.081	565.453	508.644	460.633	445.610	67.669.300
Leeds—Paris	612	783	49.051	46.604	49.110	61.148	63.306	64.103	6.425.681
Manchester—Paris	605	780	630.886	568.366	478.819	507.100	503.387	470.466	82.646.066
Sheffield—Paris	570	726	41.697	48.431	:	:	:	:	5.003.640
Vienna—Warsaw	557	664	222.350	205.889	205.113	208.178	207.535	207.423	28.016.037
Vienna—Budapest	215	264	103.411	100.208	104.901	107.372	104.075	106.144	4.239.851
Vienna—Zagreb	268	481	164.100	160.326	163.421	163.748	158.326	163.022	8.533.200

(continued)

(continued)

Route	Distance in a straight line (km)	Distance on rail network (km)	Passengers 2017	Passengers 2016	Passengers 2015	Passengers 2014	Passengers 2013	Passengers 2012	CO_2 produced in 2017 (Kg)
Brussels—Paris	261	315	199.065	203.444	205.727	170.833	156.712	132.912	9.754.161
Brussels—Lyon	560	742	245.347	216.856	239.002	240.540	221.730	226.416	32.140.392
Brussels—Amsterdam	180	219	259.780	219.141	243.416	241.883	223.166	213.768	12.729.220
Brussels—Birmingham	466	488	152.911	133.353	144.297	120.092	107.179	105.405	16.208.566
Brussels—Manchester	535	698	397.564	376.620	385.279	332.714	287.574	262.263	48.321.378
Brussels—London	318	378	651.306	572.195	722.546	673.362	568.996	547.948	48.196.607
Antwerp—London	316	433	50.229	54.422	58.319	54.043	70.810	71.962	3.114.198
Copenhagen—Goteborg	226	339	289.570	331.470	298.020	336.134	306.589	327.313	11.872.350
Budapest—Prague	440	609	109.502	78.296	:	:	:	75.708	12.264.224
Riga—Warsaw	958	676	115.588	95.197	86.196	80.683	65.635	66.016	15.373.138
Riga—Vilnius	265	347	162.870	165.797	165.048	175.214	181.108	178.129	8.794.953
Warsaw – Vilnius	390	568	167.559	121.675	101.614	100.675	94.416	72.707	15.247.869
Barcelona—Lyon	530	680	294.658	308.922	302.408	282.955	294.310	212.719	36.832.250
Barcelona—Marseille	335	580	78.118	78.985	:	:	:	:	5.833.050
Barcelona—Bordeaux	443	678	162.115	120.481	77.669	68.707	61.595	67.767	17.670.480
Amsterdam—Paris	428	522	1.422.740	1.324.720	1.154.023	1.158.098	1.127.698	1.141.903	125.997.340
Paris—Turin	580	731	171.647	174.458	181.807	178.319	178.723	189.469	22.314.045
Warsaw—Prague	517	747	212.637	196.005	179.672	162.341	153.787	160.386	26.366.988

France: national medium-distance routes

Route	Distance in a straight line (km)	Distance on rail network (km)	Distance on road network (km)	Passengers 2017	Passengers 2016	Passengers 2015	Passengers 2014	Passengers 2013	Passengers 2012	CO_2 produced in 2017 (Kg)
Paris–Toulouse	586	816	680	3.254.630	3.258.878	3.214.776	3.169.217	3.192.829	3.154.149	450.067.524
Bordeaux—Lyon	437	985	556	532.009	508.416	474.885	459.698	450.582	451.107	59.052.999

Germany: national medium-distance routes

Route	Distance in a straight line (km)	Distance on rail network (km)	Distance on road network (km)	Passengers 2017	Passengers 2016	Passengers 2015	Passengers 2014	Passengers 2013	Passengers 2012	CO_2 produced in 2017 (Kg)
Munich—Hamburg	610	825	775	1.739.043	1.805.203	1.811.381	1.762.070	1.721.782	1.721.358	269.551.665

Italy: national medium-distance routes

Route	Distance in a straight line (km)	Distance on rail network (km)	Distance on road network (km)	Passengers 2017	Passengers 2016	Passengers 2015	Passengers 2014	Passengers 2013	Passengers 2012	CO_2 produced in 2017 (Kg)
Naples—Genoa	590	923	700	79.120	48.233	55.321	28.979	58.654	78.390	10.681.200

Spain: national medium-distance routes

Route	Distance in a straight line (km)	Distance on rail network (km)	Distance on road network (km)	Passengers 2017	Passengers 2016	Passengers 2015	Passengers 2014	Passengers 2013	Passengers 2012	CO_2 produced in 2017 (Kg)
Valencia—Seville	539	862	654	182.431	113.126	125.082	100.509	151.100	200.946	21.891.720

European Union: international medium-distance routes

Route	Distance in a straight line (km)	Distance on rail network (km)	Distance on road network (km)	Passengers 2017	Passengers 2016	Passengers 2015	Passengers 2014	Passengers 2013	Passengers 2012	CO_2 produced in 2017 (Kg)
Bremen—Paris	650	1042	787	71.977	81.496	94.566	111.216	93.261	86.607	10.220.734
Hannover—Paris	650	832	773	162.607	155.232	154.388	152.446	160.392	155.251	23.252.801
Dresden—Amsterdam	626	935	734	40.541	26.969					5.473.035
Leipzig—Vienna	586	906	582	55.397	50.189	53.549	55.844	53.747	50.139	6.093.670
Nurnberg—Paris	637	808	759	169.229	157.601	161.757	159.913	147.525	146.125	23.861.219
Duesseldorf—Birmingham	611	1054	783	270.567	241.274	219.054	198.654	184.347	175.607	36.797.044
Berlin—Copenhagen	365	803	750	629.851	679.784	634.634	632.246	579.484	525.643	46.356.508
Cologne—Copenhagen	645	948	750	110.729	110.874	:	:	:	:	15.280.602
Frankfurt—Turin	565	819	738	192.261	186.392	206.678	168.617	202.560	189.370	23.648.103
Frankfurt—Wroclaw	600	844	725	148.028	138.459	140.732	108.570	105.793	100.426	20.427.795
Munich—Brussels	602	835	780	403.195	349.794	347.172	351.434	363.450	365.824	54.431.325
Munich—Lyon	570	902	731	186.797	188.556	190.066	202.512	222.971	211.898	25.030.731
Munich—Wroclaw	512	1005	720	137.124	138.119	126.466	126.359	128.033	122.759	15.220.764
Duesseldorf—Prague	554	942	724	234.022	192.126	166.656	178.444	164.417	154.948	29.252.750
Duesseldorf—Copenhagen	625	931	728	312.164	378.972	380.323	379.853	386.314	326.087	44.327.217
Duesseldorf—Lyon	620	1317	759	78.996	:	:	76.649	131.316	137.295	11.296.428
Hamburg—Gdansk	572	767	858	68.010	43.745	:	:	:	:	8.705.280
Zagreb—Stuttgart	610	818	780	55.176	54.103	50.675	50.082	50.270	51.362	10.593.792

(continued)

(continued)

Route	Distance in a straight line (km)	Distance on rail network (km)	Distance on road network (km)	Passengers 2017	Passengers 2016	Passengers 2015	Passengers 2014	Passengers 2013	Passengers 2012	CO_2 produced in 2017 (Kg)
Sofia—Athens	526	850	790	201.971	146.979	95.600	96.364	89.203	96.117	24.640.401
Budapest to Warsaw	550	867	781	238.254	212.363	196.648	172.417	181.743	115.697	30.496.448
Madrid—Lisbon	503	1324	629	1.427.516	1.296.808	1.173.975	1.027.638	974.675	1.109.369	166.970.934
Madrid—Toulouse	552	702	1022	366.401	201.864	154.094	160.710	187.944	206.766	47.265.793
Madrid—Bordeaux	554	686	849	103.995	:	:	46.408	43.146	95.782	13.311.360
Madrid—Porto	422	562	856	615.261	533.548	484.125	395.307	362.452	398.131	63.987.196
Malaga- Lisbon	470	1863	650	88.069	:	:	:	:	:	8.962.026
Seville—Lisbon	312	1790	433	95.730	:	:	:	:	:	6.318.906

References

Albalate D, Bel G (2012) The economics and politics of high-speed rail. Rowman and Littlefield Publishers (Lexington Books), Lanham, MA, Lessons from Experiences Abroad

Alonso G, Benito A, Lonza L, Kousoulidou M (2014) Investigations on the distribution of air transport traffic and CO_2 emissions within the European Union. J Air Transp Manag 36:85–93

Anger A (2010) Including aviation in the European emissions trading scheme: impacts on the industry, CO_2 emissions and macroeconomic activity in the EU. J Air Transp Manag 16:100–105

CAPA (2018) Aircraft fleets: Western v Eastern/Central Europe. Airbus leads orders [Online]. Available at https://centreforaviation.com/analysis/reports/aircraft-fleets-western-v-easterncentral-europe-airbus-leads-orders-410122 [Accessed July 2019]

Chèze B, Chevallier J, Gastineau P (2013) Will technological progress be sufficient to stabilize CO_2 emissions from air transport in the mid-term? Transp Res Part D 18:91–96

EASA (2019) Emissions [Online]. Available at https://www.easa.europa.eu/eaer/topics/overview-aviation-sector/emissions [Accessed July 2019]

EEA (2019) EEA greenhouse gas–data viewer [Online]. Available at https://www.eea.europa.eu/data-and-maps/data/data-viewers/greenhouse-gases-viewer [Accessed July 2019]

EEA (2018) Total greenhouse gas emission trends and projections [Online]. Available at https://www.eea.europa.eu/data-and-maps/indicators/greenhouse-gas-emission-trends-6/assessment-2 [Accessed July 2019]

European Commission (2020) Report from the commission to the european parliament and the council—updated analysis of the non-CO_2 climate impacts of aviation and potential policy measures pursuant to EU Emissions. Trading Syst Direct Article 30(4). [Online]. Available at https://eur-lex.europa.eu/legal-content/EN/TXT/?uri=SWD:2020:277:FIN [Accessed December 2020]

European Commission (2019) Report verified emissions for 2018 [Online]. Available at https://ec.europa.eu/clima/sites/clima/files/ets/registry/docs/verified_emissions_2018_en.xlsx [Accessed July 2019]

European Commission (2017) Annual analyses of the EU air transport market 2016. [Online]. Available at https://ec.europa.eu/transport/sites/transport/files/2016_eu_air_transport_industry_analyses_report.pdf [Accessed July 2019]

Eurostat (2019) Greenhouse gas emission statistics—emission inventories [Online]. Available at https://ec.europa.eu/eurostat/statistics-explained/pdfscache/1180.pdf [Accessed July 2019]

Lee DS, Fahey DW, Skowron A, Allen MR, Burkhardt U, Chen Q, Doherty SJ, Freeman S, Forster PM, Fuglestvedt J, Gettelman A, De León RR, Lim LL, Lund MT, Millar RJ, Owen B, Penner JE, Pitari G, Prather MJ, Sausen R, Wilcox LJ (2021) The contribution of global aviation to anthropogenic climate forcing for 2000 to 2018. Atmos Environ 244. [Online]. Available at https://www.sciencedirect.com/science/article/pii/S1352231020305689?via%3Dihub [Accessed December 2020]

Lufthansa (2017) Lufthansa Group airlines to offer many new destinations worldwide in winter 2017/18 [Online]. Available at https://newsroom.lufthansagroup.com/english/newsroom/lufthansa-group-airlines-to-offer-many-new-destinations-worldwide-in-winter-2017-18/s/7ded97ba-a414-4e04-8c8a-a9fa906b63d4 [Accessed July 2019]

Mendes LMZ, Santos G (2008) Using economic instruments to address emissions from air transport in the European Union. Environ Plan A 40:189–209

Morrell P (2007) An evaluation of possible EU air transport emissions trading scheme allocation methods. Energy Policy 35:5562–5570

Prussi M, Lonza L (2018) Passenger aviation and high speed rail: a comparison of emissions profiles on selected European routes. J Adv Transport [Online] Volume 2018 Available at https://www.hindawi.com/journals/jat/2018/6205714/ [Accessed July 2020]

Ryanair (2019) Our network [Online]. Available at https://corporate.ryanair.com/network/?market=it [Accessed July 2019]

Tardivo A, Sánchez Martín C, Carrillo Zanuy A (2020) Covid-19 impact in Transport, an essay from the Railways' system research perspective [Online] Available at https://advance.sagepub.com/articles/Covid-19_impact_in_Transport_an_essay_from_the_Railways_system_research_perspective/12204836 [Accessed July 2020]

Competition on the Domestic Rail Passenger Transport Market Under Public Service Obligation in Some Selected European Countries and Slovak Republic

Anna Dolinayova, Lenka Cerna, and Igor Domeny

Abstract Sustainable rail transport is an important aim within European transport policy. One way to achieve this is to open-up the rail passenger market to competition. The White paper 2011 of the European Commission (EC) stated that: a competitive and resource efficient transport system presents the need towards opening domestic rail passenger market to competition, including the mandatory award of public service contracts through competitive tendering as a first initiative within the single European transport area. The open access to competition is realised by two different approaches in many European countries; 1) as a public service provision through public tender and 2) on a commercial basis—though open access. Markets for domestic rail passenger services through public tenders have not been opened in many EU countries. The study looks at the rail passenger transport market situation in selected European countries (Austria, Italy, Hungary, Germany, etc.) and places a particular focus on the transport market situation in Slovak Republic. It describes different mechanisms to realise public tenders for the operation of rail passenger transport services under Public Service Obligation (PSO), analysis process of public tenders in the Slovak Republic, quantifies the risk of unsuccessful public tenders and suggest the criteria to achieve successful public tenders for providing rail passenger transport services under PSO.

Keywords Competition · Criteria · Public tenders · Rail passenger transport · Risk

1 Introduction

In recent years, the European railway system has become a subject of extensive reforms. These reforms have different forms in individual countries, but all of them have a common goal—to reverse the development of the European railways. In most countries, services with regard to rail passenger transport and rail freight transport were provided by incumbents or undertakings, (national rail service carriers), which

A. Dolinayova (✉) · L. Cerna · I. Domeny
University of Zilina, Univerzitna 8215/1, Zilina 010 26, Slovakia
e-mail: anna.dolinayova@fpedas.uniza.sk

M. Marinov and J. Piip (eds.), *Sustainable Rail Transport 4*, Lecture Notes in Mobility,
https://doi.org/10.1007/978-3-030-82095-4_12

were unable to react flexibly to constantly changing social-economic conditions and passenger requirements. In Europe, mainly in Western-European countries, there was a diversion of heavy industry and dynamic development regarding automotive passenger transport. Owing to these changes, the transport performance of freight and passenger transport decreased with the current increase in indebtedness of state-owned undertakings. Over the last few years, the railway system in many European countries has come under increasing criticism within the media, and many people are complaining about capacity issues, delays, cost, etc. (Fraszczyk et al 2016). These were the main reasons for implementing the reforms, in order to revitalise the railways and increase their performance and competitiveness in the European transport market. The main aims of these reforms are liberalisation of the European railways and open competition in the market (Janeckova 2015). The aim (whether open access services or rail passenger transport under public services obligation) is to achieve sustainable rail transport (Abramovic et al 2018). There are two different approaches in the field of competition. The first one is based on public tenders for the operation of PSO (public service obligations) services, and the second is based on open access services, when several carriers operate trains in direct competition with each other, competing for passengers.

In relation to railway transport, the open access approach gives better identification to effective competition. Rail carriers are under pressure to reduce costs and to use market opportunities better. In most EU countries, where railway undertakings have started to provide open access services, there were effects like price reduction, improvement in the quality of provided services and an increase in the frequency of train connections. Noticeably, open access is also accompanied by a number of problems. Railway transport operation takes place in a network, where individual lines with different profitabilities exist side by side. There could be cross-financing of individual lines during the network operation. Due to the high amount of fixed costs and the entry of new carriers, total costs have been increasing and it is questionable, if the investments of new carriers in railway technology would be cost-effective. Open access often leads to fragmentation of transport flows between more subjects, which can endanger the achievement of economies of density. By definition, 'economies of density are cost savings resulting from spatial proximity of suppliers or providers'. In the context of fare price reductions, open access can trigger price 'fights' between rail undertakings, which eventually cause operational losses. This type of open competition could be classified as 'cut-throat' competition (Tomeš 2013).

The second way of providing inland rail passenger transport services in the EU is public tender for PSO (public service obligation) services. PSO services have a 90% share of the total operational performance in the EU and are mainly related to regional and commuter trains, but also to some long-distance trains. According to the 4th railway package Directives, these services have to be awarded through public tenders (Tomeš and Jandová 2018).

Rail passenger transport markets with services provided based on public tenders are still closed in most EU countries. For example, markets for inland passenger transport services have been opened in Germany, Sweden, Italy, the United Kingdom,

Austria and Czech Republic. The rail passenger transport market is still closed for public tender in Lithuania, Luxembourg, Slovenia etc. (Záhumenská 2018). Experiences from these open markets indicate an improvement in service quality and availability, as well as an increase in passenger satisfaction. In some cases, the passenger numbers have grown by over 50% in 10 years. In other cases, public tenders for PSO services brought savings of 20–30%, which could be reinvested in improving the service quality. Nowadays, following Regulation (EC) No 1370 /2007 rules, the railway sector is excluded from an obligation to award PSO contracts through public tender, whilst the Council Directive 91/440/EEC does not include the general right of access to rail infrastructure around Europe for inland rail transport operation—as a result, the railways can remain national monopolies. According to Regulation (EC) No 1370/2007, national authorities in member states may decide to award PSO contracts in railway transport directly or through public tender (Transport and tourism policy 2020).

The European Commission (EC) suggested changing and supplementing Regulation (EC) No 1370/2007 on public passenger transport services by rail and by road, in order to introduce compulsory public tender for awarding PSO services in rail transport. The EC also suggested changing and supplementing Directive 2012/34/EU in order to allow open access in inland rail transport, with the possibility of limiting this access in case of a threat to the economic balance of PSO contracts (The 4th railway package 2016).

PSO services are services where carriers have an obligation to transport specified groups of the population, in terms of relevant pricing legislation, free of charge or with social discount (Regulation (EC) No 1370/2007). A competent authority regulates fares (on selected routes and trains). Losses that arise to carriers from passenger transport operation, are compensated. Public interest is realised indirectly by covering the loss from operating unprofitable routes. PSO services mainly include services that are accessible to every citizen regardless of their financial position and physical possibilities, anywhere in the member state territory (Act No. 514/ 2009).

PSO services in rail passenger transport are very important for the provision of transport service for all EU residents, because there are many areas, where public transport services would not be operated by railway undertakings on a commercial basis, or would not be operated in the necessary range. PSO services are reimbursed by resources that are provided by the contracting authority (Strategy for the development of public passenger and non-motorised transport in the Slovak Republic until 2020).

With the arrival of the 4th railway package in 2016, the question arises regarding opening the domestic passenger rail services market. Regulation (EU) 2016/2338 deals with the method of awarding PSO contracts and there is a requirement to implement a system for compulsory public tenders with regard to these contracts. After the set deadline, directly awarding PSO contracts will no longer be possible.

The 4th railway package is a sum of six legal norms for the Community railway development in the EU, particularly the uniform railway market. The main aims of the 4th railway package are rail transport resumption and increase rail transport competitiveness compared to other transport modes. It consists of two pillars—the technical pillar and the market pillar. The market pillar's role is to complete the process of gradually rail transport market opening, which began with the 1st railway package. The market pillar lays down the general legislation for railway undertakings established in one of the Member states, for the operation of all types of passenger transport services in any other EU state. It also lays down rules aimed at improving the impartiality of railway infrastructure management as well as preventing discrimination and introduces the compulsory public tendering principle for PSO services in passenger rail transport. Competition on the passenger rail transport market will encourage infrastructure managers to respond to customer requirements better and improve the quality of services and cost effectiveness. Fair public tender for PSO contracts can achieve savings in public finances. It is expected that the market pillar will bring more choice and better quality in the services provided for EU citizens, which are the main aims (The 4th railway package 2016).

Based on the 4th railway package and amendments to the EU laws, railway undertakings will be able to offer inland passenger rail transport services throughout the entire EU. Railway undertakings have two options: to operate services in open access or to take part in public tender for PSO contracts.

On the basis of the EU Commission's proposal, public tender can only be applied for contracts exceeding certain thresholds. Organising public tenders in the case of small volume contracts would not make sense, because expected cost savings for public authorities would be low. The maximum limit value for PSO contracts shall be determinate taking into account the operational performance (train-kms) or the percentage of total volume of PSO services for each member state will be available on the market to ensure that suitable tenderers are able to submit a tender. In order to avoid a disproportionable effect on existing directly awarded contracts, temporary measures will be adopted in connection with the introduction of public tender (European Commission Notice 2013).

In Slovakia, there have been several public tenders for operation of PSO services in recent years, on several routes such as Bratislava—Komárno, Žilina—Rajec, etc. They all ended in failure and without a final provider for these services. The aim of the paper is trifold: 1) to conduct a detailed analysis regarding the current state of public tenders in selected countries and in Slovakia in paricular, 2) to analyse the reasons for failure of public tenders and 3) to draft procedures for the public tendering process in the Slovak republic.

2 Literature Review

The problem of competition in the transport sector has been solved in many scientific papers and studies. Many authors have researched intermodal competition from different viewpoints. Bergantino and Madio (2020) studied whether and to what extent the introduction of HSR can lead to inter-modal substitution and they identified the demand segments which are less captive to the currently chosen mode. Raturi and Verma (2020) applied a game-theoretic framework in order to model the competition scenario between high-speed rail and airlines in the Indian context and assess the impact of speed and passenger characteristics on the equilibrium of the game. Many other studies have addressed competition/cooperation between high-speed rail and airlines such as Wang et al (2020), Zhang et al (2019), Tsunoda (2018), Cadarzo et al. (2017), D'Alfonso et al (2016), Capozza (2016) etc. Beria and Bertolin (2019) dealt with the main long-distance transport characteristics in a liberalised environment (Italy) in terms of market concentration levels, fares and product specialisation. They considered rail, coach and carpooling services, which are not fully separate markets and interact in terms of supply, prices and customer groups. Fearnley et al (2018) researched the competition and substitution between public transport modes on the basis of 50 empirically estimated cross elasticities between public transport modes from over 20 sources collected from Australia, Europe and USA.

The aim of competition in rail passenger transport is to improve the quality and efficiency of this railway transport segment. While the quality of provided services has increased in almost all countries where competition has been introduced, there is no clear evidence of efficiency gains. Dias and Graca (2016) conducted the first thorough cross-country comparison of European expectations in relation to railway competition effects by using a multilevel factor model. The results of the Bougna and Crozet (2016) study showed that competition tendering improves productive efficiency, whereas free entry delivers the opposite result. Their results also showed that competition and overall liberalisation have no impact on productive efficiency. Tomeš (2017) performed analysis on 27 European countries in the period between 1995 and 2013 and showed that there is no evidence that vertical separation and competition entry increase European railway modal shares. Nash et al (2019) studied the experience of Europe's three most liberalised railways—Sweden, Germany and Britain—in opening-up rail passenger services to competition by competitive tendering, and considered the wider experience in terms of market competition for commercial services. Jandová and Paleta (2019) researched the budgetary impacts and development of public spending efficiency in the context of on-track competition in the Czech Republic. Their general results indicate that the situation regarding the budgetary impact is more complex. In Poland almost all entries to the long-distance commercial market in Poland (including the Interregio case) were made from another market segment (regional public services). Interestingly, by publicly-owned operators (Król et al 2019).

Competitive Tendering has become in the last decade the prevailing approach for procuring Public Transport passenger services both at urban and rural/intercity level (Papaioannou et al 2020). Market liberalization is crucial in terms of speeding up the procedure regarding the conclusion of the public service obligation contract (Humic and Abramovic 2019). Contracting is the main trend, which can indeed take many forms, as many kinds of relationships are possible between transport authorities and transport operators, the way authorities divide regulatory powers upon public transport, the way public transport funding is organised, the type of regulatory regime and so forth (Pedro and Macário 2016). Poliak et al (2018) identified the importance of public passenger transport and to define the effective method to increase public transport competitiveness. At present, public transport tendering and contracting is used mainly in public passenger road transport. Many authors researched how competitive tendering in the public transport by road affect this segment of transport in term of costs, economic efficiency, quality of services and other aspects such as Hensher and Stanley (2003), Rojo et al (2015), Kavanagh (2016), Valkama et al (2018), Iossa and Waterson (2019), etc.

There are a few studies that deals with rail passenger transport under public service obligations. Wegelin and von Arx (2016) dealt with addressing the transaction costs of different governance forms in regional public rail transport by comparing Germany's competitive contract awarding model with Switzerland's direct contract awarding model. Avenali et al (2020) estimated the standard cost in Italian regional public rail passenger transport services (LPTR), depending on service characteristics and proposed regulatory adjustments to accomplish policy targets regarding the fair allocation of public LPTR funds to Regions and Local Authorities and a more efficient use of (scarce) local and national public resources. Nash and Smith (2020) researched public transport procurement in Britain in the bus and rail sector.

Our literature review showed there are a many articles which deal with competition in the rail passenger transport but only a few of them deal with rail passenger transport services in the public interest.

3 Mechanisms to Realise Public Tenders for the Operation of Rail Passenger Transport Services Under PSO

In almost every country within the European Community, the incumbent has predominance in the rail passenger market. Most performances in countries are services operated under PSO contracts concluded with a certain administrative authority, e.g. ministry or regional authority, unless this authority provides the services itself. Public interest means the provision of transport services and other public services at the contracting authority's expense. The PSO contract can be concluded through direct award or through a fairer public tender (Regulation (EC) No 1370/2007).

Under Regulation (EU) 2016/2338 of the European parliament and of the Council from December 14th, 2016 amending Regulation (EC) No 1370/2007 concerning opening the market for domestic passenger transport services by rail. If the contracting authority is not also the PSO services provider, it may provide these services by concluding a contract with another legally different entity through direct award, if the range of services is only within the agglomeration, or through public tender for railway undertakings that are not internal and the range of services exceeds agglomerations (Regulation (EU) 2016/2338).

3.1 *Direct Award*

A direct award is a way of ordering public services or a method of concluding the PSO contract between the contracting authority and the selected railway undertaking. This method has its advantages and disadvantages.

Direct award advantages:

- speeding-up the contracting process,
- conditions in the contract are adapted to the possibilities of the railway undertaking,
- reducing the risk of the inability to fulfil obligations under the contract,
- there is always a railway undertaking (e.g. incumbent),
- planned performances specified in the contract can be converted into public tender.

Direct award disadvantages:

- insufficient market competition,
- an incumbent is often preferred,
- reducing the quality of services,
- it is not approved by EU law, except selected routes,
- services are provided even for higher economically justified costs.

Under Regulation (EU) 2016/2338 of the European parliament and of the Council from December 14th, 2016 amending Regulation (EC) No 1370/2007 concerning opening the market for domestic passenger transport services by rail, direct awarding means concluding the contract with a pre-selected railway undertaking, which is chosen by the contracting authority for public services. The EU is simultaneously pushing for a reduction in the number of contracts concluded through direct award. Contracts can be awarded directly by the contracting authority, if the contract's annual value in passenger rail transport is less than € 7,500,000 or the volume of services is less than 500,000 train-kms per year. Contracts can also be awarded directly if it is justified by geographical and social conditions or the directly awarded contract would lead to higher quality and efficiency in the provision of these services. Contracts can also be awarded directly if the railway undertaking provides services on its own infrastructure. If the operation of PSO services is interrupted, the contracting authority may temporarily enter into a contract with another railway undertaking, for a maximum of two years (Regulation (EU) 2016/2338).

3.2 *Public Tender*

Another method of awarding PSO services is a public tender. Public tender is announced by a state administration authority, and railway undertakings with a valid licence can participate here. It is another form of transport market liberalisation, which can replace direct award. The result of the tender is that a successful tenderer is able to conclude a PSO contract with the contracting authority. The tenderer´s success is decided by an independent commission. The commission will evaluate the tenderer's individual possibilities of meeting the conditions specified in the tender documents, published by the contracting authority of these services. The public tender will facilitate the entry of the private sector, whereas this does not presuppose the preference regarding the incumbent, but the railway undertaking which has more suitable opportunities to provide services on tendering route. The contracting authority of PSO services in Slovakia is the Slovak Republic Ministry of Transport and Construction, and individual regions also have the opportunity to announce public tenders (Act No. 514/ 2009).

Public tender advantages:

- there is no preference of incumbent,
- higher probability that the contract will be conclude with a private railway undertaking,
- increasing competition in the passenger rail market,
- higher quality services provided,
- conditions in the tender documents should be adapted to the specific route.

Public tender disadvantages:

- long process duration,
- poorly set conditions in the tender documents, the risk of public tender failures,
- insufficient number of railway undertakings on the market,
- the risk of inability to meet the conditions after concluding the contract.

Under Regulation (EU) 2016/2338, public tender is a procurement for the provision of PSO services, announced by the contracting authority of these services, and the participants are transport undertakings. The tendering procedure under this Regulation must always be non-discriminatory, fair and transparent. Before the tender has started, the Regulation allows (if national law also allows) to temporarily award the contract through direct award. Such award must be reported to the Commission. The validity period is up to five years and the contract must be published. Before announcing the public tender, the contracting authority must check the possibilities of access to rolling stock for tenderers (Regulation (EU) 2016/2338).

For the purpose of initiation of a public tender, competent authorities assess if measures are needed to ensure effective and non-discriminatory access to suitable rolling stock. This assessment shall take into account the fact, whether there are rolling stock leasing companies or other market players providing rolling stock rental. An evaluation report will be published. Competent authorities can decide in accordance with national law and with State aid rules, that they will accept appropriate measures to ensuring effective and non-discriminatory access to appropriate rolling stock (Regulation (EU) 2016/2338).

The contracting authority is required to submit a tender notice in accordance with this Regulation, where it indicates the volume of services, service type, information in relation to rolling stock and the contract duration, in the Official European Union Journal. It has to do so at least one year before the tender starts. The contracting authority does not have to provide this information, if the services provided in the contract are lower than 50,000 train-kms per year. If the contracting authority requires compliance with social and quality standards, so in that case they must be a part of the tender documents and also as an annexe to the contract. Tender documents must indicate the volume and possibility of subcontracting.

Potential railway undertakings have the opportunity to express their interest within 60 days of the public tender announcement. If only one tenderer has been notified by this time, and will be able to fulfil the obligations under the contract, or if there has not been a better alternative, then the contracting authority can conclude a contract with this tenderer. If the railway undertaking already has more than one contract and the aim is to increase the competition and to attract other railway undertakings into tenders, the contracting authority may limit the number of contracts for this undertaking.

According to Regulation (EU) 2016/2338, the Commission will decide with regard to the provider of PSO services in a public tender and if the contract is awarded directly, then the contracting authority will decide itself. Thereafter, the railway undertaking will provide rail passenger services and other obligations under the contract, while the contracting authority will reimburse it for these obligations. The railway undertaking must state: "It is necessary for the railway undertaking to provide information to the competent authority in the PSO contract while ensuring the legitimate protection of confidential business information. Contracting authorities make relevant information for preparing the tender available to all interested parties, while ensuring the legitimate protection of confidential business information concerning passenger demand, fares, costs and revenues related to passenger public transport, included in the tender and detailed information in relation to infrastructure specifications, which are important for the operation of rolling stock, in order to allow shareholders to develop well-founded business plans. Infrastructure Managers assist the contracting authorities in provision of all relevant infrastructure specifications " (Regulation (EC) No 1370/2007).

The methods for calculating compensation shall be set in a transparent manner, in order to avoid overpayment. Operating costs, social costs, as well as vehicle repair and maintenance costs are compensated. Sharing the railway undertaking´s revenues will be specified in the contract. Compensation is controlled by the commission on the basis of a summary report on public service obligations, which is published by the contracting authority at least once a year. This report must be published on the authority's website and also in public transport policy documents. This information may not be published, if the volume of services provided is lower than 50,000 train-kms per year (Regulation (EC) No 1370/2007).

The validity of contracts for passenger rail transport awarded through public tender is 15 years, and this period may be extended by half that time, if the railway undertaking uses its assets to develop its obligations under the contract (but the contract must not be concluded between 03.12.2019 and 24.12.2023). It is also possible to extend the contract by 50%, if services are operated in more remote areas. Contracts awarded directly are valid for a maximum of 10 years. With an integrated transport system, the contract can be extended by 50%, if the share of rail services in integrated transport system is at least 50%—If the PSO contract has been awarded through public tender, it can only be extended for a period longer than 50% of the validity period with the commission's consent. The validity period of PSO contracts awarded between 3.12.2019 and 24.12.2023 can be a maximum of 10 years (Regulation (EC) No 1370/2007).

Regulation (EC) No 1371/2007 of the European Parliament and of the Council from October 23rd, 2007 on rail passenger´s rights and obligations, was created for the purpose of establishing the same conditions of carriage for passengers within the Community's territory. This Regulation also applies to passengers and railway undertakings operating PSO services on the basis of public tender or direct award, and also applies to passengers and railway undertakings operating their services in open access. Public authorities are governed by this Regulation and they have to ensure, that the participating subjects rights and obligations will not be breached, and they should also ensure that the services will be operated (Regulation (EC) No 1371/2007).

The main topics of Regulation (EC) No 1371/2007 are:

- carriage contracts and their conclusion, providing public information, transport documents;
- responsibility of railway undertakings for passengers and their luggage;
- train delay or cancellation, missed connection and compensation for such cases;
- transporting people with disabilities or reduced mobility;
- safety;
- quality of the operated services and complaints;
- informing passengers of their rights and ways of enforcing them (Regulation (EC) No 1371/2007).

Based on these rights and obligations, the conditions in the tender documents are specified. The currently or contractually valid PSO contracts also refer to these conditions. These conditions must be complied by the railway undertakings and they must not contradict them in their own transport regulations.

4 Rail Passenger Transport Market Services in Selected European Countries

The process of gradually opening the railway market in the EU began in 1991, when the Council Directive 91/440/EEC was issued on the development of the Community´s railways. The Directive introduces a certain degree of market liberalisation and calls on railway undertakings to increase performance rates and competitiveness. Until then, railway undertakings in individual countries had the function of unitary railways, which means they were operating railway transport and simultaneously managing the railway infrastructure. As part of the forthcoming liberalisation of the railway market within the EU, it was unacceptable for a railway undertaking to perform both of these functions, especially in terms of non-discriminatory access of railway undertakings to rail infrastructure. Therefore, the Directive requires at least accounting separation of these two subjects (EU 1991).

4.1 Liberalisation of Rail Passenger Transport Services

Czech Republic

Transformation of railway undertakings in Czech Republic began in 1993, when České Dráhy s.o. was founded after a division in the Czechoslovak State Railways. The vertical separation occurred in 2003, when České Dráhy s.o. was split into two independent companies—České Dráhy, a.s. as a carrier and SŽDC as an infrastructure manager. Transformation has been inspired by the French model, where ownership of station buildings and the traffic management are provided by the carrier, and infrastructure manager operates the rest of the rail infrastructure. This model was a barrier for full market opening and entry of new undertakings on the tracks. This problem was solved in 2011, when the ownership of stations and traffic management including operating staff, have been transferred under SŽDC administration. Horizontal transformation was completed in 2007, when the ČD Cargo subsidiary operating freight transport services, was founded by spin-off from České Dráhy (Tomeš 2013).

The first private railway undertaking, that has entered on the passenger rail market in open access, was Regio Jet in final months of 2011. Regio Jet started operating its services on the main route between the capital Prague and Ostrava in the east of the country. At the end of 2012, a private company Leo Express entered on the same route. Further expansion of open access services came in December 2016, when Regio Jet entered on a route between Prague and Brno (the second largest city in the Czech Republic) with four pairs of connections per day. After a year, connections on this route was expanded to 12 pairs of trains a day, while four of them have a start/destination station in Vienna and eight of them in Bratislava (Tomeš and Jandová 2018).

In 2013, another carrier entered a partially deregulated market—Arriva, with eight daily connections on the route Praha Masarykovo nádraží—Kralupy nad Vltavou. However, operation of these trains ended in December of the same year. Following that year, in 2016 Arriva entered on the Benešov u Prahy—Praha route, where 14 commercial connections operate on working days. In the same year, Arriva began operating international trains from Prague to Nitra in Slovakia, with two pairs daily connections.

In response to the entry of new railway undertakings in open access between Prague and Ostrava, since the end of 2011, the Czech Republic Ministry of Transport has exempted the trains from the national carrier České Dráhy, a.s. on this route outside of the PSO contract. These trains have been operating since the timetable period in 2011/2012 as open access services. Super City trains are also operating in open access on the Prague—Ostrava route—state border with Slovakia. The main route Prague—Ostrava has become fully deregulated and open to competition. The Czech Republic is also one of the EU countries, where the incumbent has a smaller market share in the open access segment than private railway undertakings (Annual Report of the Czech Railways Group 2011).

In response to the enactment of the 4th railway package, several public tenders for PSO services have been announced in the Czech Republic, both in regional and long-distance services. From the viewpoint of the valid system of competencies for ordering PSO services, tenders for long-distance services were announced by the Czech Republic Ministry of Transport and tenders for regional services were announced by regional authorities. The largest number of private railway undertakings entered the market of PSO services at the end of 2019—companies Regiojet, Leoexpress, Arriva, AŽD Praha and die Länderbahn. Tenders for long-distance services ordered by the Ministry of Transport were won by Regiojet (on the Brno—Bohumin route) and Arriva (on the Praha—Turnov—Tanvald, Kolín—Nový Bor, Praha—Rakovník and Praha—Příbam—České Budějovice routes) (Sůra 2019).

Before that, private company GW Train regio entered the market of PSO services in Czech Republic, which have operated their services on the long-distance R25 Plzeň—Most routes are based on direct award by Ministry of Transport after the cancelled tender in 2014, which was one of the first public tenders in this country. In the same year, a public tender for operating PSO services on three regional routes, was announced by the South Bohemian region. The private company GW Train region won this tender and has been providing services on these regional routes since 2017.

Poland

Transformation of the railway market in Poland began in 1995, when the state-owned company Polskie Koleje Państwowe (PKP) was divided into several accounting units. In 1998, these units were also separated institutionally, including their management. Then in 2001 they were transformed into separate subsidiaries, which became part of the holding PKP S.A. with 4 main divisions—infrastructure PKP Polskie Linie Kolejowe, long-distance passenger transport PKP Intercity, regional passenger transport Przewozy Regionalne and freight transport PKP Cargo (Amos 2005).

In 2007–2008, the concept of the transfer of ordering the regional railway transport services to regional authorities was developed. For this reason, the new company Przewozy Regionalne was subsequently spun-off from the holding PKP. Company shares in this new company have been transferred to 16 regional authorities in the country (OECD 2013).

The monopoly of the domestic incumbent was already broken in 2005, when Koleje Mazowiecke S.A. owned by the region of Mazovia was established. This company has been providing PSO services in regional rail transport based on a public tender. In addition to Przewozy Regionalne, several other regional companies are operating the PSO services in regional rail transport, while public tenders are preferred for this type of services.

The long-distance passenger rail transport market in Poland was opened from a legislative viewpoint in 2010, in response to the 3rd railway package published by the European Union. However, for many years, the domestic incumbent PKP InterCity was the only railway undertaking operating long-distance and in the international passenger rail market in Poland. The majority of long-distance services have been awarded directly since 2005, except trains EIC (Expres Inter City), EIP (Expres Inter City Premium) and TLK (Tvoje linie kolajowe) on the route between Lodz and Warsaw, which are operated on an open access basis, without subsidies. The first private company Leo Express entered the international rail passenger transport market in 2018 with one pair of trains twice a week on the Prague to Krakow route. In 2019, the company received a licence to operate open access services on the route from Prague to the Belarus borders (Terespol station) and to the Ukrainian borders (Medyka station). The minority of services are provided by private regional undertakings (market share 0–5%) also operated in open access (IRG-Rail 2020).

Hungary

Implementation of the first railway package began in Hungary in 2003, when the unitary Hungarian railways Magyar Álom Vasutak (MÁV) were divided into 5 separated accounting units—traction, passenger transport, freight transport, infrastructure and engineering services. In 2005, MÁV was transformed into a holding, from which 16 subsidiaries have been gradually separated, like freight transport in 2006, passenger transport MÁV Start in 2007, traction in 2008 and independent infrastructure manager MÁV Pályavasút in 2010. The subsidiary MÁV Cargo operating freight rail transport was in 2008 privatised by Rail Cargo Austria (OECD 2013).

Legal liberalisation of the international passenger in the rail transport market in Hungary began in 2009, however no private railway undertaking has yet entered the market in open access (IRG-rail 2020).

Hungary has had problems with distortion of competition and the incumbent's abuse of a dominant position since the beginning of implementing EU legislation. The situation on the passenger rail transport market has been stagnant for several years and there are still only two incumbents on the market—MÁV and GYSEV. All rail services are provided under PSO contracts, which are awarded directly by the state authority. Legal entry of new railway undertakings is possible by law, but conditions for their entry, such as licencing or infrastructure allocation system are not favourable and mainly favour domestic incumbents. For this reason, Hungary has been facing lawsuits from the European Commission relating to abuse of a dominant position by the domestic incumbent (OECD 2013).

Nowadays, Czech private company RegioJet is applying for a licence for providing open access passenger rail transport services in Hungary, on the route Prague—Vienna—Budapest. Its success would be a milestone in process of the deregulating the passenger rail transport market in Hungary.

Austria

Railway transformation in Austria was based on the Federal Railway Structure Act of 2003, which responded to the Council Directive 91/440/EEC from July 29th, 1991 on the development of the Community's railways. Due to the required accounting division of infrastructure management from transport operation, on January 1st, 2005 were the state-owned unitary railways Österreichische Bundesbahnen (ÖBB) transformed into a holding with several subsidiaries. Main subsidiaries of the ÖBB holding are ÖBB-Personenverkehr AG, which operates the passenger rail transport, Rail Cargo Austria AG operating the freight rail transport, and ÖBB -Infrastruktur AG as the infrastructure manager. 100% of company shares in this holding are owned by the state through the Federal Ministry of Transport, Innovation and Technology.

The second domestic incumbent in Austria from 1872 is Die Raaberbahn AG (Raab-Ödenburg-Ebenfurter Eisenbahn), known as GySEV in Hungary. The share of the Federal republic of Austria in the Raaberbahn company shares is 33.3%. The rest of company shares are divided between Hungary and a private stakeholder. Raaberbahn have their own railway infrastructure in the area of Austria and Hungary border (Raaberbahn AG 2020).

Legal implementation of the 3rd railway package in the field of deregulation of the international passenger rail transport market occurred in Austria in 2010. It was a matter of opening the market de jure, which means that legislative conditions for licencing and entering the rail infrastructure for railway undertakings in open access were created. The first private company entered the inland passenger rail transport market in 2011—WESTBahn Management GmbH (a subsidiary of RAIL Holding AG), which is still operating trains on the main route in Austria between the capital Vienna and Salzburg. In its first year of operation in the market Westbahn operated 13 pairs of connections daily at two-hour intervals. Westbahn has already had problems with allocating railway infrastructure capacity in its efforts to acquire it, due to the way of vertical separation of the infrastructure manager and domestic incumbent. At the end of 2018, Westbahn doubled its offer of connections between Vienna and Salzburg from 15 to 29 pairs of trains a day. At present, there has been a temporary reduction in connections of more than 50%, due to investment of rolling stock and waiting for new ones to be delivered (Tomeš and Jandová 2018).

Open competition on the market of inland passenger rail transport also takes place on the Vienna—Airport Vienna route. There are 2 companies operating trains—domestic incumbent ÖBB and private company (although partly owned by the incumbent) City Air Terminal Betriebs GmbH (CAT). The majority of operational performances (95%) on mentioned two routes is operated in open access (IRG-rail 2020).

The first private company entered the market for international passenger rail transport in 2017, when Czech company RegioJet has started operating 4 pairs of trains daily between Prague and Vienna, as an open access service. The same range of connections have been maintained to this day (Sůra 2017).

Ordering PSO services in regional passenger rail transport in Austria is in the competence of regional authorities (same as the Czech Republic), which are the individual federal republics. Austria has not had much experience with public tender for PSO contracts yet and most of the PSO services were awarded directly to incumbent, due to its strong position and quality of services, unlike countries like Slovakia, Czech Republic or Poland, where the competition can force the incumbent to increase the unsatisfactory quality of services (IRG-rail 2020).

Germany

The process regarding the transformation of railway undertakings in Germany began in 1994, when the state-owned company Deutsche Bahn A.G was created by merging two companies—Deutsche Bundesbahn (DB) and Deutsche Reichsbahn (DR), after the fall of the Iron Curtain and the unification of Germany.

In the second phase of the railway reforms in 1999 there were 4 independent divisions spun-off as separate subsidiaries of the DB A.G. holding. Completion of the transformation cycle in 2008 was the formation of the group DB, which was created by the merger of the DB A.G. holding company and the newly established holding company DB Mobility Logistics A.G. There were six subsidiaries as a part of DB Mobility Logistics holding: DB Regional (passenger transport under PSO contracts), DB Long-Distance (long-distance passenger transport), DB Schenker Rail (freight transport), DB Urban (regional bus transport), DB Schenker Logistics (logistics services), DB Services (property management, including station buildings) and since 2010 also DB Arriva (Deville and Verduyn 2012).

The legal environment for open access in the passenger rail transport market in Germany was created in 1994. There were several public tenders for PSO contracts for regional transport services in the country at the same time. In 2006, the market share of the private companies in the regional passenger transport market reached a value of 15.2% of operational performance in train km, then in 2017 increased to 26%. There was a stagnation in 2016 after the previous sharp increase. Awarding PSO services in Germany is like most European Union countries, based on a decentralised system—they are in the competence of federal republics as regional authorities. The common regulatory authority in Germany is the Federal agency for infrastructure (Bundesnetzagentur Jahresbericht 2018).

Long-distance markets (both inland and international) passenger transport in Germany is characterised by open competition. All long-distance trains, including trains operated by incumbent DB, are operated on an open access basis. Nevertheless, there is full deregulation on this market, the incumbent (IRG-rail 2020) still has the biggest market share (more than 90% of operational performance in train km).

Transdev Germany was the first private company entering the inland long-distance passenger rail transport market in Germany on an open access basis in 2002, when it began operating occasional connections several times a week on the Berlin—Leipzig route. After that, in 2007 the company Locomore, a founding member of Hamburg-Köln-Express GmbH was founded and in 2012 began operating connections between Hamburg and Köln in competition with Deutsche Bahn ICE trains. Locomore later moved to the Berlin—Stuttgart route. But the company did not survive the competition in the market, and in May 2017 declared bankruptcy for insolvency. Subsequently, the Czech private railway undertaking Leo Express showed an interest in the Locomore company shares. Nowadays, Leo Express is operating under the brand Filxtrain one pair of train a day on the Berlin—Stuttgart route and one pair of trains per day between Hamburg and Köln (LEO Express to relaunch the Locomore 2017 service).

Several dozen railway undertakings are providing their services on the open long-distance passenger rail transport market in Germany, but their market share is minimal in comparison with the incumbent (0–5% of the operational performance in train km). Some examples of these railway undertakings on the international market are Thalys International on the Köln—Brussels route, SNCF on the Paris route or OBB Nightjet and Snälltåget (night train services). On the domestic market it is f.e. Flixtrain, DB

Regio and several dozen small private companies operating services on local and tourist railways (IRG-rail 2020).

Italy

The first hint of railway reform in Italy dates back to 1985, when the state-owned railway undertaking Ferrovie dello Stato (FS) as a part of the Ministry of Transport, became an independent company. A few years later in 1998, in response to European legislation, there was a vertical separation and two new subjects were founded—the infrastructure manager and railway infrastructure capacity allocator Rete Ferroviaria Italiana (RFI), and Trenitalia operating passenger rail transport services. Both of them are a part of Ferrovie dello Stato holdings owned by state (Desmaris 2016).

The international and long-distance passenger rail transport market in Italy has been opened legal (de jure) for open access since 2001, what is a long time before market opening and was ordered by EU legislation within 3rd railway package. Italy also became the first European country that opened the market of services provided on the High-Speed Railways (HSR) (Desmaris and Croccolo 2018).

Despite the earlier creation of a legal environment for open access on Italian long-distance passenger rail market, the first private railway undertaking obtained a licence from the regulatory authority after more than ten years in 2012, when Nuovo Trasporto Viaggiatori (NTV) entered the HSR market in open access. NTV began train operation under the Italo brand in April 2012 on the Torino—Salerno and Venice—Salerno routes, which connect the north and south of the country. The scope of operation began with 21 trains a day and increased to 98 connections a day on average. In 2019 the share of NTV's operational performance on the HSR market in Italy reached value of 35%. As well as NTV trains, trains operation by Trenitalia are also on the HSR market. Since NTV entered the market, Trenitalia has increased its HSR operational performance by 110% at the expense of more than 40% reduction of performance on conventional long-distance routes (Bergantino 2017).

Besides the domestic incumbent Trenitalia, foreign incumbents are also operating services on open HSR market in Italy, like French SNCF on the Torino—Milano route through SNCF voyages Italia and Trenord owned by the DB and OBB corporations, operating international trains to Austria and Germany. Their market share reaches low values (0–5% of the operational performance in train km), compared to 70% share of Trenitalia and more than 20% share of NTV (IRG-rail 2020).

Regional passenger rail transport services and long-distance services on conventional routes are in Italy under PSO contracts. PSO services are awarded by regional authorities. Since 1999 these authorities have had the opportunity to decide if they want to award PSO contracts directly or through public tender. Public tenders have been used by only 4 Italian regions so far. These tenders took place between 2003 and 2007, accompanied by difficulties and unsatisfactory results (Desmaris and Croccolo 2018).

Sweden

Sweden was the first country in the EU that made the vertical separation of the subject responsible for infrastructure management and railway undertaking providing transport services, this was prior to the Council Directive 91/440/EEC being published. The next part of the transformation process in Sweden was the divide of SJ in 2001 into several new divisions—division of the passenger transport SJ Ltd, freight transport division Green Cargo, division Jerhusen for property management and divisions EuroMain and Swemaint providing vehicle maintenance. The 3rd railway package brought with it the obligation to create conditions for open access in the international passenger rail transport market and was gradually implemented in Sweden between 2009 and 2011. Prior to that, in 2007, night trains were exempt from PSO contracts. Since July 2007, they have been transferred from PSO contracts to open access services also SJ's long-distance connections during weekends and from October 2007 as well as all international trains. All long-distance passenger transport services have gradually become exempt from PSO contracts and began operating on an open access basis since 2011, which has made the long-distance passenger rail transport market in Sweden fully deregulated. Then on the basis of the new Public Transport Act in 2012, Sweden became the first country in the EU, that also allows open access de jure on the regional passenger rail transport market, but this market is still heavily subsidised and PSO contracts are there awarded through public tenders (Alexandersson and Rigas 2012).

Skandinaviska Jernbanor was the first private railway undertaking, which entered the passenger rail transport market in open access under the brand Blå Tåget (Blue train) in 2011. This company is still operating trains on the Göteborg to Upssala route, but it is not a direct competitor of the incumbent SJ, because of specific concept of "historical and luxurious train experience " and also due to the travel time on these trains being two hours longer. In April 2013, private railway undertaking MTR applied for a licence and for an allocation of the railway infrastructure capacity on the route between Stockholm and Göteborg (the main railway corridor in Sweden). MTR based in Hong Kong has become the first railway undertaking based outside Europe, which is operating passenger rail transport services in a European country. MTR has operated its trains since 2015, in range of seven pairs of connections a day. For the comparison, SJ is operating 18 pairs of trains a day on the same route. In terms of market share the domestic incumbent, similar to Germany, still has a dominant position on this fully deregulated market (Vigren 2017).

The Netherlands

In the Netherlands, there was only decentralisation of individual services. Open access in general in the market of the inland passenger transport does not exist. Exclusive rights for operating inland passenger transport in the Netherlands network has concession holders. The only case when a private railway undertaking obtained a concession, was the operation of trains by company Lovers Rail from 1997 to 1999. After that Lovers Rail terminated its operation prematurely due to unprofitability.

The market for international passenger rail transport has been deregulated of law since 2010, in response to the 3rd railway package (Deville and Verduyn 2012).

Abroad, data differs from country to country, its history, territory, or the extent of the railway network and public interest of the state. Ordering services under PSO contract is only very sporadic through tendering in EU countries. Most countries are trying to extend contracts with existing railway undertakings, or use direct award.

4.2 Declared Public Tender

The current state of PSO services operation in selected European union countries has been analysed in two ways. One of them is finding currently announced public tenders, which have been published on the European Union Official Journal website. The second way is a questionnaire survey for ascertaining the state of public tenders. Questionnaires were sent to all of European Union (or EC) countries by e-mails due to the current pandemic situation across Europe. Personal and telephone inquiries could not be carried out because individual ministries in the EU worked in home office.

The status and way of announcing public tenders in the EU countries was detected on the basis of mandatory publication "Tender announcements ". We focused mainly on information concerning tender subject, languages, time ranges, technical competence etc.

As of March 11th, 2020, these seven public tenders have already been announced in the EU countries (Supplement to the European Union Official Journal 2020):

- Czech-Pilsen region—public tender for PSO contract on lines: P4 Plzeň—Žihle, P13 Plzeň—Bezdružice, P14 Planá u Mariánských Lázní—Tachov, P21 Rokycany—Nezvěstice, P22 Plzeň—Radnice, P31 Nýřany—Heřmanova Huť,
- Germany—Dresden—trains on the non-electrified routes in Dresden region on lines: (RB 33 Dresden—Königsbrück, RB 34 Dresden—Kamenz, RB 71 Pirna—Neustadt / Sa.—Sebnitz, RB 72 Heidenau—Altenberg and RE 19 Dresden—Heidena—Altenberg),
- Germany—Munchen—PSO services in Munchen region, (München Hbf—Passau, München Hbf—Plattling/Regensburg, (Erding—Schwaigerloh)—München Flughafen Terminal—Regensburg Hbf, (München Hbf—Landshut Hbf),
- Norway—regional and local trains under PSO contract, (L1 Spikkestad—Lillestrøm, L2 Stabekk—Ski, L21 Stabekk—Moss, L22 Skøyen—Mysen (–Rakkestad), R20 Oslo S—Halden, L3 Oslo S—Jaren, R30 Oslo S—Gjøvik),
- Poland-Subcarpathian Voivodeship—passenger rail transport services in the agglomerations of the Subcarpathian Voivodeship,
- Italy-Bari—regional passenger rail transport services in Bari region,
- Italy-La Spieza—passenger rail transport services in La Spieza and connections with cities Marittima, Migliarina a Santo Stefano di Magra.

In making recommendations to support public tenders in Slovakia, we used the search for the notifications of the announcement of public tenders.

4.3 Survey on Ways of Ordering Services in the Public Interest

Analysis of the current situation in service provisions under PSO contracts was carried out by a questionnaire survey at the beginning of 2020. Questionnaires were sent to ministries of transport of individual EU countries. This survey has been addressed to 25 countries and upon concluding the case study 6 responses were received. The survey contained the following questions:

1. *Which entity is the purchaser of public service obligations in the railway sector in Your country?*
- State/region/district/ another territorial unit
2. *What is the percentage proportion of railway passenger transport performance depending on public service obligations and beyond?*
- Public service obligations / Outside public service obligations: %
3. *How are public service obligations ordered in Your country?*
- Tendering only/Direct award only/ Tendering as well as direct award /other:
4. *What is the percentage (indicative/approximate number) of public service contracts which are ordered by tendering or direct award?*
- Direct award/Tendering/otherwise: %
5. *Can you name examples of carriers that provide these services on the basis of a successful tendering (only if the tendering in your country has been succeeded)*?
6. *How is railway passenger transport financed in the public interest in your country?*
- train-kms/passenger-kms/ subsidy in some volume package /otherwise
7. *What is Your opinion regarding the* 4th*railway package measure that pushes on realising tendering within the public contract ordering in public interest?*
8. *How is the plans or measures with implementation of Regulation XYZ that pushes on realising public tenders within the PSO contracts?*
9. *In the existing public service contracts framework, do the carriers consider the appropriate profit by cost calculation?*
- Yes/No/Public sector only/otherwise:

The realised survey results are listed by the countries.

Czech Republic

In the Czech Republic, PSO services are ordered by the state and individual regional authorities, while about 90% of these services are under PSO contracts and the remaining 10% of services are commercial. PSO contracts in the Czech Republic are awarded both directly and based on success in public tender, which can be announced by Ministry of Transport or regional authority. Currently 75% of services under PSO contract have been awarded directly and 25% through public tenders. Carriers that have achieved success in public tender are for example České dráhy a.s.; GW Train Regio a.s., etc. The opinion of Ministry of Transport on public tenders is positive and it sees it as an objective comparison of costs and revenues items, which are affected by qualitative parameters of services provided by individual carriers. The Ministry of Transport is ready to increase the number of public tenders. PSO services are financed based on operational performances in train-kms and carriers have right to reach an appropriate profit.

Republic of Poland

Services under PSO contracts in Poland are ordered by state or regional authorities (voivode ships), while these services represent approximately 90% of all passenger railway transport performances. Contracts can be awarded directly (90% of them) or through public tender. Contracts awarded in Poland based on the success in public tender represent around 10% of all PSO performances. These tenders can be announced by the state or by voivode ships. The Polish opinion regarding the 4th railway package is that public tenders can increase competitiveness on the railway market. Services under PSO contracts are financed based on set values of operational performance in train-kms and carriers can reach an appropriate profit.

Kingdom of Denmark

Entities responsible for ordering PSO services in Denmark are state and regions, while 98% of all performances are under PSO contracts and 2% are commercial. Contracts are then awarded directly or through public tenders. 90% of all PSO performances are operated based on direct award and 10% based on success in public tender—these services are provided by the private railway undertaking Arriva. Performances under PSO contracts are financed based on some volume package and carriers can reach an appropriate profit.

Republic of Finland

In Finland, PSO services are ordered by state and by regional authorities. Currently, concluded contracts have been awarded based on direct award, but there are currently some public tenders ongoing. Performances under PSO contracts are financed by a combination of several indicators, like train-kms, passenger-kms, financial packages, etc., but the Ministry of Transport did not give an exact specification for them. Carriers can also reach an appropriate profit.

Republic of Estonia

The authority responsible for ordering services under PSO contracts in Estonia is the state and all performances are operated based on PSO contracts. Contracts are awarded only directly, however there is a plan to organise public tenders in the future. The Ministry of Transport is currently worried that public tenders could cause an increase in prices for operation transport services in railway transport. Performances under PSO contracts are financed based on train-kms and carriers can reach an appropriate profit.

Republic of Lithuania

The competent authority for ordering services under PSO contracts in Lithuania is, or will be the state, because Lithuania is just about to conclude the first PSO directly awarded contract ever. Currently all performances in Lithuania are outside of a PSO contract. The Ministry of Transport's opinion on public tenders is positive, they see them as an opportunity to increase competitiveness and quality of provided services. Performances under PSO contracts will be financed based on financial package and carriers will not be able to reach an appropriate profit.

The results of our survey can be summarized in several points as following:

- the purchaser of public service obligations in the railway sector is state and Regional authorities (only in Estonia Lithuania the purchaser is only state),
- only in three countries is commercial performances (Czech Republic, Poland and Denmark—in all the same 10%) and PSO are ordered both—tendering as well as direct award,
- railway passenger transport is most often financed by train km or combination train km and passenger km,
- the carriers can consider the appropriate profit by cost calculation (beside Lithuania).

5 Rail Passenger Transport Market in the Slovak Republic

The transformation process for railway undertakings in Slovakia began on 1.1.1993, when Železnice Slovenskej republiky (ŽSR) was founded after the division of the Czechoslovak State Railways, based on a decision by the Slovak Republic Government. On 30.9.1993 has been issued the Act No. 258/1993 about ŽSR. This Act defined ŽSR as a state-owned company with the application of elements of commercial and public administration, one of a kind. The Slovak Republic Government approved with Resolution no. 830 of 2000 for the restructuring and transformation project and then as a result, two new subjects were created from the former ŽSR company: Železničná spoločnosť, a.s. as a carrier in both passenger and freight transport and Železnice slovenskej republiky as an infrastructure manager. The transformation process was completed in 2005, when Železničná spoločnosť, a.s was divided into 2 independent companies: Železničná spoločnosť Slovensko, a.s (ZSSK) as a passenger transport carrier and Železničná spoločnosť Cargo Slovakia, a.s as a freight transport carrier (History of railways 2007).

The Slovak Republic has been applying a separate model since 2002, which means that the infrastructure manager is not only accounting but also institutional separated from the national carrier (incumbent). This fact allows more open and less discriminatory access for railway undertakings to railway infrastructure and their entry on the market. First, licencing for operation of open access services was granted in 2011 to domestic incumbent ZSSK, that has removed since 2012 its IC trains on the main route Bratislava—Košice from PSO contract. The first private carrier that entered the international passenger rail transport market in open access was Regio Jet in 2014, when this company extended its line from Prague—Ostrava to Košice. Subsequently, this carrier also started operating an open access services on the inline route Bratislava—Košice, however this line's operation was terminated in 2016 (Pečený et al. 2016). In 2016, they began operating their services on the Praha—Košice route as well as another Czech private railway undertaking Leo Express. The 4th railway undertaking operating passenger rail transport services on the partially deregulated market is Arriva, who have been operating trains on the route between Prague and Nitra since 2016. There are currently 2 pairs of trains a day operating on this route by Arriva.

In terms of passenger rail transport services under PSO contracts, there are currently two PSO contracts concluded in Slovakia (both long-distance and regional transport), between the Slovak Republic Ministry of Transport and Construction as a competent authority responsible for ordering services in public interest and carriers ZSSK (national incumbent) and Regio Jet (private railway undertaking). The first PSO contract directly awarded to a private railway undertaking was concluded in 2010 between Ministry of Transport and Regio Jet for a 10 year' period. This contract is related to operation of services on the Bratislava—Komárno route to the agreed volume of 1,246,451 train-kms per year. The first public tenders for regional service operation in the public interest were tenders on the Žilina—Rajec and Bratislava—Komárno routes, announced at the end of 2019. The first and so far, the only one public tender for long-distance transport on the Bratislava—Banská Bystrica route was cancelled after failure (Public service contract ZSSK, a.s., Regiojet 2010).

5.1 Regulation and Financing of Rail Passenger Transport Services

Ensuring the mobility of people cannot be implemented effectively on market principles. The need for regulation results from the economic character of transport and objective action of factors, which can cause a market failure. Significant factors in ensuring the mobility of people is passenger transport financing. Public finances are not only involved in financing infrastructure for passenger transport, but also in financing vehicles for passenger transport and in compensation for PSO services operated by public transport undertakings (Nedeliaková et al. 2016).

Because public tenders are made for PSO services, for which price is regulated, therefore price regulation must be based on Act No. 18 of 1996 of the Slovak Republic National Council on prices. This Act modifies and defines the rights and obligations of legal and natural persons and state administration authorities, in the application, negotiation and controlling prices of products, services, labour costs, rental or real estate prices and measures to prevent unwanted price distortions and market failures. This act also deals with price regulation, which is one of the measures for preventing unwanted price developments. Price regulation is the price determination or the method of determining price and the conditions of price regulation by the regulatory price authority, which in railway transport is the Transport Authority (Act No. 18/1996).

Regulation of passenger transport is caused by three basic factors, which are:

- objective causes of market failure,
- enforcing the subjective government aims in other areas of social life through transport,
- guarantee of fundamental civil rights by the state.

The optimal functioning of transport in the social system is conditioned by adopting transport policy measures, which respect the conditions and characteristics of society in a certain development period. The EU Transport policy is one of the starting points for implementing passenger transport regulation at national level towards ensuring adequate, sustainable and safe mobility and quality of life of society. It is implemented on the basis of law. In the EU Member States, the duality of law also applies to transport (Transport and tourism policy—rail transport 2020).

Price authority can regulate prices if:

- there is an extraordinary situation affecting the prices of goods and services on the domestic market,
- the market threat is caused by an underdeveloped market environment (determined by the Antimonopoly Office),
- it is the protection of a consumer or a service in public interest (subsidised goods and services),
- there will be a change in the exchange rate, tax, customs or subsidy regulations,
- the goods are delivered or provided through network industries,
- the goods are delivered or provided through the universal postal service or postal payment service (Act No. 18/1996).

Price regulation according to Act No. 18 of 1996 of the Slovak Republic National Council on prices as amended is implemented by (Act No. 18/1996):

- Official pricing:

Official pricing is determining the price of defined goods by the price authorities as maximum or fixed. The maximum price is a price that cannot to be exceeded. A fixed price is a price that cannot be changed.

- Factual price guidance:

Factual price guidance means determining the conditions for price negotiation by the price authority. These conditions are the maximum extent of the allowable price increase in a certain period or a binding pricing procedure.

- A combination of the previous two methods.

Price regulation in public passenger rail transport fulfils several transport policy objectives in the economic, regional, social and environmental fields. These include the availability of public transport services, cohesion, the redistribution of demand in favour of public transport in order to reduce the negative environmental impact of transport and reduce congestion (Strategic plan for the development of transport in the Slovak Republic until 2030, 2016).

Free travel

Within services operated under PSO contracts, the Slovak Republic Ministry of Transport and Construction established free fares for selected groups of passengers. Each railway undertaking must respect this fact and provide its services to selected passengers free of charge. At present, they are the ZSSK, a.s. and RJSK, a.s. railway undertakings. Free fares were introduced on November 17th, 2014 in order to increase interest in rail transport and also for the state social policy (Chúpek 2015).

Registration for free fares has been possible since October 27th, 2014 in the following sales channels: in the personal cash of ZSSK, a.s. with KVC—registration and issuance of the card is immediately, in ZSSK, a.s. cash without KVC—registration and issuance of the card will take place within three working days, and on the internet (students with a contactless smart card and pensioners over 62).

Through registration, a customer account was created in the ZSSK system for every passenger who have the right to free train travel. After creating this account, a card with a 2D code is issued for the passenger. For further purchases, the passenger is obliged to prove themselves with this card, or to prove themselves with this card on the train. By loading the 2D code when the ticket is issued, the passenger's expedition and the ticket's automatic assignment to the customer's account is ensured.

The possibility of travelling with a free ticket is not to apply to journeys by InterCity trains, which have a special tariff. Nevertheless, fares for passengers entitled to free transport, are advantageous over other passengers on IC trains as follows:

- children under 15 years of age, while up to 6 years of age travel without any ticket and card. From 6 to 15 years of age, they must already have a ticket for free transport together with a card issued by the railway undertaking.
- pupils and full-time students at universities and high schools up to 26 years of age, with citizenship in the Slovak Republic or citizenship or permanent residence of an EU member state. Pupils / students prove themselves with a ticket for free transport together with a student card or a card issued by the railway undertaking.
- persons over the age of 62 and pensioners. They prove themselves with a free transport ticket together with a card issued by the railway undertaking (Public service contract 2010).

For category R trains—fast trains and selected Ex—express non-commercial trains, a contingent is set, i.e. a limit on the number of passengers for free transport on a certain route.

Positive effects of free travel on passenger transport are (Chúpek 2015):

- increase in demand for passenger rail transport,

- resumption of passenger train operations on routes where operation has been suspended or cancelled,
- increase in the number of connections,
- transfer of services from road to rail,
- attracting new railway undertakings to the passenger rail market.

Negative effects of free travel on passenger transport are (Chúpek 2015):

- overcrowded trains,
- reduction in the quality of provided services,
- decrease in revenues volume and increase in compensation,
- complaints from paying passengers,
- discrimination against people for whom railway transport is unavailable,
- limiting the number of non-paying passengers in selected train categories.

With the introduction of free transport, it was necessary to take steps to keep travel quality standards for paying passengers while complying with the above-mentioned Government Resolutions. For this reason, in the first phase, single and multi-section limited numbers were introduced in SC, EC, Ex and R trains. However, the introduction of limits prolongs the issue of travel documents.

In suburban trains, no limits have been introduced, but capacity has increased, and traffic performance has been strengthened during peak periods. Following the above facts, in terms of regular changes to the timetable, the management of individual trains was optimised with regard to the actual use—according to passenger demand for transport. At the same time, ZSSK operatively streamlined the capacity of individual suburban transport trains (News ZSSK 2020a, 2020b).

During the crisis situation caused by the CORONA virus pandemic, the measure of free travel for the period from 1.4.2020 to 1.6.2020 was interrupted for the pupils and students. From May 2020, free travel was resumed for all categories of passengers (News ZSSK 2020a, 2020b).

5.2 *Rules for Providing Rail Passenger Transport Services Under PSO*

All elements related to the conclusion of the PSO contracts are regulated by Act 514/2009 on transport on the Railways of the Slovak Republic National Council entering into force on October 28th, 2009 with the latest amendment valid from January 1st, 2020. Act 514/2009 of the Slovak Republic National Council regulates the conditions for the provision of transport services in railway transport, tram transport, trolleybus transport and transport on special rails and cableways. It also regulates public administration and state authorities in the area of railways, driver certifications and the rights and obligations of the carrier as a provider of transport services, the passengers and customers in freight transport.

Based on carriage and transport obligation, a transport service plan must be drawn-up. This plan is also the basis for drafting the timetable and concluding PSO contracts. The contract has written form and includes information resulting from the Act:

- definition of obligations, that should be meet by the undertaking,
- territory or routes, where the undertaking has to fulfil its obligations,
- range or possibility regarding subcontracting transport services,
- tariff conditions,
- exclusive rights,
- demonstrable loss or payment schedule calculation method,
- controlling the fulfilment of obligations and recognition of incurred costs,
- measures and sanctions for non-compliance with obligation,
- contract duration,
- other information, as tariff conditions for transporting people in the armed forces and etc. (Act No. 514/2009).

The role of the PSO contract (hereinafter only contract) is to ensure suitable transport services in the area, taking into account the social and environmental factors. The contract is concluded between contracting authority of PSO services (in railway transport f. e. state) and the provider of these services (carrier) in cases, if the transport undertaking does not provide these services at all or provides them to a lesser range than necessary, and also in a worse quality and higher price, due to the economic disadvantage. A contract can be awarded directly or through a public tender, in order to ensure inland and international transport to the state border for trains, that have at least one stop in the Slovak Republic territory. A contract can be also applied to an integrated transport system, where the share of railway transport will be determined. Railway undertakings can use funds from the contracting authority, which are determined by the PSO contract within an integrated transport system. For determining the duration of awarded contracts, the Act refers to the Regulation (EC) No 1370/2007 and its amendment—the Regulation (EU) 2016/2338 (Regulation (EU) 2016/2338).

The public tender is announced by the competent authority by notice on its website, where it determines conditions for participation and the deadline for submitting applications to participate in the tender, more than 30 days from publication, or 15 days if justifies it as urgent (Act No. 514/2009). Tenderers that have demonstrated that the conditions are met and have met the deadline, will receive a call to submit their offer. A selected tenderer can submit only one offer alone or as a consortium of undertakings. Deadline for submitting offers is more than 30 days from the call's dispatch date. The call contains:

- a link to published tender notice,
- a website with tender documents, if it is unable to publish them then the tender documents are attached to the call,
- a deadline, address and language (or languages), in which offers can be submitted,
- a list of documents needed,
- the weight of the criteria or descending order of the criteria's importance, if not included in tender documents or the notice,

- and other (Act No. 514/2009).

Basic and special fares for PSO services must be regulated. It is regulated by the railway transport regulatory authority, which is the Transport Authority. The regulatory authority sets the maximum fare amount, that is based on economically justified costs, by generally applicable legislation. This legislation will be published in the Slovak Republic Ministry of Transport and Construction Journal (Regulation ÚRŽD No. 1/2010).

By law, the payment of obligations from a PSO contract which was awarded directly, is compensation of a demonstrable loss from the fulfilment of the obligation arising from the contract. In the case of a contract concluded by public tender, the reimbursed amount is the amount per unit of operational performance, based on the tender's conditions. The demonstrable loss represents the difference between the economically justified costs, including an appropriate profit, and revenues from regulated fares and other revenues from obligations specified in the contract.

In addition to compensation for appropriate loss the railway undertaking may also request a subsidy from the contracting authority from its budget for the purchase and modernisation of rolling stock, that will be used for operation of the services under PSO contract. Accident and emergency costs, environmental costs and infrastructure costs caused by a competing transport mode and exceeding the equivalent costs of the railway undertaking, may be also subsidised. A railway undertaking can only acquire subsidies in order to increase the quality of services provided (Public service contract ZSSK, a.s., Regiojet 2010).

5.3 The Current Status of Rail Passenger Transport Services Under PSO Ordering

The Slovak Republic is a country in which PSO services in passenger rail transport have only been awarded so far on the basis of direct awarding. Public tenders, that have taken place so far in the Slovak Republic have not been successful or have been cancelled. Approximately 90% of the performances are services under PSO contracts and the other 10% are outside of PSO contracts (open access services). The contracting authorities in the Slovak Republic are the Slovak Republic Ministry of Transport and Construction or regional authorities—higher territorial units.

At present, the Slovak Republic Ministry of Transport and Construction has concluded two PSO contracts. One of the contracts has been concluded with Železničná spoločnosť Slovensko, a.s. (hereinafter as ZSSK, a.s.) This contract is network-wide and concluded by direct award. The second PSO contract has been concluded with private railway undertaking RegioJet a.s. (hereinafter as RJSK). Currently, the validity of these concluded contracts is coming to an end. Selected routes on a network-wide contract with ZSSK, a.s. will be or are announced for public tender. The route from the contract with RJSK, a.s. is also the subject of public tender.

5.3.1 Contract with ZSSK, a.S.

This contract was established for the needs of providing public passenger rail transport, which can be national or international with at least one stop in Slovak Republic territory. It is a network-wide contract. The contract has been awarded directly, between the Slovak Republic Ministry of Transport and Construction state authority and the national railway undertaking ZSSK, a.s., The consortium guarantees that it will provide these services in accordance with conditions arising from this contract and other annexes or appendices, which are an integral part of the contract. The contract enters into force 01.01. 2011 for a 10 year' period, which may be extended by a maximum of 5 years. For each year of the contract's validity, the annual operational performance is ordered and the amount of its compensations is stated (outside the base year it is stated in the appendices) (Public service contract ZSSK, a.s. 2010).

Based on this contract, the contracting authority for PSO services guarantees that it will pay the railway undertaking compensation for the provision of these services, but the railway undertaking may not misuse them for other purposes. The Ministry is also a public administration body in the field of rail transport, but it also has the rights of a ZSSK, a.s. shareholder, whereby the sole shareholder is the Slovak Republic. Based on the Act on the State Budget, the amount of expenditures that will be reimbursed in the relevant year is determined.

The Railway undertaking has three general obligations:

- Transport obligation—the railway undertaking must receive and transport each passenger on specified routes, on specified connections, according to a valid timetable. It must carry passengers on the basis of the established tariff, transport conditions and on the basis of meeting the conditions arising from the contract and other legal regulations.
- Operational obligation—the railway undertaking guarantees that it will provide services with a licence and use vehicles suitable for the carriage of passengers. The railway undertaking has to maintain, repair or renew these vehicles, which it uses for PSO services.
- Tariff obligation—the fare should be set by the railway undertaking so as not to exceed the maximum amount set by the decree of the regulatory body. For unregulated services, prices are set by the contracting authority. These fares may differ according to the category of passengers. If the railway undertaking agrees with the contracting authority, it can transport passengers for lower fares (Public service contract ZSSK, a.s. 2010).

The railway undertaking also has other obligations. It should ensure, that the services provided meet the requirements for vehicle quality, capacity and cleanliness. It must also ensure that the operation is safe, smooth, accurate and regular. The contracting parties will establish a quality management system. The railway undertaking must separate the services, which provides in public interest from other services in a commercial way, on an accounting basis. The contracting authority may order an audit of the obligated railway undertaking.

The contracting authority's obligation is that it will pay the railway undertaking compensation for the ordered services, which have a range on the basis of the Act on the State Budget. Compensation for the contract of Ministry of Transport with ZSSK, a.s. is calculated according to the following steps:

The initial calculation of compensation parameters is done once, before the contract has entered into force. Contract compensation is based on economically justified costs, which are divided into 6 categories such as:

- costs related to vehicles (capital costs),
- costs related to distance (labour costs, maintenance costs, spare parts costs…),
- costs related to time (personnel costs in operation),
- costs related to train weight and performed kms (fuel, electricity),
- selected actual costs (infrastructure charges, service charges, cross-border services…)
- other costs (overhead and others) (Public service contract ZSSK, a.s. 2010).
  The indexation of compensation parameters is performed at the beginning of each contract year. Indexed can be costs for:
- oil fuel;
- electricity;
- wages and social insurance;
- interest costs;
- other costs, inflation (Public service contract ZSSK, a.s. 2010).

Determination of planned compensation and advance payments shall take place at least 6 months before the year of validity of the supplement. Advance payments divide the calculated value of compensation into 12 equal part, which are paid monthly. The calculation is:

$$Compensation = (PP * CP) + AP - PR \quad (1)$$

where

PP planned performances
CP compensation parameters
AP appropriate profit
PR planned revenues.

The adjustment of compensation parameters takes place at the beginning of each contract year at either party's request, when the cost amount changes. The contracting authority also has an obligation concerning the quality management system (submission of proposals) to provide the necessary information to the railway undertaking and to assist in investments related to performance in the public interest.

Fig. 1 The Bratislava main station—Banská Bystrica route. *Source* Author, via ŽSR

5.3.2 Contract with Regiojet, a.S.

The contract for the provision of PSO services was concluded by direct award on the basis of the provision of transport services by public passenger railway transport on the route Bratislava hlavná stanica—Bratislava Nové Mesto—Dunajská Streda—Komárno (ŽSR no. 131 and 132 routes). The contract has been concluded between the contracting authority—the Slovak Republic Ministry of Transport and Construction, and private railway undertaking RJSK, and has entered into force 30.09.2011. The contract's validity will end before the timetable change in 2020. Operation of passenger trains on this route began on 04.03.2012. Based on this contract, the contracting authority must pay compensation for services from the state budget (Public service contract Regiojet, a.s. 2010).

Annexe 1 of this contract contains the factual and time definition of trains for the first period of operation and the timetable period 2011/2012. For subsequent years, the volume of services is adjusted by an addendum to the contract. As in the contract with ZSSK, a.s. the contract includes an order for services for each valid year. Total volume of operational performance may change based on external factors, such as the different number of days in a year, so that they deviate from the original order by a maximum of 5%. Trains must operate on the basis of the timetable compiled by the infrastructure manager ŽSR in cooperation with the contracting authority. This contract also defines the railway undertaking's general obligations to ensure the transport service on this line (Public service contract Regiojet, a.s. 2010).

The provider, RJSK, guarantees that it will operate trains that comply with the contracting authority's quality requirements under this contract. Deviations are only permissible with the contracting authority's consent but must be in writing and are only valid for a certain period. Based on this contract, operation of services is ensured by the use of vehicles Bombardier Talent, or vehicles with similar attributes as:

- the vehicle is approved for operation and has a design speed of at least 120 km / h,
- the seats must be upholstered with headrests,
- allow for transporting pushchairs and bicycles,
- a toilet for people with limited mobility or orientation,
- optical and acoustic information system,
- air-conditioned vehicle interior,
- at least 8 seats in the first class (Public service contract Regiojet, a.s. 2010). Under this contract, the provision of services is evaluated according to operational quality indicators as:
- clean interior,
- functional acoustic device providing train position information,
- the provision of route timetables free of charge at any stop or station where the train stops,
- staff wear uniforms,
- customer line in operation at least during working days from 08:00 to 18:00,
- emergency telephone lines in each vehicle,
- free Wi-Fi,
- other qualitative indicators agreed between the railway undertaking and the contracting authority (Public service contract Regiojet, a.s. 2010).

The railway undertaking RJSK must carry out a passenger satisfaction survey at least twice annually, whereby they will take corrective measures in cooperation with the contracting authority. The contracting authority must pay compensation to the railway undertaking for the provision of services, which may not be exceeded, based on the Formula (2).

$$Compensation = TP * P_{tkm} * (1 + i) + TP * P_i \tag{2}$$

where

TP—total operational performance in train-kms for the timetable's validity period, without unrealised and erroneous performances

P_{tkm}—price per one train-km

i—yoY index of consumer price increase.

P_i—price for the use of railway infrastructure per one train-km.

Compensation is paid by the contracting authority in monthly advance payments. The contracting authority should cooperate with the infrastructure manager to ensure a quality, safe and efficient transport infrastructure and traffic management also for the private railway undertaking. Based on this, a triple agreement has been concluded with RJSK, ŽSR and the Ministry of Transport (Public service contract Regiojet, a.s. 2010).

5.4 Quality Standards of Rail Passenger Transport Services Under PSO

Obligations concerning the quality of services provided in the public interest are also included in the contract. These are often parameters, which must already be proven by the railway undertakings during the application for the contract. During the operation of services, compliance with these parameters is checked.

These parameters are described in supplement No. 10 of the RJSK, a.s. PSO contract and in supplement No. 9 of the ZSSK, a.s. PSO contract (2010). The monitored parameters are identical for both railway undertakings and are only approximately specified in this study. Individual parameters are:

- *Accuracy and fulfilment of the timetable*—The contracting authority requires a minimum accuracy of trains of 94%. If it is not complied with, the contracting authority can ask for a penalty of € 16,000 for each full percentage. However, the most common delay is caused by traffic reasons, by the Infrastructure Manager. It is a delay beyond the control of the railway undertaking, and therefore this delay is not included in the accuracy of trains. If the trains are accurate to at least 96%, the contracting authority will pay a bonus of € 16,000 for each 0.2% increase in accuracy.
- *Train composition*—The train composition must be realised according to the passenger trains composition plan. In the event of non-compliance with the standards (e.g. use of lower-class wagons), the railway undertaking will be sanctioned.
- *Equipment and functionality of wagons*—Wagons should be fully equipped when leaving the home station or maintenance station and without visible defects in look and functionality. Only new and original parts must be used to repair wagons. This standard focuses on toilets, exterior and interior doors, passage doors, windows, seats, shelves, interior walls, lighting, baskets, floor, Wi-Fi, sockets and the possibility of transporting bicycles.
- *Interior cleanliness*—Every first departing train must have a clean interior that is accessible to the public and must also be free of objects that are not part of the wagon. If the train leaves a station where there is no technical and hygienic maintenance, it must be cleaned at least every second day or at a station where it stays longer than 45 min.
- *Exterior of wagons and locomotives*—Graphite-free cleanliness, clean and translucent window panes without traces of dried detergents are required. The logo of the railway undertaking should be visible on wagons and locomotives, unless the wagon is leased from another company. Advertising paints are also approved.
- *Interior comfort*—The temperature in wagons intended for the public must not be lower than 18 °C. At temperatures of 30 °C and higher, the temperature in air-conditioned wagons should be at least 5 °C lower. At least 80% of the lamps should be lit in the wagon and the dimming device should be in permanent operation.

There should be no odours in the carriage and the ride should be quiet and without shocks.

- *Train reliability and safety*—The contract specifies the maximum number of trains that railway undertaking can cancel or delay by 30 min or more at the destination station, if the delay is caused by the carrier's fault. This number will decrease each year during the contract's validity. The railway undertaking must ensure that the train runs safely and that no incidents or accidents occur.
- *Information*—The railway undertaking must publish the train's location via the Internet and, if the train has a delay of more than 30 min, also the development of its delay. Seating, couchette and sleeping wagons must be marked with either a notice board or an electronic optical board. The wagon must also be marked with direction signs.
- *Travel documents*—The railway undertaking can only accept the documents that it issues, or documents approved by the contracting authority, unless it is in an exceptional situation. The railway undertaking should not allow carriage without travel documents or must collect a penalty if a passenger does not meet the conditions. Passengers should have their documents checked within 30 min of boarding.
- *Calculation of justified costs*—The railway undertaking can only calculate costs related to the contract, otherwise it would be in breach of the agreed contractual conditions and would be sanctioned for it.
- *Repeated deficiencies or defects*—These deficiencies or defects must not be detected repeatedly on the same wagon within 1 month. The railway undertaking guarantees that it will not neglect maintenance and repair (Amendments No. 5—Public service contracts 2010).

5.5 Analysis of Contract Performance Under PSO

Fulfilment of the PSO contract was analysed on the basis of data on planned and realised operational performances and the amount of planned compensation for demonstrable loss. Table 1 shows the fulfilment of the ZSSK contract, Table 2 shows the RJSK contract.

As can be seen in Table 1, planned and realised transport services of (ZSSK) had a declining trend until 2014. In 2012, the private railway undertaking RJSK entered the Slovak passenger rail transport market, but the services provided by it only slightly affected ZSSK, as the performance on the taken-over route was not provided to a large extent by incumbent. However, the contracting authority's expectations from RJSK were to use the potential of this route and therefore for the first year of operation it was given almost 1 million train km.

In 2014, a small increase in ordered transport performance can be observed for both railway undertakings. Along with transport performance, the amounts of planned compensations also increased. That year, the contracts were amended in

Table 1 Fulfilment of the PSO contract concluded with ZSSK, a. s

Year	Planned performances (mil. train km)	Realised performances (mil. train km)	Planned reimbursement of provable loss (mil. €)
2011	31.4	31.252	205.000
2012	29.35	29.358	199.342
2013	29.104	29.121	197.559
2014	29.604	29.594	197.559
2015	31.304	31.169	209.559
2016	31.304	31.418	209.559
2017	31.304	31.399	209.559
2018	32.37	32.347	224.559
2019	33.114	33.161	218.709
2020	33.54	–	218.709

Source ŽSR Annual Reports and data from the Slovak Republic Ministry of Transport and Construction

Table 2 Fulfilment of the PSO contract concluded with RJSK, a. s

Year	Planned performances (thousand train km)	Realised performances (thousand train km)	Planned reimbursement of provable loss (mil. €)
2012	947.743	1,014.783	5.658
2013	1,196.422	1,682.087	7.222
2014	1,196.526	1,169.838	7.223
2015	1,197.248	1,200.106	8.357
2016	1,197.236	1,202.350	8.417
2017	1,197.368	1,197.453	8.438
2018	1,211.094	1,207.859	8.664
2019	1,226.706	–	10.593
2020	1,176.218	–	11.408

Source ŽSR Annual Reports and data from the Slovak Republic Ministry of Transport and Construction

such a way that free (unpaid) transport for selected passenger groups was introduced, binding for both undertakings. Free transport has affected the demand for passenger rail services so far.

The realised transport performance is mainly affected by extraordinary events as accidents, replacement bus services or construction works. In Western Slovakia, the realised transport performance is also affected by the cancellation of some passenger trains due to a lack of train drivers.

The ordered operational performance for both railway undertakings is higher for 2020 than in 2019, but the realised operational performance is expected to be lower, due to the COVID-19 pandemic. Due to this pandemic, railway undertakings provided services only to a minimal extent (limited number of connections during the working week to the number of connections dispatched on weekends or school holidays, cancellation of IC trains, etc.).

5.6 *Public Tenders of Providing Rail Passenger Transport Services Under PSO*

Public tenders within passenger rail transport in Slovakia are still at the beginning and the issue of public tenders is insufficiently addressed. The Slovak Republic Ministry of Transport and Construction has a list of 35 lines that can be put out to tender. The first attempt at a public tender in Slovakia was on the Bratislava—Zvolen—Banská Bystrica route, but the tender was cancelled. In the Slovak Republic, one public tender is currently announced, specifically on the Košice—Veľká Ida—Moldava nad Bodvou-mesto route, but it is only in the preparatory phase. The other 3 public tenders have already ended. They took place on the Žilina—Rajecké Teplice—Rajec, Bratislava—Dunajská Streda—Komárno and Bratislava—Plavecké Podhradie routes. More public tenders will probably be announced in the future.

The public tender process in Slovakia is in accordance with the legislation and is as follows:

- Selection of the route for the tender and determining the conditions,
- Publication of a Preliminary Notice of PSO contracts,
- Publication of a notice regarding the tender announcement,
- Registration of tenderers for the tender, submission of applications,
- Selection of tenderers fulfilling the conditions to participate in the tender,
- Making tender documents available to tenderers,
- Submission of offers by tenderers,
- Evaluation of offers,
- Decision on success,
- Conclusion of the contract, possibly temporary direct award (Liberalization of passenger rail transport 2020).

Public tender Bratislava—Banská Bystrica

The tender was announced by the Ministry of Transport in 2015 and it was the first public tender for the operation of passenger rail services in Slovakia. The route of the tender was Bratislava main station—Levice—Zvolen—Banská Bystrica (Fig. 1).

Fig. 2 Route Žilina—Rajec. *Source* Author, via ŽSR

However, this tender was cancelled in 2016, by a decision from the Slovak Republic Minister of Transport and Construction. The reason was insufficient preparation concerning the public tender and the need for better preparation of the public tender. Trains that would run on this route would be unlike other public tenders, category R—Fast train. It is possible, that the tender for the provision of services on this route will be repeated in the future (Záhumenská 2018).

Public tender Žilina—Rajec

This tender was announced for the operation of regional passenger rail transport services on route no. 126 from Žilina to Rajec (Fig. 2).

The start of the tender on this route was announced through the publication of the Tender Notice on August 7th, 2019. Previously, the tender was announced by the Preliminary Notice of PSO contracts on March 15th, 2018. The announcer of this tender was the Slovak Republic Ministry of Transport and Construction, and the validity period of the contract is 119 months. The contract was to be concluded on February 1st, 2020 (Prior notice of invitation to tender Žilina—Rajec 2018).

Tender documents are not available to the public. The deadline for the submission of offers was November 8th, 2019 at 15:30. The deadline for opening the submitted tenderer´s offers prepared on the basis of the tender documents was set for November 22nd, 2019, from twelve o'clock (12:00), at the office of the Slovak Republic Ministry of Transport and Construction (Notice of invitation to tender Žilina—Rajec 2020a, 2020b).

Tender offers were submitted by only one railway undertaking ZSSK (incumbent), which has operated services on this line as the original provider. The Commission decided, that the tenderer did not meet the requirements for the conclusion of PSO contract. A public tender on the Žilina—Rajec route does not have a successful tenderer. The result was published on December 16th, 2019. No further information

Fig. 3 Route Bratislava—Komárno. *Source* Author, via ŽSR

is available about the future of the tender or the operation of services on this route (Notice of the result of the evaluation of tenders Žilina—Rajec 2020a, 2020b).

Public tender Bratislava main station– Komárno

This tender was announced for the provision of regional passenger rail transport services on the route No. 136 Bratislava main station—Dunajská Streda—Komárno (Fig. 3).

The start of the tender on this route was announced through the publication of the Tender Notice on October 10th, 2019. Previously, the tender was announced by the Preliminary Notice of PSO contracts on June 29th, 2018. The announcer of this tender was the Slovak Republic Ministry of Transport and Construction, and the validity period of the contract is 120 months. The contract should be concluded on December 13th, 2020 (Notice of invitation to tender Bratislava—Komárno 2019).

The tender documents have been made available for the purposes of this case study, and an analysis of the failure of public tenders in Slovakia will be made on the basis of them. The deadline for submission of offers was February 14th, 2020 at 15:30. The deadline for opening the submitted tenderer´s offers prepared on the basis of the tender documents was set for February 17th, 2020, from ten o'clock (10:00), at the office of the Slovak Republic Ministry of Transport and Construction (Tender documents Bratislava—Komárno 2019; Explanation of tender documents Bratislava– Komárno I.,II. 2020a, 2020b, 2020c).

Five railway undertakings were admitted to submit their offers, but the envelope with tender offers was only submitted by RJSK, which is also the original provider of railway passenger transport services on this route. In the given tender envelope, the tender documents were not filled in, but only comments on the course of the public tender were added. The Commission decided that the tenderer RJSK, a. s. did not meet the requirements for concluding the PSO contract. The public tender on the

Fig. 4 The Bratislava main station—Plavecké Podhradie route. *Source* Author, via ŽSR

Bratislava main station—Komárno route does not have a successful tenderer. The result was published on March 6th, 2020.

Public tender Bratislava main station– Plavecké Podhradie

This public tender takes place for the seasonal provision of rail passenger transport services on route No. 110 and No. 112 between Bratislava main station, Zohor and Plavecké Podhradie (Fig. 4). It is the provision of rail passenger transport services seasonally, in the summer, by means of a cycling train "Záhoráčik ", and it will therefore be a contract for the provision of services of a local and recreational character. The announcer of this tender is Bratislava self-governing region.

The start of the tender on this route was announced through the publication of the Tender Notice on February 7th, 2020. The contract is scheduled to start on May 1st, 2020 and end on October 9th, 2022. This public tender also accepts Czech language (Notice of invitation to tender Bratislava—Plavecké Podhradie 2020).

The deadline for submission of offers was March 11th, 2020 at 10:00. The deadline for opening the submitted tenderer´s offers prepared on the basis of the tender documents was set for March 11th, 2020, from eleven o'clock (11:00), at the Office of the Bratislava Self-Governing Region. The main evaluation criterion was the cost of operating the cycling train (Tender documents Bratislava—Plavecké Podhradie 2020).

Tender documents were submitted electronically. Only 1 tenderer submitted tender documents—ZSSK, a. s. According to the information we received from the tenderer's employee, the tenderer did not succeed in their tender. The announcer considered their offer to be disproportionately high and also cancelled the public tender (Answers to public procurement questions Bratislava—Plavecké Podhradie 2020).

Public tender Košice—Moldava nad Bodvou-mesto

Fig. 5 The Košice—Moldava nad Bodvou-mesto route. *Source* Author, via ŽSR

This public tender takes place for the provision of regional rail passenger transport services on route No. 160 and also partially on No. 168 Košice—Moldava nad Bodvou—Moldava nad Bodvou-mesto (Fig. 5), where the regional passenger transport terminal is located. The announcer of this tender is Ministry of Transport and Construction of the Slovak republic.

The execution of the public tender on this route has so far been announced only by the Preliminary Notice of PSO contract, which was published on May 9th, 2018. The contract is expected to start on January 1st, 2022 and will be valid for 120 months, 10 years. The tender will be announced after the completion of electrification of line 160. So far, the tender is only in the preparatory phase (Prior notice of invitation to tender Košice—Moldava nad Bodvou 2018).

5.7 *Survey of Public Tender Contentment*

In order to fulfil the case study's aim, a satisfaction survey on the course of public tenders was carried out in the months of January–February 2020 , on the basis of the opinion of railway undertakings. We ascertained this opinion through questionnaires sent to passenger railway transport undertakings in Slovakia and Czech Republic. As a part of the survey, employees who worked in managerial positions with carriers were contacted.

Respondents, railway undertakings, stated in the questionnaire whether they have experience with the provision of PSO services and, if they were providers of such services, how they concluded PSO contracts. They should also state satisfaction / dissatisfaction with the course of the public tender, express the area of objection to the public tender and evaluate the appropriateness of the method of the contract's conclusion.

Železničná spoločnosť Slovensko, a. s.

ZSSK is the Slovak incumbent, which mostly provides services on the basis of a PSO contract. These obligations have been so far awarded to ZSSK, a. s. only through direct award. This railway undertaking was also a tenderer in a public tender, the course of which it was dissatisfied and plans to object to the course of the tender. ZSSK points out incorrectly set tender conditions (unfulfillable and incomprehensible) and also on the timeliness of publication of tender documents. In this case, the railway undertaking does not see an error in the concept of the tenders, but in the conditions set by the announcer. ZSSK, a. s. prefers direct award to public tender. The individual risks resulting from the analysis of the failure of public tenders are elaborated in details in Chapter no. 5.8.

České dráhy, a. s.

České dráhy, a. s. is an incumbent in the Czech Republic, which also provides services on the basis of a PSO contract. These obligations have been so far awarded to České dráhy, a. s. both through direct award and public tender. The railway undertaking was less satisfied with the course of the public tender and points out the non-acceptance of comments by the announcer. České dráhy, a. s. prefers direct award to public tender.

Arriva

Arriva is a German railway undertaking, which provides services in Czech Republic as well as in Slovakia. In the Czech Republic, Arriva provides these services on the basis of a direct awarded PSO contract. Arriva was a tenderer in public tender, with which it was less satisfied but did not provide a reason of this dissatisfaction. Arriva prefers public tenders to direct award.

Leo Expres, a.s., (Leo Expres Tenders s.r.o.)

Leo Expres, a.s. is a railway undertaking operating mainly in the Czech Republic, where it also provides services under PSO contract. They operate commercial trains in Slovakia on the Prešov—Praha route. The questionnaire was filled in by the subsidiary LE Tenders s.r.o., which deals with the preparation of the LE railway undertaking for public tenders. The carrier has acquired PSO performance through direct award. The carrier was also a tenderer in the public tender. The carrier was satisfied with the course of the public tenders in the Czech Republic, but no with the public tender in the Slovak Republic. In its opinion, there is a very short time for the preparation of tender documents, or offers and there is also a very short time between the results announcement and the start of operation. The carrier refers to Germany, where the time between the announcement of the results and the start of operations is around 3–4 years.

The carrier proposes to be able to choose the timetables according to transport demand. Also proposes, that requirements for rolling stock should be given only in a framework, as it is difficult to procure vehicles on the current market in such a short period of time. The carrier recommends that the tenders be announced at least 2 years in advance, and that the contracting authority should consult the details of the contract with the railway undertaking.

5.8 'Non–success' Public Tender Risk

Finished public tenders announced by the Slovak Republic Ministry of Transport and Construction and by the Bratislava Self-governing Region took place without a successful tenderer. The specified conditions of participation, and the tender documents were unable to be met by tenderers or they did not even go to the public tender. The offer as such was not submitted in any of the tenders announced by the Ministry. Only objections were submitted in the envelopes (Pribula 2020).

All public tenders in Slovakia have been in accordance with international and national legislation. But this is not enough for the tender to be successful, if participation conditions, tenders and future contracts awarded by the announcer, are not rational or feasible. Based on the analysis of public tenders in Slovakia and abroad, we have determined the most common factors that affect the failure of public tenders in the Slovak Republic. The COVID-19 pandemic did not affect the course of public tenders in Slovakia.

In the analysis of individual risks, we based objections to the course of public tenders, outputs from questionnaires, current legislation, the current state of foreign public tenders and other. For the purposes of this study, we have identified 12 risks, which are important bases for drafting recommendations for a more effective announcement of public tenders in the Slovak Republic.

Risk 1: Public tender time range

There is often a longer time between the announcement of a tender by the Preliminary Notice and announcement of a tender by Tender Notice, but there is a very short time from its announcement to the end of the deadline for the offer's submission, and tenderers often fail to meet the conditions set-out in the tender documents. During the course of a public tender, the tender documents may also be modified on the basis of the tenderers' objections. There is also a very short period between the end of the public tender and the contract's entry into force.

Time intervals between the publication of the Tender Notice and the deadline for submission of offers in the Slovak Republic is:

- Žilina—Rajec: approximately 4 months
- Bratislava main station—Komárno: approximately 4 months
- Bratislava main station—Plavecké Podhradie: approximately 1 month (Liberalization of passenger rail transport 2020).

It follows on from the above-mentioned time periods, that the tenderers do not have even half a year to prepare a tender offer in the tenders that have been announced in the Slovak Republic.

The most obvious factor affecting the course of public tenders in Slovakia was time. Railway undertakings have complained about this time period both in the objections of the public contract and in the questionnaires. If the conditions in tender documents were rationally and appropriately set, it is very likely that tenderers would be unable to submit properly drafted offers.

Risk 2: Estimated value of services per unit of operational performance—economically justified costs

Estimated values of costs per unit of operational performance, which the contracting authority has entered as a precondition and the indicative value, are too low for the railway undertaking. Railway undertakings agreed in the survey, that the contracting authority does not include all cost items in the calculation of unit costs. The contracting authority estimated this value for the Bratislava—Komárno route in the amount of €7.47 / train km and for the Žilina—Rajec route €7.87 / train km (Tender documents Bratislava—Komárno 2019).

Railway undertakings object to the value of unit costs expected by the announcer as unattainable, when the conditions for the contract and the tender are met.

Risk 3: The amount of the bank guarantee

The announcer of the public tender requests from the tenderer is a very high value of the bank guarantee to be applied in case the tenderer/railway undertaking is unable to provide the obligations arising from the PSO contract. In such a case, the bank guarantee is intended to cover outstanding obligations under the contract. Such a situation is unlikely, if the railway undertaking provides these services on the basis of success in the tender, which has been fairly decided by a commission appointed by the announcer to assess whether the railway undertaking will be able to meet these obligations.

In case of the public tender on the Bratislava—Komárno route, the announcer requests a bank guarantee in the amount of € 8,977,317.36, which represents almost 45 times the bank guarantee from the previous contract, awarded on this route directly (€ 200,000). The value of the bank guarantee has been calculated by the product of the initial annual operational performance and the maximum unit value of the service provided. Before the first modification of the tender documents, this value was 10 times higher (Tender documents Bratislava—Komárno 2019).

In the Czech Republic in the Pardubice Region, the amount of the bank guarantee for the annual transport performance of a regional trains of 572,000 train km was approximately € 0.5 mil. Similarly, for long-distance trains connecting Brno with Bohumín, with an operational performance of approx. 2.1 mil. train km per year, the amount of the bank guarantee was set at approximately € 1.2 mil. For the performance of transport services in the Pilsen Region, in the amount of 25.5 mil. train km per year, the region requires a bank guarantee in the amount of € 2 mil. (Explanation of tender documentation trains—Plzeňsko 2020 and Binding text of the draft Treaty—Plzeňsko 2020).

Railway undertakings see discrimination in the amount of this bank guarantee against small businesses, whereas previous contracts required a bank guarantee outside the calculation of the product of annual operational performance and unit costs. The determined value of the bank guarantee outside this calculation was much lower than in the case of the usual calculation of the Slovak Republic Ministry of Transport and Construction.

Risk 4: Technical competence

"Minimum level of standards required: List of services provided of the same or similar character, as is the subject of the contract for the previous 5 years preceding the date of submission of tender offers, supplemented by certificates of the quality of service provision, stating prices, delivery times and purchasers " (Notice of invitation to tender Bratislava-Komárno 2019).

The contracting authority requests, that tenderers have at least several years long experience of providing similar services, whereas the minimum volume of operational performance of the provision of services was similar and a higher volume of performance on the tendered route. Long experience was different for every public tender in Slovak Republic. Fulfilment of these conditions is verified by the contracting authority, and the tenderer must declare this in the annexes to the application for participation in a public tender.

The problem is, that these conditions worsen the access of newly founded railway undertakings to the passenger rail transport services market, and also the access of railway undertakings, which have been on the market for a long time, but have not yet had long enough experience in providing passenger rail transport services. New railway undertakings are unable to obtain the possibility of providing PSO services on currently tendering routes. It is difficult for new railway undertakings to obtain sufficient experience in providing services via commercial trains, as there are not many suitable routes where this would be effective.

Risk 5: Vehicle fleet

The announcer does not provide information on the availability of rail vehicles on the market, or rental or purchase options. The announcer does not also ensure the vehicles by itself. This is recommended by the Regulation (EU) 2016/2338, in article No. 5a. The market for required rolling stock is not sufficiently saturated, and railway undertakings often have a problem with the availability of required vehicles that fully meet the contracting authority's requirements.

Risk 6: Language

Much of the content of the Tender Notice is in the national language. It is also published in the national language preliminary notices, tender documents for public tenders in the European Union Official Journal. All documents and other documentation related to the public tender are also in the national language.

Acceptance of only one language in the public tender makes access to the passenger rail transport market more difficult for foreign railway undertakings. If the application for the tender or the tender offer is in a language other than the national language, participation in public tender may be invalid. In the Slovak Republic, participation was conditioned only by the Slovak language, except for the tender of a cycling train, where Czech language was also accepted. The language barrier creates obstacles in the transparency of public tenders to foreign countries.

Risk 7: Timetables

The announcer creates preliminary timetables in cooperation with the infrastructure manager, where the individual travel times are set. Due to the different traction characteristics of different vehicles, it is not always be possible to comply with these timetables with each vehicle, although these vehicles would meet other requirements. The infrastructure manager always has the final word in the creation of timetable, and possible train connections and other traffic on the tendered route must be accepted when creating them.

The disadvantage of the preliminary creation of the timetable is its non-adaptation to most of the possible vehicles that could drive on the tendered route, which also restricts tenderers to offer these vehicles. These timetables are considered binding by the announcer.

Risk 8: Capacity of vehicles

According to the tenderers, the required capacities of rolling stock by the announcer are exaggerated. When determining the minimum capacity of vehicles, the announcer does not allow standing passengers and does not take the required capacity for peak hours or saddle hours into account. The announcer also does not take into account that selected train connections will not be dispatched on the entire tendered route.

When a train is running, it is not only expected that passengers will board at stations or stops, but that they will also disembark. The tenderers consider that the required vehicle capacity could only be used during peak hours, if the train were to run on the whole route. In this way, operating costs increase and the problem of vehicle availability on the market also increases.

Risk 9: Vehicle quality standards

Strict vehicle quality standards are set in tenders. They are set so, that they cannot be ensured until a contract enters into force. Most vehicles are not equipped with devices to ensure passenger comfort or other quality parameters.

Railway undertakings do not have time to ensure, that all vehicles meet these standards (such as air conditioning etc.) until the contract enters into force. Their implementation takes place after the tender ends, until the beginning of contract's validity. In the public tender Bratislava—Komárno, this should be 50% of the vehicles offered. However, the railway undertaking had less than a year to equip them.

Risk 10: Requirements for the use of vehicles

The announcer in tender documents requires that vehicles used for contract services will be used only for services covered by the contract. However, this can affect the creation of optimal circulations of rail vehicles and reduces the efficiency of their economic use.

Risk 11: Electronic cash register

The announcer requests too detailed requirements for electronic cash registers for the issuance and registration of travel documents. In our opinion, as well as some tenderer´s, these requirements are irrelevant to the operation of PSO services. For example, these requirements are the screen size of the device and the number of identical sensors, their brand, etc.

Risk 12: Information in the PSO contract's Preliminary Notice

These notices are often published more than a year before the public tender starts but contain only little relevant information about the future tender. This is information as the subject of the tender, who the announcer is and the expected date of the contract's entry into force.

The specific conditions of participation are only in the Tender Notice, and they are published very late, before the start of the tender. The information contained in the Preliminary Notice do not provide sufficient information to assist in a railway undertaking's preparation for public tender.

5.9 Draft Criteria for Selecting Providers of Rail Passenger Transport Services Under PSO

Based on the analysis of the risks of public tenders in the Slovak Republic and abroad, the current state of ordering services in the public interest in the Slovak Republic and abroad and on the basis of a satisfaction survey among selected railway undertakings, we have proposed a set of criteria to support the effectiveness and success of public tenders in Slovakia. The main aim of the proposed criteria is to create partial recommendations, that are to streamline the course of public tenders in Slovakia and ensure their successful completion. The individual criteria have been elaborated with regard to the performed analysis and questionnaire surveys, and simultaneously take into account the valid national and international legislation (Pribula 2020).

The individual criteria follow the analysis of the risks of public tender failure in the Slovak Republic, and are as follows:

Criterion 1: Time range of the public tender

We propose to extend the total time range of the public tender as follows:

- Period from the Preliminary Notice of PSO contract until the beginning of the contract's validity—at least 3 years.
- Separate the time of submitting applications to the public tender from the time of the public tender duration (elaboration of the offer).
- Time from the publication of tender documents to the deadline for submitting tender offers—at least half a year.

- Time from the public tender evaluation to the beginning of the contract—at least one year.

In our opinion, this is one of the most important measures, that could increase the chances of a successful course of public tender in Slovakia and possibly abroad, because in current public tenders announced by the Slovak Republic Ministry of Transport and Construction, the tenderers were unable manage to prepare tender documents. There would also be more time for preparing the public tender and the possible modification of the tender documents.

Criterion 2: Estimated cost value per unit of operational performance

We propose that the estimated value of unit costs not be stated in the public tender. The aim of public tenders is to ensure a more efficient provision of rail passenger transport services, which can be achieved primarily by offering better services to customers. The resulting effect is a higher use of passenger rail transport and higher revenues, which ultimately reduces the requirements for non-investment subsidies (depending on the contract conditions).

In case the public tender announcer insists on the maximum unit value of the service, we propose that this value will be estimated by independent experts.

Criterion 3: Bank guarantee amount

Our proposal is a reduction of the bank guarantee amount. Railway undertakings and experts do not consider it correct to calculate it from the service's annual value, but the set fixed and rational value. Alternatively, it is necessary to re-evaluate the methodology for calculating the bank guarantee amount, so that its value is rational.

In the Bratislava—Komárno public tender, tenderers would be willing to provide a bank guarantee amount of 3 times the value of the bank guarantee required in the previous contract, which is € 3,200,000.

The benefit of this measure is a reduction in unit operating costs due to lower interest rates and the prevention of problems in obtaining such a high bank guarantee.

We have proposed a formula (3) to calculate the increased bank guarantee amount for undertakings, that do not have sufficient experience. In this formula, the original bank guarantee would be increased taking into account the difference in the years of presence of the undertaking operating the passenger rail transport services on the market, from the required presence.

$$BG_i = BG_b * (PE_r - PE_t) * \left(\frac{OP_t}{OP_s}\right) \tag{3}$$

where

BG_i—increased bank guarantee in €

BG_b—basic bank guarantee in € (stated in the tender documents),

PE_r—required period of experience with services similar or the same as in the announced tender, in years.

PE_t—period of experience of the tenderer, in years (a minimum value of 0 years, and if the undertaking has more than required years' experience, then the PE_r value is used),

OP_t—average operational performance on the tendered route, in train km/year.

OP_s—average operational performance of the tenderer's service, in train km/year (the maximum value used in the calculation is equal to OPt).

Maximum value of the ratio of the average expected annual operational performance on the tendered route to the tenderer's realised average annual performance, would be fixed and would be used by railway undertakings, that had not carried out any passenger rail transport services until the public tender has been announced.

Application of the proposed formula for the undertaking, who has been operating on the passenger rail transport market less and with lower operational performance than in the announced public tender.

The railway undertaking has operated passenger rail transport services for 3 years, while providing a service worth 30,000 train km per year. The announcer, in compliance with the basic bank guarantee amount, required the railway undertakings 5 years' experience with a minimum value of the service 53,000 train km per year. The values used in the calculation are:

- BG_b = € 125,000
- PE_r = 5 years
- PE_t = 3 years
- OP_t = 53,000 train km
- OP_s = 30,000 train km
- BGi = 125,000 * (5–3 + 1)*(53/30) = € 662,500

As the tenderer did not meet the basic conditions, they must expect an increase in their bank guarantee to € 662,500.

Application of the proposed formula for undertaking, which has been operating on the passenger rail transport market for a sufficiently long time, but with lower operational performance than in the announced public tender.

The railway undertaking is able to declare a sufficient number of years' experience in operating passenger rail transport services, while providing a service in the amount of 100,000 train km per year. The announcer, in compliance with the basic bank guarantee amount, required the railway undertakings 4 years' experience with a minimum value of the service 120,000 train km per year. The values used in the calculation are:

- BG_b = € 150,000
- PE_r = 4 years
- PE_t = 8 years
- OP_t = 120,000 train km
- OP_s = 100,000 train km
- BG_i = 150,000 * (4–4 + 1)*(120/100) = € 180,000

The railway undertaking had a sufficient number of years' experience in operating passenger rail transport services but did not provide a service in sufficient annual operational performance. Therefore, its bank guarantee was increased from € 150,000 to € 180,000.

With this proposed criterion, we would increase the possibilities of participation in the public tender for all undertakings, that meet the conditions for operating rail transport (have received a licence for operating passenger rail transport services).

Criterion 4: Language

When examining the Tender Notice in Slovakia, but also in various countries, we found that language is also a significant limiting factor, especially for foreign railway companies. Our proposal is to introduce compulsory acceptance of at least 2 languages within the course of the public contract, namely the national language and at least English. In the Slovak Republic, we recommend full acceptance of Czech language due to its similarities with Slovak language. Selected languages will be listed in tender documents.

For the ministry, we propose to issue tender documents in both / more languages, and also acceptance of applications to the public tender and submitted tender offers in these languages. Publishable documents in the national language and translated into English should be made available on the Internet, at least in the European Union Official Journal.

The aim of this proposed criterion is making the conditions transparent to all potential tenderers on the declared route, a possible increase in interest in service provisions by foreign railway undertakings, more efficient selection and, ultimately, more efficient provision of the PSO services.

Criterion 5: Requirements for the use of vehicles

We propose that the railway undertaking should be able to use the rolling stock necessary for the services resulting from the specific contract for other purposes, other than those purposes arising from this contract.

The benefit of this measure is the efficient use of the carrier's vehicle fleet and therefore, a reduction in the carrier's operating costs associated with parking vehicles, empty runs and other factors. It would also reduce the capital intensity of procuring a fleet for a specific route.

Criterion 6: Timetables

In public tenders, we propose to set the train connection frequency at peak times and saddles on a given line / route, instead of timetables. Binding timetables would only be created after the tender's successful completion, which would take into account the tenderer's offered vehicles. The railway undertaking would consult these timetables with the infrastructure manager and as well as with the contracting authority.

The timetables proposed by the tenderer would be proposed according to its available vehicles, and therefore, there would be no restriction on the choice of vehicle fleet for the tender on the basis of strict adherence to travel times, which have only been calculated for specific vehicle parameters based on the announcer's assumption.

Criterion 7: Capacity of vehicles

We propose that when determining the capacity of rail vehicles, the public tender announcer also takes standing passengers in regional or suburban trains into account. Taking into account the total capacity of railway vehicles, they will be used more efficiently, and therefore reducing the capital costs of the railway undertaking.

Criterion 8: Electronic cash register

We propose that the requirements for the electronic cash register are not too detailed, and that the announcer will state only functions, which the device is intended to ensure, and the necessary compatibility with public authorities.

Criterion 9: Information in the Preliminary Notice of PSO contract

The information on the future tender in the Preliminary Notice should be more detailed, similar to the information provided in the Tender Notice. The information could be extended to requirements for public tender participation. Preliminary notices are often published well in advance before the tender, therefore more detailed information in them would help railway undertakings to prepare to participate in the tender.

6 Conclusions

Transport infrastructure, public passenger transport and other transport services are an integral part of the people's daily life. At the same time, they condition the achievement of economic growth, increasing society's competitiveness and prosperity. They contribute to increasing employment opportunities and are a key factor in the inflow of foreign investment, development of tourism, and help to reduce disparities between regions. For the transport sector, the European Commission has set a condition in the form of providing comprehensive plans for transport infrastructure development, including plans for the sustainable development of urban, suburban and regional transport. These plans should be based on a thorough analysis of the needs in the sector and the subsequent identification of key bottlenecks (e.g. missing sections, unsatisfactory infrastructure parameters, etc.) and potential development factors, whereby implementation will make a significant contribution to improving the existing situation, whether from a transport, economic or environmental viewpoint, or their combination (Strategic plan for transport development in the Slovak Republic until 2030 2016).

Transport infrastructure, quality and accessibility of public passenger transport, have a significant impact on the people's transport behaviour. Shortening travel times increases the number of journeys made, e.g. while commuting to work. Quality public passenger transport leads to an increase in its use and a reduction in individual passenger transport. As a result, the Slovak Republic's modernised transport system will lead to overall economic growth, reduce economic disparities between regions, lead to the inflow of foreign investment, reduce unemployment, and tourism development. Market openness and non-discriminatory conditions for all undertakings interested in providing passenger rail transport services, can also help to ensure quality public passenger rail transport.

Liberalisation measures on European railways can lead to significant changes, that allow new railway undertakings to enter the market, not only through public tender for the exclusive right to operate a certain route, but also in open access. Such conditions must be created for the business of rail passenger transport, which enable railway undertakings better access to the provision of passenger rail transport services in the public interest. These conditions are of a special nature and need to be met in order to be entitled to operate rail services.

The opening a domestic passenger rail services market has created a number of challenges that will need to be solved such as quality of rail passenger transport, sector efficiency, socio-economic efficiency, environmental aspects. It will be necessary to examine the impact of the opening of the rail passenger market on socio-economic efficiency using econometric models, the DEA method and other models and tools in order to effectively PSO tendering and increasing competitiveness of rail passenger transport in the transport market.

Acknowledgements The paper was supported by the VEGA Agency, Grant No. 1/0379/20 "Socio-economics aspects of rail transport market services deregulation in the context of EU legal framework and sustainable mobility", at Faculty of Operations and Economics of Transport and Communication, University of Žilina, Slovakia.

Literature

Abramovic B, Sipus D, Leko M (2018) The analysis of the organisation of railway passenger transport on the liberalised market. In: Road and rail infrastructure V, 5th international conference on road and rail infrastructure (CETRA). Zadar, Croatia. May 17–19, 2018, p. 847–853

Act No. 18/1996 about prices (1996) (Online) Available https://www.epi.sk/zz/1996-18 Accessed May 2020

Act No. 514/2009 of transport on track (2009) (Online) Available https://www.epi.sk/zz/2009-514 Accessed October 2019

Alexandersson G, Rigas K (2012) Rail Liberalisation in Sweden. Policy development in a European context. Res Transport Business and Management 6:88–98. ISSN 2210–5395

Amos P (2005) Reform, Commercialization and Private Sector Participation in Railways in Eastern Europe and Central Asia: Transport paper series. Washington, DC.: The World Bank. No. 4 (Online) Available <https://openknowledge.worldbank.org/bitstream/handle/10986/17406/33906a10tp141eca1rail1reform.pdf?sequence=1&isAllowed=y>. Accessed June 2020

Annual Report 2011 of the Czech Railways Group (2012) (Online) Available http://www.ceskedrahy.cz/assets/skupina-cd/fakta-a-cisla/vyrocni-zpravy/vyrocni-zprava-skupiny-cd-2011.pdf Accessed February 2020

Answers to questions Bratislava—Plavecké Podhradie (2020) (Online) Available https://www.uvo.gov.sk/vyhladavanie-dokumentov/detail/3081929 Accessed January 2020

Avenali A, Boitani A, Catalano G, Matteucci G, Monticini A (2020) Standard costs of regional public rail passenger transport: evidence from Italy. Appl Econ 52(15):1704–1717

Bergantino AS, Madio L (2020) Intermodal competition and substitution. HSR versus air transport: Understanding the socio-economic determinants of modal choice. Res Transport Econom. 79:SI, Article Number UNSP 100823

Bergantino AS (2017) Open access passenger rail competition: the case of Italy. Italy, University of Bari (Online) Available <https://www.autorite-transports.fr/wp-content/uploads/2017/07/Session-1_Pre%CC%81sentation_Angela-Stefania-Bergantino.pdf>

Beria P, Bertolin A (2019) Evolving long-distance passenger services. market concentration, fares and specialisation patterns in Italy. Res Transp Econ 74:77–92

Binding text of the draft Treaty—Plzeň (2020) (Online) Available https://ezak.cnpk.cz/document_download_67527.html Accessed April 2020

Bougna E, Crozet Y (2016) Towards a liberalised European rail transport: analysing and modelling the impact of competition on productive efficiency. Res Transp Econ 59:358–367

Bundesnetzagentur Jahresbericht (2018) (Online) Available https://www.bundesnetzagentur.de/SharedDocs/Downloads/DE/Allgemeines/Bundesnetzagentur/Publikationen/Berichte/2019/JB2018.pdf;jsessionid=F4CA7DC2E9401F22EF1BF1254F2A08B1?__blob=publicationFile&v=6 Accessed June 2020

Cadarso L, Vaze V, Barnhart C, Marín Á (2017) Integrated airline scheduling: considering competition effects and the entry of the high speed rail. Transp Sci 51(1):132–154

Capozza C (2016) The effect of rail travel time on airline fares: first evidence from the Italian passenger market. Econ Transp 6:18–24

Chúpek M (2015) Strengthening rail passenger transport in line with the public interest. Proceedings of the professional seminar—logistics monitor (Legislative changes in public rail passenger transport 2015) (Online) Available https://www.logistickymonitor.sk/images/prispevky/zbornik-seminar-2015.pdf Accessed April 2020

Council Directive of 29 July 1991 on the development of the Community's railways (91/440/EEC) (1991) (Online) Available https://www.rail.sk/arp/slovakia/leg/1991-440.htm Accessed January 2020

D'Alfonso T, Jiang CM, Bracaglia V (2016) Air transport and high-speed rail competition: environ implications and mitigation strategies. Transport Res Part A—Policy and Practise 82:261–276

Desmaris Ch (2016) High speed rail competition in italy—a major railway reform with a "Win-win Game?" Discusion Paper [Online]. Paris: International Transport Forum. No. 2016/11 (Online) Available https://www.oecd-ilibrary.org/docserver/6da6e5ca-en.pdf?expires=1589894440&id=id&accname=guest&checksum=4C684B0F55A1035BA9AE65D6DF548C68 Accessed May 2020

Desmaris Ch, Croccolo F (2018) The HSR competition in Italy: How are the regulatory design and practices concerned? Res Transport Econ 69:290–299

Deville, X, Verduyn F (2012) Implementation of EU legislation on rail liberalisation in Belgium, France, Germany and The Netherlands: working paper document [online]. Brusel, National Bank of Belgium. No. 221 (Online) Available https://www.nbb.be/doc/ts/publications/wp/wp221en.pdf Accessed February 2020

Dias J, Graca T (2016) The Europeans' expectations of competition effects in passenger rail transport: a cross-national multilevel analysis. Soc Indic Res 129(3):1383–1399

Directive 2012/34/EU of the European Parliament and of the Council of 21 November 2012 establishing a single European railway area (Online) Available https://eur-lex.europa.eu/legal-content/EN/ALL/?uri=CELEX:32012L0034 Accessed January 2020

European Commision Notice (2013). (Online) Available https://ec.europa.eu/commission/presscorner/detail/en/MEMO_13_45 Accessed April 2020

Explanation of tender documentation trains—Pilsen region (2020) (Online) Available https://ezak.cnpk.cz/document_download_68444.html Accessed April 2020

Explanation of tender documents Bratislava—Komárno 2 (2020) (Online) Available https://www.mindop.sk/ministerstvo-1/doprava-3/zeleznicna-doprava/liberalizacia-osobnej-zeleznicnej-dopravy/bratislava-komarno Accessed April 2020

Explanation of tender documents Bratislava–Komárno (2020) (Online) Available https://www.mindop.sk/ministerstvo-1/doprava-3/zeleznicna-doprava/liberalizacia-osobnej-zeleznicnej-dopravy/bratislava-komarno Accessed April 2020

Fearnley N, Currie G, Flügel S, Gregersen FA, Killi M, Toner J, Wardman M (2018) Competition and substitution between public transport mode. Res Transp Econ 69:51–58

Fraszczyk A, Lamb T, Marinov M (2016) Are railways really that bad? an evaluation of rail systems performance in Europe with a focus on passenger rail. Transport Res Part A: Policy and Practice 94:573–591

Hensher DA, Stanley J (2003) Performance-based quality contracts in bus service provision. Transport Res Part A: Policy and Practise 37(6):519–538

History of railways (2007) (Online) Available https://www.zsr.sk/o-nas/historia-zeleznic/od-roku-1993/od-roku-1993.html Accessed February 2020

Humic R, Abramovic B (2019) Criteria for the quality of services of public interest organized by train operators. In: 13th International scientific conference on sustainable, modern and safe transport (TRANSCOM 2019), High Tatras, Novy Smokovec—Grand Hotel Bellevue, Slovak Republic, May 29–31, 201, pp 259–264

Independent Regulators´Group—Rail (IRG-Rail) (2020) Eighth annual market monitoring working document (Online) Available https://www.irgrail.eu/irg/documents/market-monitoring/260,2020.html Accessed April 2020

Invitation notice of the announcement of the public tender Bratislava——Plavecké Podhradie (2020) (Online) Available https://ted.europa.eu/udl?uri=TED:NOTICE:70062-2020:TEXT:SK:HTML&src=0 Accessed March 2020

Invitation notice of the tender of Košice—Moldava nad Bodvou (2018) (Online) Available https://www.mindop.sk/ministerstvo-1/doprava-3/zeleznicna-doprava/liberalizacia-osobnej-zeleznicnej-dopravy/kosice-moldava-nad-bodvou Accessed April 2020

Invitation notice of the tender Žilina—Rajec (2018) (Online) Available https://www.mindop.sk/ministerstvo-1/doprava-3/zeleznicna-doprava/liberalizacia-osobnej-zeleznicnej-dopravy/oznamenia-o-vyhlaseni-verejnej-sutaze. Accessed February 2020

Iossa E, Waterson M (2019) Maintaining competition in recurrent procurement contracts: a case study on the London bus market. Transp Policy 75:141–149

Jandová M, Paleta T (2019) Impact of on-track competition on public finances—the case of the Czech Republic. J Rail Transp Plann Manage 12(Article Number UNSP 100145)

Janečková B (2015) Experiences with passenger railway open access competition. Diploma thesis. (Online) Available https://is.muni.cz/th/f42vw/OPEN_ACCESS_Janeckova.pdf Accessed June 2020

Kavanagh P (2016) A case for negotiated performance-based contracting rather than competitive tendering in government public transport (bus) service procurement. Res Transp Econ 59:313–322

Król M, Taczanowski J, Jarecki S, Kołoś A (2019) Publicly-owned operators can also challenge incumbents. New cases of open-access passenger rail competition in Poland. J Rail Transp Plann Manag 12(Article Number UNSP 100150)

Latest posts Raaberbahn (2020) (Online) Available <https://www.raaberbahn.at/das-unternehmen/die-raaberbahn-ag/> Accessed February 2020

LEO Express to relaunch Locomore service (2017) (Online) Available https://www.railwaygazette.com/news/passenger/single-view/view/leo-express-to-relauch-locomore-service.html Accessed February 2020

Liberalization of passenger rail transport (2020) (Online) Available https://www.mindop.sk/ministerstvo-1/doprava-3/zeleznicna-doprava/liberalizacia-osobnej-zeleznicnej-dopravy Accessed March 2020

Nash Ch, Smith A (2020) Public transport procurement in Britain. Res Transp Econ 81(SI, Article Number UNSP 100847)

Nash Ch, Smith A, Crozet Y, Link H, Nilsson JE (2019) How to liberalise rail passenger services? Lessons from European Exper Transp Policy 79:11–20

Nedeliakova E, Harmanova D, Panak M, Kudlac S (2016) Selected problems of risk management in the company of railway infrastructure manager. In: Transport means—proceedings of the international conference. October 05–07, Juodkrante, Lithuania. pp 877–880

News ZSSK (2020) Free Transport (Online) Available https://www.zssk.sk/bezplatna-preprava/ Accessed March 2020

News ZSSK (2020) Free transport for children, pupils and students will be resumed (Online) Available https://www.zssk.sk/aktuality/bezplatna-preprava-pre-deti-ziakov-a-studentov-bude-obnovena/ Accessed June 2020

Notice of invitation to tender Žilina—Rajec (2020) (Online) Available https://www.mindop.sk/ministerstvo-1/doprava-3/zeleznicna-doprava/liberalizacia-osobnej-zeleznicnej-dopravy/oznamenia-o-vyhlaseni-verejnej-sutaze Accessed March 2020

Notice of the announcement of the public tender Bratislava—Komárno (2019) (Online) Available https://www.mindop.sk/ministerstvo-1/doprava-3/zeleznicna-doprava/liberalizacia-osobnej-zeleznicnej-dopravy/bratislava-komarno Accessed March 2020

Notice of the result of the evaluation of offers Žilina—Rajec (2020) (Online) Available https://www.mindop.sk/ministerstvo-1/doprava-3/zeleznicna-doprava/liberalizacia-osobnej-zeleznicnej-dopravy/oznamenia-o-vyhlaseni-verejnej-sutaze Accessed February 2020

OECD (2013) Recent Developments in Rail Transportation Services (Online) Available http://www.oecd.org/daf/competition/Rail-transportation-Services-2013.pdf Accessed January 2020

Papaioannou P, Georgiadis G, Nikolaidou A, Politis I (2020) Public transport tendering and contracting arrangements in countries under regulatory transition: the case of Cyprus. Research in Transportation Economics. In press, corrected proof

Peceny L, Gasparik J, Gaborova, V (2016) Development of the public tender of paths ordering in long-distance rail transport in the slovak republic. In: 3rd International conference on traffic and transport engineering (ICTTE), Assoc Italiana Ingn Traffico Trasporti Res Ctr, Belgrade, SERBIA, November 24–25, pp 462–466

Pedro MJG, Rosário M (2016) A review of general practice in contracting public transport services and transfer to BRT systems. Res Transp Econ 59:94–106

Pribula D (2020) Analysis of the conditions of tender for the provision of transport services in the framework of public services. Bachelor thesis. University of Žilina. (Online) Available https://opac.crzp.sk/?fn=detailBiblioForm&sid=2DE4D0DBB221E90D1D8BFABEC259&seo=CRZP-detail-kniha

Poliak M, Poliaková A, Mrníková M, Šimurková P, Jaśkiewicz M, Jurecki R (2018) The competitiveness of public transport. J Competitiveness 9(3):81–97

Public service contract, Regiojet (2010) (Online) Available https://www.mindop.sk/ministerstvo-1/doprava-3/zeleznicna-doprava/zmluvy-o-dopravnych-sluzbach-vo-verejnom-zaujme-pri-prevadzkovani-osobnej-dopravy-na-drahe/zmluva-o-dopravnych-sluzbach-vo-verejnom-zaujme-pri-prevadzkovani-osobnej-dopravy-na-zeleznicnej-trati-bratislava-dunajska-streda-komarno Accessed March 2020

Public service contract, ZSSK, a.s. (2010) (Online) Available https://www.mindop.sk/ministerstvo-1/doprava-3/zeleznicna-doprava/zmluvy-o-dopravnych-sluzbach-vo-verejnom-zaujme-pri-prevadzkovani-osobnej-dopravy-na-drahe/zmluva-o-dopravnych-sluzbach-vo-verejnom-zaujme Accessed March 2020

Raturi V, Verma A (2020) A game-theoretic approach to analyse inter-modal competition between high-speed rail and airlines in the Indian context. Transp Plan Technol 43(1):20–47

Regulation (EC) No 1370/2007 of the European Parliament and of the Council of 23 October 2007 on public passenger transport services by rail and by road and repealing Council Regulations (EEC) Nos 1191/69 and 1107/70 (2007) (Online) Available https://eur-lex.europa.eu/legal-content/SK/ALL/?uri=CELEX%3A32007R1370 Accessed May 2020

Regulation (EC) No 1371/2007 of the European Parliament and of the Council of 23 October 2007 on rail passengers' rights and obligations (2007) (Online) Available https://eur-lex.europa.eu/legal-content/SK/ALL/?uri=CELEX:32007R1371 Accessed June 2020

Regulation (EU) 2016/2338 of the European Parliament and of the Council of 14 December 2016 amending Regulation (EC) No 1370/2007 concerning the opening of the market for domestic passenger transport services by rail (Text with EEA relevance) (2016) (Online) Available https://eur-lex.europa.eu/legal-content/SK/TXT/?uri=CELEX%3A32016R2338 Accessed May 2020
Regulation of ÚRŽD No. 1/2010 on the requisites of the proposal for the determination of fares and on the procedure for the regulation of fares in railway transport (2010) (Online) Available http://drahy.nsat.sk/regulacia-na-zeleznicnych-drahach/osobna-doprava-regulacia-cestovneho/vydane-vynosy/ Accessed January 2020
Rojo M, dell'Olio L, Gonzalo-Orden H, Ibeas Á (2015) Inclusion of quality criteria in public bus service contracts in metropolitan areas. Transp Policy 42:52–63
Strategic plan for the development of transport in the Slovak Republic until 2030 (2016) (Online) Available file:///C:/Users/lenka/Downloads/Strategicky_plan_2030.pdf Accessed June 2020
Strategy for the development of public passenger and non-motorized transport in the Slovak Republic until 2020 (2013) (Online) Available file:///C:/Users/lenka/Downloads/strategia_vond_2013.pdf Accessed May 2020
Supplement No. V to the Contract on Transport Services in the Public Interest, ZSSK, a.s. (2010) (Online) Available file:///C:/Users/lenka/Downloads/Pr5.pdf Accessed June 2020
Supplement to the Official Journal of the EU (Online) Available https://ted.europa.eu/TED/browse/browseByBO.do Accessed April 2020
Sůra J (2017) RegioJet for the first time in Vienna. They just came to lick the cream, says the head of the ÖBB (Online) Available https://zdopravy.cz/regiojet-poprve-ve-vidni-prisli-jen-sliznout-smetanu-rika-sef-obb-5030/ Accessed February 2020
Sůra J (2019) One month to a large rotation of carriers. Let's go, we have people, they promise (Online) Available https://zdopravy.cz/mesic-do-velkeho-stridani-dopravcu-vyjedeme-lidi-mame-slibuji37283/?fbclid=IwAR2PEKY47fkH9oNpl_PXZZ0EqmrTKp76L1Li3l_gNxMQZwdqKR7UfREvUCo Accessed February 2020
Tender documents Bratislava—Komarno (2019) (Online) Available https://www.mindop.sk/ministerstvo-1/doprava-3/zeleznicna-doprava/liberalizacia-osobnej-zeleznicnej-dopravy/bratislava-komarno Accessed March 2020
Tender documents Bratislava—Swimming Podhradie (2020) (Online) Available https://www.uvo.gov.sk/vyhladavanie-dokumentov/detail/3079366 Accessed April 2020
The 4th railway package (2016) (Online) Available https://ec.europa.eu/transport/modes/rail/packages/2013_sk Accessed February 2020
Tomeš Z (2013) Open access in passenger railway transport—review of foreign experiences and model results. Conf Regul Unregul Compet Rails, Telc 2013:192–220
Tomeš Z (2017) Do European reforms increase modal shares of railways? Transp Policy 60:143–151
Tomeš Z, Jandová M (2018) Open access passenger rail services in central Europe. Res Transp Econ 72:74–81
Transport and tourism policy (rail transport) (2020) (Online) Available https://www.europarl.europa.eu/ftu/pdf/sk/FTU_3.4.5.pdf Accessed June 2020
Tsunoda Y (2018) Transportation policy for high-speed rail competing with airlines. Transp Res Part A: Policy and Practice 116:350–360
Valkama P, Kankaanpää J, Anttiroiko A-V (2018) Financial and structural impacts of quasi-marketization of the Helsinki Metropolitan Area's bus services. Case Stud Transp Policy. 6:246–256
Vigren A (2017) Competition in Swedish passenger railway: entry in an open access market and its effect on prices. Econ Transport 11–12:49–59. ISSN 2212–0122
Wang J, Huang J, Jing Y (2020) Competition between high-speed trains and air travel in China: from a spatial to spatiotemporal perspective. Transp Res Part A: Policy and Practice 133:62–78
Wegelin P, von Arx W (2016) The impact of alternative governance forms of regional public rail transport on transaction costs. case evidence from Germany and Switzerland. Res Transp Econ 59:133–142

Záhumenská Z, Vojtek M, Gasparik J (2018) Position of railway passenger transport companies on current liberalized transport market. In: 10th international scientific conference horizons of railway transport, Strecno, Slovakia, 11 October 2018 through 12 October 2018, Code 14230
Zhang A, Wan Y, Yang H (2019) Impacts of high-speed rail on airlines, airports and regional economies: a survey of recent research. Transp Policy 81:A1–A19

Challenges of Dangerous Goods Transport by Rail in Thailand

Mayurachat Watcharejyothin, Koji Nishimura, and Marin Marinov

Abstract For the transport of dangerous goods (DGs), trucks are more commonly used in Thailand. However, transportation by truck is often prone to accidents and even environmental damages. Rail freight transport is an environmentally friendly mode of transport, which is considered safe and reliable, though it has shown the lowest market share amongst all the freight transport modes in Thailand. This is because of a lack of logistics facilities, infrastructure, locomotive units, supported regulations and efficient operational practice. This situation led to the increase of risks of possible damages on inland logistics in the country. This paper carries out a study of the challenges of DGs transport by rail in Thailand. A comparison of legislations for DGs transport by rail between Europe, Japan and Thailand is organised to provide appropriate guidelines for the development of regulations applicable to the situation in Thailand. Results of the study indicate that DGs transport by rail in Thailand requires regulations, controlled operational procedures, and training provisions covering direct and block train operations, reliable maintenance of trains and shipping. With proper guidelines in place, the market share of DGs transport by rail in the country would increase and thus promote the railway transport mode as the backbone of an environmentally friendly transportation logistics component in the overall development of Thailand.

Keywords Rail freight · Transportation · Logistics · Dangerous goods · Rules and regulations

M. Watcharejyothin (✉)
PTT Group Logistics Management Department, PTT Public Company Limited, Bangkok, Thailand
e-mail: mayurachat.w@pttplc.com

K. Nishimura
Japan Freight Railway Company (JRF), Tokyo, Japan

M. Marinov
Department of Engineering Systems and Supply Chain Management (ESSCM), Aston University, Aston Triangle, Birmingham, UK

M. Marinov and J. Piip (eds.), *Sustainable Rail Transport 4*, Lecture Notes in Mobility,
https://doi.org/10.1007/978-3-030-82095-4_13

1 Introduction

Rail freight transport is a safe, reliable and environmentally friendly mode for transporting commodities and different types of goods to considerable distances. Remote railway routes coupled with reliable timetables and schedules provide a certain degree of safe transport of Dangerous Goods (DGs) like flammables, explosives, and toxic materials. In 2017, the market share of the rail freight transport was 17.3% of the total inland freight transport in the European Union (Eurostat 2017). That number is relatively high compared to Thailand, which is eight times less.

Overall, the total domestic freight transport of goods in Thailand stood at 616 million tons in 2018. Of this, rail transport accounted for only 1.7% of the total freight transport market in the country, followed by waterways at 19.7% and road at 78.6% (NESDB 2019) as shown in Fig. 1. According to (WHO 2017) Thailand is ranked as the world's second-highest rate of road fatalities per capita. This is because of the intensive use of the roads and highways to transport people and goods.

The State Railway of Thailand (SRT) is the only rail operator in Thailand. It is a government sector under the Ministry of Transport with over 100 years of history. However, both passenger and freight transport by rail still have low utilisation rates in Thailand.

Currently, the major commodities transported by rail in Thailand are the general goods in containers, fuel (liquid petroleum gas and crude oil), petroleum products and cement (see Table 1). Most containers are transported between the Laem Chabang port and the Lat Krabang Inland Container Depot (ICD), the latter is located near the Suvarnabhumi Airport. Fuel is transported to the inland area from the coastal area by a special freight wagon. General goods transported in containers have the largest share of total rail freight transport in Thailand accounting for 45%, followed by cement products (19%), and petroleum products (18%). The 118 km route from Ladkrabang Inland Container Depot (ICD) to Laem Cha Bang port is a major route accounting for 72% of total rail freight transport. There are no goods in ISO container tank transported by rail. Petroleum products contained in 26 × 30,000-L tanks fixed with bogies that are transported to/from the oil terminal to the North and the Northeast regions are the only case of DGs transported by rail in Thailand. For over a 500 km

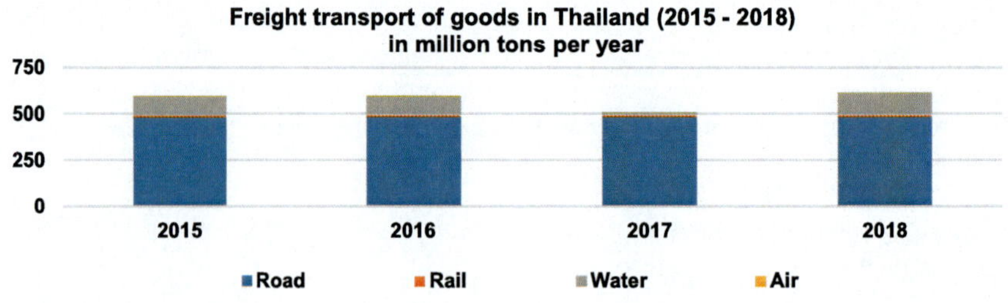

Fig. 1 Transport of goods in Thailand (2015–2018). *Source* NESDB (2019)

Table 1 Transport revenue by major commodities

Commodities fiscal year 2018	Ton carried Tons ('000)	%	Revenue Baht ('000)	%	Average Kms/Ton
Liquid petroleum gas	631	6.2	206,779	12.8	457
Crude oil	803	7.9	296,018	18.4	543
Petroleum products	549	5.4	181,079	11.2	491
Baggage cement	310	3.0	47,176	2.9	372
Bulk cement	1,510	14.8	189,042	11.7	267
Rice products	0	0.0	0	0.0	0
Gypsum	0	0.0	0	0.0	0
Container	0	0.0	0	0.0	0
East Line	3941	38.7	276,386	17.2	118
Land Bridge	360	3.5	52,193	3.2	245
Others	1,538	15.1	301,054	18.7	412
Military Effects	8	0.1	2,627	0.2	467
Miscellaneous	522	5.1	58,585	3.6	119
Total	**10,172**	100.0	**1,610,939**	**100.0**	**272**

Source SRT (2019)

distance, the transportation cost by rail is about 20% lower than by the transportation cost trucks in the country.

In many countries, coal is the main commodity transported by rail. However, because Thailand does not use a lot of coal, it is deemed uneconomic for it to be transported by rail. Additionally, the box wagon transportation, which is used in many countries around the world, was found inconvenient for Thailand compared to the use of trucks and as a result its use was abolished in the country.

Transportation of Dangerous Goods (TDGs)

Handling of DGs, when done improperly could have devastating impacts on living organisums, environment, social security, etc., therefore it requires strict rules and requirements in place. As it can be seen in Fig. 2, the transport of DGs by road (truck) in Thailand has the largest market share (of 90%), 43.8 million tons per year (GTZ 2002). The transportation by truck has been dominant in Thailand for the last 50 years. The growth of the TDGs by truck has been supported by the construction of extensive highways and road networks in the country, offering less investment cost for the road transporters and door-to-door services to customers. By comparison, rail freight transport has less developed infrastructure and support facilities. As a result, truck transport has a more competitive edge than that of the rail transport in Thailand.

In the transportation of general goods, cost is regarded as important. There are many shippers deciding on a transport mode only by a cost comparison. However, it is not appropriate to decide on the DGs transport mode based on a cost comparison alone. In transporting DGs, safety must be paramount. The damages are far

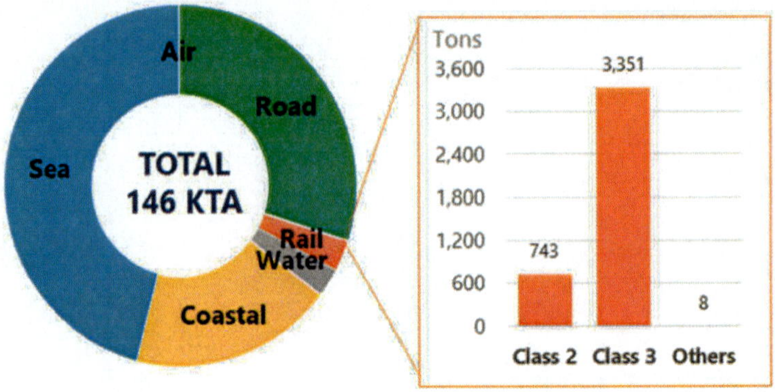

Fig. 2 TDGs by mode in Thailand in thousand tons per year. *Source* GTZ (2002)

more serious than that of general goods especially when an accident occurs because of a failure during the transportation of DGs. In accidents involving transport of DGs, fire and toxic material leakage often occur, thus putting people's lives at risks. Therefore, safety should be given priority than cost when choosing the DGs transport mode. Consideration of the risk control and the social responsibility of the shipping company should also be given. As rail transport is considered safer than the truck, therefore TDGs by rail should be chosen and promoted.

The Thailand government aims to increase transport modal shift from road to rail, as it views the shift to rail as more energy efficient and environmentally friendly. A mega project of 400,000 million baht to double rail lines and build a 3000 km track extension has been carried out. As it can be seen in Fig. 3, it is envisaged for the railway network to become the backbone transportation network of the country in the next 20 years (OTP 2018). Due to doubling the railway lines, an increased share of passenger rail transport is also targeted. The market share target of goods transport by rail is projected to reach 10% of the total goods transport in next five years in the country. To achieve these targets, there are three critical aspects that rail freight in the country should strengthen; (i) Loading and unloading facilities, (ii) Readiness and sufficient number of movable carriers e.g., locomotives, bogies and wagon containers and (iii) Appropriate regulatory framework and operation practice to support multimodal and intermodal freight transport.

A regulation of inland TDGs in Thailand under the Royal Thai Government Gazette B.E. 2545 and DG Act B.E. 2535 is in place. The acts govern the TDGs by truck and any other carriers but it excludes rail transport (Royal Thai Government Gazette, 2545). TDGs by rail in Thailand therefore is still limited though it is governed by the State Railway of Thailand Act B.E. 2494 and the Rail and Highway Management Act B.E. 2464. The Asian Development Bank (ADB) also suggested that the SRT or a Thai regulatory agency should first establish rail safety and environmental compliance guidelines for the TDGs by rail (ADB 2013). The Railway Department (DRT), a national body of rail regulation compliance and legislation, was

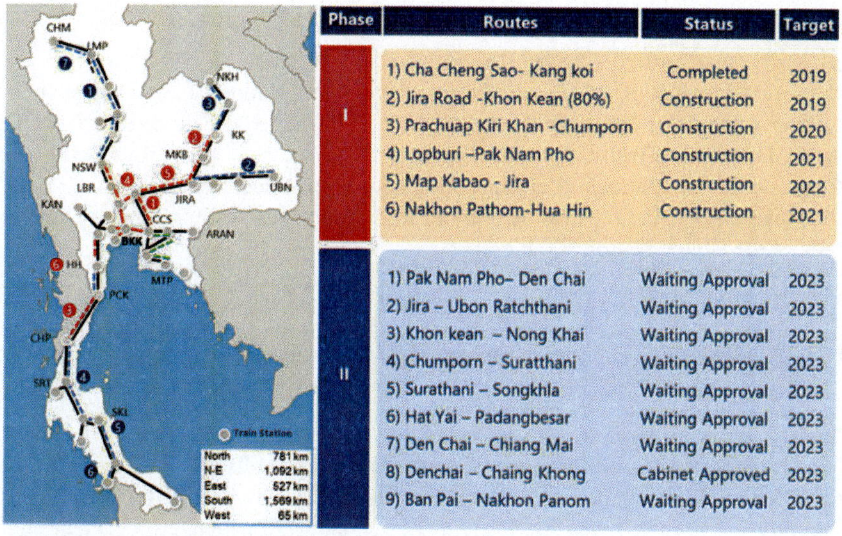

Fig. 3 Double track railways project in Thailand. *Source* SRT (2019)

established in 2019 with a main mandate to formulate national rail safety standards and regulations.

TDGs by rail is common in many developed countries. A good example of transporting chemicals in ISO tanks is the continental container transport along the Rhine River. There are several studies proposing improvement of safety in TDGs by rail. Studies of risk assessment tools of TDGs by rail are used to evaluate factors of risk, and prioritization of risk mitigation methodologies. Gheorghe et al. (2005) proposed Master Logical Diagram and developed a software platform to determine a loss of containment frequency. Macciotta et al. (2018) proposed a hazard-ranking tool for DGs transport in the Canadian railway corridors. Khanmohamadi (2018) evaluated route selection and less critical areas of Chlorine transport along railway route along Texas-Illinois corridor in order to avoid terrorist attack occurrence. Huang et al. (2020) and Huang (2019) analysed relationships between risk factors and accidents for DGs transport by railway. A study of Nowacki et al. presented realization of risk management system to improve safety of TDGs by any modes of transport in Poland and the EU.

Meanwhile, in Thailand, a study of the challenges of TDGs by rail in the country was carried out which proposed a regulatory framework followed by its proper implementation. Reviewing of related legislations of TDGs by rail in Thailand with other jurisdictions at international level would provide appropriate guidelines and safety control procedures suitable for the situation in Thailand. This would extend up to date legislations in compliance with current transportation related laws.

2 Regulations for TDGs Around the World

International regulations of DGs are issued by the United Nations (UN). They include the International Air Transport Association (IATA), Dangerous Goods Regulations, the ICAO Technical Instructions, the International Carriage of Dangerous Goods by Road, and the International Maritime Dangerous Goods (IMDG) Code. The IMDG Code divides DGs in nine classes. Further information can be found in the Orange Book of the United Nations (Table 2).

2.1 *Regulations for TDGs by Rail in Europe*

The TDGs by rail in Europe is governed by the Appendix C of the Convention Concerning International Carriage by Rail (COTIF). Appendix C covers the uniform regulations concerning the International Carriage of DGs by Rail (RID) established since 1999, with forty-seven states having been contracted with it. RID launches its revision every two years. It has seven parts as shown in Fig. 4. RID classifies DGs under IMDG code.

2.2 *Regulations for TDGs by Rail in Japan*

TDGs by rail in Japan is governed by national laws, technical and working standards. DGs control is governed by four national Acts; (i) the Explosive Control Act, (ii) the High-Pressure Gas Safety Act, (iii) the Fire Service Act, and (iv) Poisonous and Deleterious Substances Control Act. Applying a similar concept with IMDG code, the Fire Service Act classifies DGs into six classes as shown in Table 3.

Table 2 Classification of DGs by IMDG code

Class 1	Explosives
Class 2	Gases
Class 3	Flammable liquids
Class 4	Solid flammables, spontaneous combustion, substances and dangerous when wet
Class 5	Oxidising substances and organic peroxides
Class 6	Toxic substances
Class 7	Radioactive materials
Class 8	Corrosive substances
Class 9	Miscellaneous dangerous substances and articles

Source IMDG Code

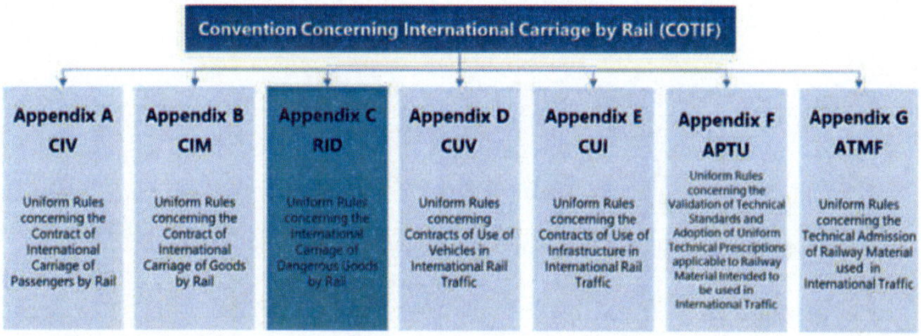

Fig. 4 Convention concerning international carriage by rail. *Source* COTIF (2020) and Galieriková et al. (2018)

Table 3 Classification of DGs by the fire service act

Class 1	Oxidising solids	Class 4	Inflammable liquids
Class 2	Combustible Solids	Class 5	Self-reactive substances
Class 3	Spontaneous combustible/ water-reactive substances	Class 6	Oxidising liquids

However, the Japanese Acts related to DGs including the Fire Service Act do not apply for DGs carried by rail. The TDGs by rail is governed by the Railway Operation Act, a Ministerial ordinance for technical standard of railway. Regulation for railway transportation established by the Ministry of Land, Infrastructure and Transport (MLIT) and Conditions of Carriages established by the Japan Freight Railway Company (JRF) (Fig. 5). Laws (i)–(iv) are applied only when a container loaded with DGs is located in a railway container yard.

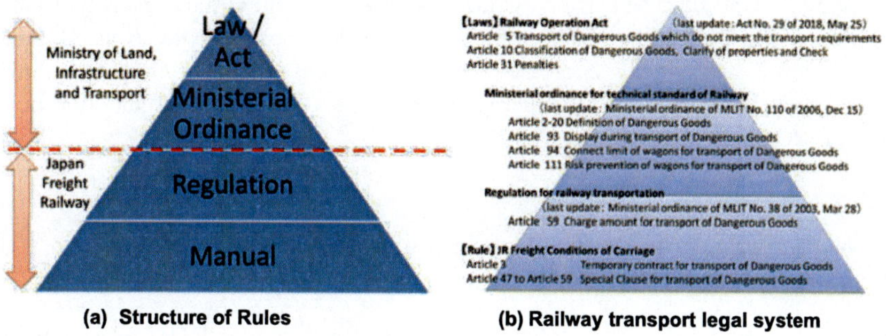

Fig. 5 Railway transport legal system in Japan. *Source* JRF (2019)

Fig. 6 Railway transport legal system in Thailand. *Source* DRT (2020)

2.3 *Regulations for TDGs in Thailand*

Legal framework for DGs logistics in Thailand is governed by Ministerial Acts and regulations by the government institutions e.g., Ministry of Interior, Ministry of Energy, Ministry of Transport and Ministry of Industry. These have played major role to control DGs and petroleum handling safety. TDGs in Thailand governed by the Thai Provisions Volume 1 (TP-1) referred to UN RTDG 11th Edition (2000) is in line with the UN Recommendations and IMDG Code. Similar to the Japanese case, Acts related to DGs do not apply for DGs carried by rail. The TDGs by rail falls under the Royal Thai Government Gazette B.E. 2545 and DGs Act B.E. 2535.

Excluding chemicals and other DGs substances, crude oil, gas oil, gasoline and LPG are the only commodities that SRT allows to be carried and transported by rail. Regulations for the transport of petroleum is governed by SRT rules and regulations as well as by the Regulations of the Ministry of Energy as shown in Fig. 6, while DGs are not allowed to be carried by rail. Under Ministerial Regulations B.E. 2558: petroleum is to be transported by tank, filled petroleum products in fixed oil tank welded, connected with bogies are delivered to petroleum depot at product unloading station. In addition, there is no any operating practice yet on product handling in a form of ISO container tank by rail in Thailand.

3 Proposals for TDGs by Rail in Thailand

There is an opportunity for TDGs by rail in Thailand because DGs and chemical substances. These products are contained in various forms: (i) Fixed tank, (ii) general container, and (iii) ISO container tank. A modern way of rail freight transport by a new form of container tank should be considered in order to promote modal shift

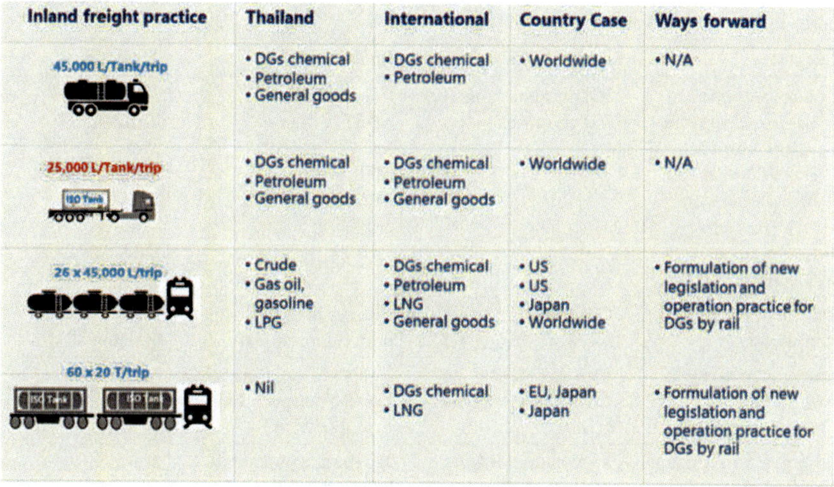

Fig. 7 Rail freight operation practice in Thailand in comparison to international practice

from road to rail in the country. However, formulation of new legislations, support facilities and best operation practices for TDGs by rail are needed (Fig. 7).

A comparison of the legislations of TDGs by rail amongst EU, Japan and Thailand cases includes three different aspects, namely; (i) General provision, (ii) Transport operation, and (3) Training needs. Similar to the case of Japan, the legislation and control of DGs transport by rail in Thailand deal with various government authorities. For example, the Railway Police Division is responsible for emergency response. In addition, research centres of railways operation in Japan and Europe contribute rail freight innovation and technology to provide risk protection and to help mitigate accidents (as shown in Table 4) while Thailand has not yet assigned any particular government agency for this part.

For Thailand, the responsibilities between the railway operator and the infrastructure manager should be clearly defined. Just like the JRF of Japan and railway transporters in Europe, railway transporters of DGs in Thailand should prepare vehicles well while the shipper should prepare loading tanks, marks and documents according to DGs control regulations. The infrastructure manager should be mandated to prepare emergency plan and access information e.g., UN number and position of DGs wagon during carriage. Table 5 lists down more details for the proposed rail transportation regulations for DGs in Thailand.

Table 6, enumerates proposals for Thailand to conduct periodical technical assessment of TDGs by a third party, which is needed to ensure that updated knowledge and techniques of works are performed by the providers of TDGs. Technical standard and guideline by individual DGs categories are set and for the staff to refresh their skills. In addition, an appointed safety advisor is set to be responsible for consigning carriage and packing of DGs just like in the case of EU.

Table 4 General provisions

General	RID	Japan	Thailand
Control authorities – Explosive substance – Petroleum – Hazardous, toxic, DGs substances – DG Transport – Environment – Regulator	RID is applicable for EU countries	Cabinet Office MLIT METI MHLW MIC MHLW	Authorities MoD MoE MoI MoT MoEnv DRT
Emergency organization to respond	Control Office of Network Rail (UK)	Fire services department police department	Railway police division
Emergency research organization	Rail Accident Investigation Branch (UK)	Railway Department, Japan Transport Safety Board of MLIT	Nil
Emergency disclosure of information	Viewable in RAIBs website (UK)	Japan transport safety board of MLIT	Nil

Remark METI—Ministry of Economy, Trade and Industry, MHLW—Ministry of Health, Labour and Welfare, MIC—Ministry of Internal Affairs and Communications of Japan
MoD—Ministry of Defense, MoE—Ministry of Energy, MoI—Ministry of Industry, MoT—Ministry of Transport, MoEnv—Ministry of Environment, DLT—Department of Railway of Thailand

4 Actions and Incentives

The future of TDGs by rail in Thailand depends on foundations of formulation of legislation, transportation guideline, operation practice, training needs and support facilities as shown in Table 7. SRT, as a sole railway operator in Thailand, must engage government authorities (e.g., DRT, MoT, MoI, MoE), products owners and shippers to formulate new regulations, ensure proper conditions of DGs carriages, and safety standard of rail freight. The SRT staff, DGs transport operators or shippers and products handlers must have well technical knowledge and skills to properly and safely transport DGs. Given that there are a number of government authorities that has to be involved, each with its own set of challenges, the formulation of this new foundation and crafting specific DGs-related transportation legislations could take a few years.

Lastly, a real case scenario was evaluated in order to identify problems and obstacles on how to transport of LNG in a new form of LNG in an ISO container tank. The SRT has been transporting crude oil and petroleum products (LPG) in a form of fixed tank welded with bogies. A new form of LNG carrier is now under discussion with SRT and related parties. Currently, there are no any regulations for railway transportation and rail freight conditions of DGs carriages. In collaboration amongst SRT, JR Freight and PTT hosted by DRT; a draft procedure of rail freight conditions of LNG carriages was produced along with a draft of ministerial regulations for

Table 5 Transportation

RID	Japan	Proposed for Thailand case
Responsibility of the carrier		
(1) Carriers i. Rail transporter: railway transport ii. Shipper: load/ unload facilities, container, placards, marks, documents	(1) Carriers i. JRF:rolling stocks, load/unload from wagon ii. Shipper: load/unload facilities, tanks, placards, marks, documents	(1) Carriers i. SRT:rolling stocks, locomotive sets ii. Shipper: load/unload facilities from truck/ wagon, tanks, placards, marks, documents
(2) Check point i. the DGs to be carried are authorized for carriage under RID. ii. all information to provide by the consignor before carriage iii. visually the wagons and loads have no obvious defects, leakages, or cracks, mission equipment iv. Deadline for next wagons test shall not expired . Wagon are not overloaded v. Placards, marks and orange plates shall be affixed. vi. Provision of information	(2) Check point i. Cargo or container size, dimension/ weight ii. Loading and position of wagon iii. Container height and transport route iv. Loading//unloading station for large scale container	(2) Check point Operation concern: i. Cargo or container size, dimension/ weight. ii. Loading and position of wagon iii. Container height and transport route Safety concern: i. the DGs to be carried are authorized for carriage ii. all information to provide by the consignor before carriage iii. Wagons and loads are ready to use iv. Placards, marks and orange plates shall be affixed v.To prepare emergency communications plan for incident/accident area
(3) The carrier shall ensure that the manager of the railway infrastructure being used is able to obtain at any time during carriage rapid and unrestricted access to the information	(3) Special provision: DG Acts including Fire Service Act shall not be applied for Railway transport. The carriage of DGs is governed by; – Railway Operation Act – Ministerial ordinance for technical standard of Railway – Regulation for railway transportation – JR Freight Conditions of Carriages	The carriage of DGs is governed by – State Railway of Thailand Act B.E. 2494 – Rail and Highway Management Act B.E. 2464 – Royal Thai Government Gazette B.E. 2545–DGs Act B.E. 2535 – Ministerial Regulations B.E. 2558: Petroleum transport by tank – Modifications of SRTs declarations and rail freight book for DGs
Responsibility of the Infrastructure manager		

(continued)

Table 5 (continued)

RID	Japan	Proposed for Thailand case
Responsibility of the carrier		
i. To prepare internal emergency plans for marshalling yards. ii. To prepare rapid and unrestricted access of information (UN number, position of each wagon on the train during carriage)	i. To prepare rapid and unrestricted access of information by IT system (Coping procedure at the accident occurrence, UN number, position of each wagon on the train during carriage)	i. To prepare emergency plan container yard. ii. To prepare rapid and unrestricted access of information (Coping procedure at the accident occurrence, UN number, position of each wagon on the train during carriage)

Source RID (2017) and JR Freight (2019)

transport of natural gas container tank. This is in line with the agenda for Thailand to serve as the LNG hub of the ASEAN Economic Community (AEC).

5 Summary

This paper carried out a study of the TDGs by rail in Thailand. Proposals for regulatory framework, safety and operation standards are presented using Japanese and European practices. It is found that SRT as a sole rail freight carrier in Thailand needs to engage with key stakeholders to established regulations and create conditions for the TDGs in the country if it is to create opportunities for the 'TDGs by rail' business in the near future.

Table 6 Training requirements for carriage of DGs

RID	Japan	Proposed for Thailand case
Groups of personnel for the individual categories • Operations personnel involved directly in the transport of DGs (drivers, marshalling staff) • Personnel responsible for technical control of wagons used for the transport of DGs (rolling stock technician). • Personnel responsible for guiding and controlling rail and marshalling services and management personnel of the railway infrastructure (manager controllers, signalers, control center personnel)	*Ordinance for technical standard of Railway* • Railway business operator shall set training on technical and knowledge for train operation and rolling stocks and facilities maintenance employees. • Railway business operator shall NOT directly engage with employees on the train operation work, unless to confirm their knowledge and techniques required for that work or to confirm their lack of knowledge and techniques required for that work	*Ordinance for technical standard of Railway* • SRT/Carriers/Shippers shall review emergency plan for petroleum transport by rail. • SRT/Carriers/Shippers shall set training on technical and knowledge for train operation and rolling stocks and facilities maintenance employees
Safety advisor	*Japan*	*Proposed for Thailand case*
• Activities of which include consigning or carriage of DGs by rail, or related packing, loading, filing or unloading shall appoint one or more safety advisers for the carriage of DGs responsible for helping to prevent risks • The adviser shall hold a vocational training certificate, valid for transport by rail. That certificate shall be issued by the competent authority or the body designed for that purpose by each RID contract state	It is not ordained by a law, but it is customary that JRF receives advice from the person in charge of the fire department when JRF makes a training program about the transport of DGs	• Activities of which include consigning or carriage of DGs by rail shall appoint one person to be a key person in charge of the carriage of DGs responsible • The person shall have sufficient training

Source RID (2017) and JR Freight (2019)

Table 7 Proposed key responsibility and authorities for the carriage of DGs by rail in Thailand

Issues	Proposed key responsibility and authorities
(1) Legislation of DGs transport by rail	
1.1 General Provisions (i) Regulation (ii) Standard and provision (iii) Emergency Research Organization	Regulation for railway transportation Rail freight conditions of DGs carriages Railway Department or Railway Safety Board
1.2 Transportation (i) Responsibility of the Carriers (iv) Products owners (v) Shippers (vi) Emergency plans and information management	SRT: rail tracks, locomotives, rolling stocks, bogies Product owner: load/unload facilities, container, tanks, placards, marks, Shippers:, container yard, tanks, placards, marks, documents Collaborations among SRT, product owner and shipper to prepare appropriate emergency plans and access of information/communication
1.3 Training needs (i) Rail freight operator (ii) Locomotive sets maintenance (iii) Rail freight scheduler	SRT: Training of DGs rail freight operation by SRT or third party SRT: Training of DGs railways and rolling stocks maintenance by SRT SRT and shippers: DGs operation manual and service guideline
(2) Support facilities 2.1 Container yard 2.2 Lift-on and lift off equipment	LSPs: Training of DGs products storage and safety control LSPs: Training of DGs products handling and safety control

Acknowledgements The authors would like to thank Mr. Takun Indarachome, Chief Container Transport Bureau, SRT and Dr. Somnuk Ngamchai, Vice President of PTT Group Logistics Management Department, PTT Plc. for their strong support and assistances on suggestions for an initiative of DGs transport by rail and a trial test of LNG in ISO container tank between SRT, JRF and PTT Plc.

References

Asian Development Bank (2013) Supporting railway sector reform final report. Thailand

Eurostat (2017). https://ec.europa.eu/eurostat/statisticsexplained/index.php?title=Freight_transport_statistics&oldid=444926#Modal_split

Galieriková A, Sosedová J, Dávid A, Bariak M (2018) Transport of dangerous goods by rail, Horizons of railway transport 2018. MATEC Web of Conf 235(2):00004. https://doi.org/10.1051/matecconf/201823500004

German Technical Cooperation Agency (2002) Dangerous goods transport statistics in Thailand

Gheorghe A, Birchmeier J, Vamanu DV, Papazoglou IA, Wolfgang K (2005) Comprehensive risk assessment for rail transportation of dangerous goods: a validated platform for decision support June 2005. Reliabil Eng Syst Safety 88(3):247–272https://doi.org/10.1016/j.ress.2004.07.017

Huang W, Zhang Y, Kou X, Yin D, Mi R, Li L (2020) Railway dangerous goods transportation system risk analysis: an interpretive structural modeling and bayesian network combining approach. Reliabil Eng Syst Safety. https://doi.org/10.1016/j.ress.2020.107220

Huang W, Shuai B, Zuo B, Yifei Xu, Antwi E (2019) A systematic railway dangerous goods transportation system risk analysis approach: the 24 model. J Loss Prev Process Ind 61:94–103

International Maritime Dangerous Goods Code (IMDG Code)

International Maritime Organisation (2016)

Intergovernmental Organisation for International Carriage by Rail (2017) Regulation concerning the international carriage of dangerous goods by rail (RID)

Japan Freight Railway (1987) "Conditions of carriage"

Japan Freight Railway (2012) "Guidance book for railway dangerous goods transport"

Khanmohamadi M, Bagheri M, Khademi N, Farid Ghannadpour S (2018) A security vulnerability analysis model for dangerous goods transportation by rail–Case study: Chlorine transportation in Texas-Illinois. Safety Science 110:230–241

Macciotta R, Robitaille S, Hendry M, Derek Martin C (2018) Hazard ranking for railway transport of dangerous goods in Canada. Case Stud Transp Policy 6(1): 43–50. https://doi.org/10.1016/j.cstp.2017.11.006

Ministry of Energy Thailand (2020) A draft of ministerial regulations: transport of natural gas container tank

Nijole B (2020) Improving safety of transportation of dangerous goods by railway transport. Infrastructures 5(7):54. https://doi.org/10.3390/infrastructures5070054

Office of the National Economic and Social Development Council (2019) Thailand Logistics Report 2018, ISSN 1906–4373 September 2019

Office of Transport Planning (2018) OTP Annual Report

World Health Organization (2017) WHO accident rate

Zbigniew ŁUKASIK, Aldona KUŚMIŃSKA-FIJAŁKOWSKA, Jacek KOZYRA (2017) Transport of dangerous goods by road from a European aspect June 2017. Scientif J Silesian Univer Technol Ser Transp 95:109−119.https://doi.org/10.20858/sjsutst.2017.95.11

The Hyperloop Concept Development, Possible Applications and Critical Analysis

Frank Michelberger, Matthias Plavec, Kevin Pyrek, and Philipp Graf

Abstract The Hyperloop was first introduced in 2013 and is a proposed mode of passenger and/or freight transportation. The system consists of a sealed tube through which a pod may travel nearly free of air resistance. Although there are no doubts concerning the technological feasibility, there is criticism regarding the concrete implementation of the Hyperloop. Critics especially question the small system capacity and estimate that the actual costs would be substantially higher than the initial cost assumptions. Numerous companies are currently working on implementing the Hyperloop concept and different Hyperloop-routes are being discussed. An implementation of the Hyperloop only makes sense as a pure 'point to point' connection. A possible first Hyperloop route would provide important insights into the actual construction and operating costs. The paper subsequently focuses on the Hyperloop concept and analyses possible applications of this new transport system.

1 Introduction

This paper is an analysis of the current state of the Hyperloop technology, containing studies for eventual development projects, scheduled for either a medium or a long term. This requires considerable research work, mainly basing on Internet sources. All the information reflects the current technological status for 2019. Literature sources for the Hyperloop are not particularly advanced yet because the subject is still being further developed by numerous different research project teams and companies currently working on this technology. Thus, the knowledge is not yet as clearly defined as is in the case of the traditional railway system. Another source of information is through the contact with various companies that are actively working on a Hyperloop solution at the moment.

F. Michelberger (✉)
St.Poelten University of Applied Sciences, Institute for Integrated Mobility Research, Matthias Corvinus-Straße 15, 3100 St. Pölten, Austria
e-mail: frank.michelberger@fhstp.ac.at

M. Plavec · K. Pyrek · P. Graf
Institute for Integrated Mobility Research, St. Pölten, Austria

M. Marinov and J. Piip (eds.), *Sustainable Rail Transport 4*, Lecture Notes in Mobility,
https://doi.org/10.1007/978-3-030-82095-4_14

The concept of Hyperloop was introduced in 2013 by Elon Musk who described it as "Maglev train". After that, some companies showed an interest in the project and began development studies and works on this technology in collaboration with research institutions and universities.

Various countries, cities and regions also demonstrated an interest in Hyperloop, promoting feasibility studies. Despite a certain amount of interest amongst the public, the project was faced with critical opinions. This mainly relates to questions concerning the technological implementation and the lack of clear plans and statements about the financing and efficiency of the Hyperloop projects.

The goal of the new means of transport remains unclear. Also, the place it will take in the general transport system and in spatial planning, in comparison to other means such as the train and the airplane, is yet to be determined. Moreover, technical and financial aspects are still not fixed. The current objective is to define the technical requirements of the Hyperloop and the positive effects on the population.

This paper proposes a solution for the development of the Hyperloop concept and analyses the concepts of the companies active in this domain. The first step is to define the requirements for reaching the goals of the project, and to carry out a feasibility study. After defining the technical implementation, the next step is to plan the operational process and try to define the economic viability of the whole project.

To define how the Hyperloop concept can be developed, we need to provide the implementation of this new transport system. It is important to reach a definition by answering the following questions: What are the advantages of this technology? Which new opportunities does it offer? How does it impact the current transport system? What are the possible applications, and where are the limits? Which problems could it cause? And last but not least, it must be analysed why this system has not been an attractive alternative so far? A constant comparison to the current traditional train system is at the heart of the analysis.

For this purpose, we focus on the general system properties of the Hyperloop with a view to developing a high-speed line. A particular focus is put on the Californian line between Los Angeles and San Francisco. In addition, a comparison between the existing railway lines and the eventual Hyperloop connection is drawn for:

- **Vienna—Bratislava**
- **Dubai—Abu Dhabi**

2 Different Types of Railways

There are many different railway systems. The first distinguishing criteria include: the design and the drive type. In his book, Filipovic (2015) defines the railway system as "lane-bound traffic systems for the transport of people and goods, where the trains are used as means of transport and are provided with their own drive". According to this definition, e.g. funiculars are not part of the umbrella term for cable cars.

However, the common technical term for a train also includes the term cable cars. In this representation, the railway group of systems is divided into four subcategories:

the conventional track railway, the Maglev, the cog railway, and the funicular, as shown in Fig. 1.

In addition to the systems shown in Fig. 1, there are also some special-type railways (e.g., monorail, suspension track). Those systems will be briefly described in the following sections of the paper.

As already mentioned, the Hyperloop is also a track-bound means of transport and semantically probably belongs to the group of magnetic levitation trains. Before the Hyperloop system can be discussed in more detail, it is necessary to take a look at the 'classic railroad'. Hence, the next step is an analysis of the magnetic levitation train, as an indirect predecessor of the Hyperloop.

Legislative Context

In Austria, the *Railway Act 1957* (BGBl. Nr. 1957) classifies railways as belonging to the categories of "Public Railways" (main, secondary and trams) and "Non-Public Railways" (connecting and material railways). Although a more precise definition regarding the system properties is not included in the legal text, it is legally and clearly defined that this law does not apply to any magnetic levitation, therefore it does not include a Hyperloop system. Other railway systems, as the funiculars, are defined in separate acts, such as the *Seilbahngesetz 2003*. The Austrian legislation does not consider any forms of magnetic levitation trains, probably due to the fact that such systems have so far not been relevant for Austria and there has therefore been no reason to include them in the laws.

The situation is quite similar in Germany: according to the *General Railway Act* Allgemeines Eisenbahngesetz (1993), the Maglev trains are not considered to be railways. In contrast to Austria, however, Germany has several separate laws and regulations for magnetic levitation trains. In most other EU member states, the laws also take into account conventional rail only. The railway law of the European Union also refers exclusively to the conventional rail and as such does not include any other alternatives.

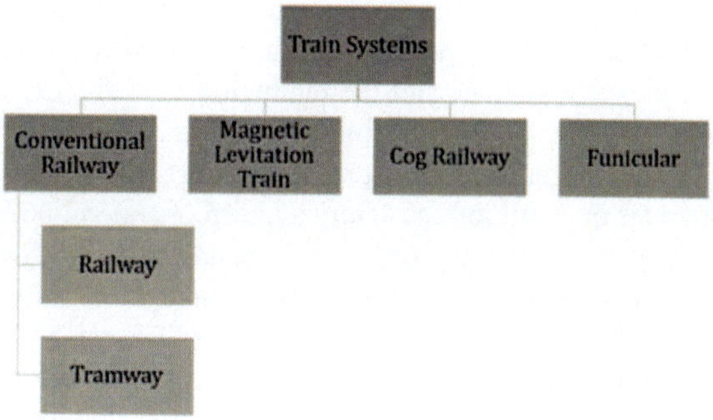

Fig. 1 Classification of the railway systems, *Source* own illustration

Conventional Railway System

The essential feature of all railways is that the transmission of forces and the guiding of the vehicle occur via wheels and rails made of steel. The low friction and rolling resistance at the contact point influence the relationship between the energy requirement and performance of the train in a positive way, which makes this system particularly suitable for the transport of large amount of goods at relatively high speed.

With regard to the energy required for performance, there are other solutions depending on the route and railways themselves. In principle, the main lines are electrified, while diesel traction is used on branch lines. Trains with steam traction are no longer relevant except for special tourist attractions. Alternative drives are currently rarely used, although this could change in the future. For example, 'Siemens is researching a battery version of the Desiro ML series (Siemens (o. J.) 2019). The Rhein-Main transport company recently ordered 27 hydrogen-powered multiple units from the French vehicle manufacturer Alstom'.

Another essential feature of the conventional railway is the extremely long braking distance. The static friction value between the wheel and the rail is about eight times lower in trains than in road vehicles. For this reason, "driving by sight" is more of an exception for full railways. Instead, the driver has to react primarily to the signalling along the route. This makes the conventional railway very complex, as security is a considerable part of traffic management. This includes signals, switches and railroad level crossings, which require a considerable contribution in terms of staff and, currently even more, in technology.

Train protection systems such as the European Train Control System (ETCS), which are among the most important components in order to significantly increase the safety of train operations, should be mentioned here as an example. In their most modern versions, these train protection systems partially take over control of the train and, initiate emergency braking if the permissible maximum speed is exceeded.

Organization of the European Railways

The railway companies are divided into infrastructure manager and transport companies (Pachl 2018). The former kind of company is responsible for the development of new railway lines and the maintenance of the existing railways, for route planning, the creation of timetables, for operations management, monitoring of the security systems, and also for transportation services. The latter kind provides transport services on a rail infrastructure and can operate in both freight and passenger transport. Such a company orders a timetable route from an infrastructure manager for a fee. A combination of both is called integrated railway company. This solution is generally applied to secondary railways. The goal of this organization is the liberalization of the railway market, by making the railway operators concurrent. Since then, companies' investments in innovations and technological advancements have increased Schienen-Control GmbH (2018), EU (2001).

In the past, the railways were developed very differently in some European countries. The effects of this can still be observed, e.g., in technical differences between the EU member states. In order to improve the competitiveness of cross-border rail transport in Europe, the EU is therefore striving to harmonize these differences (EU 2012). The differences mainly concern the following subareas (Filipović 2015):

- Gauge (broad in Spain, Portugal, Finland, standard in the rest of the EU);
- Clearance profile;
- Power system (standardization would be extremely costly);
- Catenary and pantograph;
- Track and train protection.

3 High-Speed Railway

A railway is considered a high-speed railway if the speed of the vehicle exceeds 200 km/h. Starting from this speed, other rules are applied to the operations. According to the current state of the art, driving speeds of up to at least 350 km/h are relatively easy to implement in normal operation. The current world speed record for a HS train is currently 574.8 km/h and was achieved in 2007 by a special version of the French TGV (UIC 2018) *. The high-speed lines are mainly implemented for long-distance passenger connection with a low number of stations. The biggest limitation of this system is the technical parameters of adhesion loss, air resistance, acceleration and breaking distance.*

Compared to other means of transport, the railways are considered to be very safe and environmentally friendly. In the past, both the aviation and automotive sectors introduced many improvements and showed themselves to be relatively open to innovation in their systems. This had a strong negative impact on the medium- and long-haul rail market share. In the second half of the twentieth century, some countries took measures to halt the decline of the railways and began to introduce a new approach to long-distance rail. Instead of expanding existing routes, new, dedicated high-speed routes have now been built in countries such as Japan, France and Germany. In Europe, the EU is actively supporting the construction of a trans-European rail network (Filipović 2015).

Odds...

Modern high-speed rail transport was introduced in principle as a supplement to conventional rail transport in order to increase the capacity of this means of transport. As a result, the high speeds have led to a significant reduction in travel times and significantly increased the competitiveness of the trains compared to the aircrafts on many routes. In principle, building a high-speed line only makes sense for connections for which there is high demand that is not adequately met by existing rail connections.

The speeds of up to at least 350 km/h are relatively feasible in normal operation from a technical perspective. The basic requirement for high travel speeds is the intensive and extensive maintenance of the track systems and the catenary, as well as the vehicle material itself. In the vehicles, the wheels and axles, the suspension and the pantographs are particularly affected by heavy wear. Thanks to very large radii of curvature and a high level of accuracy in the track design, tracking is ensured even at very high speeds. However, the limiting factor with regard to the maximum travel speed according to the current state of the art is the catenary and the power transmission to the vehicle through it. At very high speeds, the constant contact between the pantograph and the overhead contact line can no longer be guaranteed, which results in a loss of power.

... and Limits

The distance between cities or train stations is also a decisive factor. In principle, high-speed rail transport should connect as many cities as possible and bring them economic benefits so that they profit from better accessibility. However, more stops along a high-speed line lead to a deterioration in the real advantages of an HSR, namely lower average speed, longer travel times and lower capacities.

At present, speeds significantly higher than 350 km/h do not seem to be foreseeable for regular operation with conventional high-speed trains. The air resistance that needs to be overcome increases exponentially with higher speed, which would theoretically require very high drive power. In addition, such tensile forces can no longer be transferred to the rail by adhesion. Apart from that, the energy consumption would also be enormous and thus the ecological and economic advantages of the HSR would be significantly lowered. The braking distance for full braking is already over 4 km at 300 km/h, and also increases exponentially. In addition, acceleration times grow with higher speeds. The consequence of this is that such high travel speeds would only result in significant travel times being shortened on routes with large stations (Fiedler and Scherz 2012).

Alternatives

The authors consider that the current inefficiencies in rail transport might offer opportunities for other rail-based developments. In order to assert itself against conventional rail traffic, magnetic levitation trains or the Hyperloop would have to be the "better" or more efficient solution, at least with regard to some factors. Whether these systems actually have more advantages than disadvantages compared to the conventional 'traditional' rail systems is discussed in the following sections.

4 Other Railway Systems with Mechanical Tracking

There are also other systems classified as trains. The difference between these and the traditional system is that the contact between the vehicle and the track is not provided by a steel wheel. Various technologies such as tire trains, monorails or suspended funiculars can be considered as an alternative to the traditional railway system. In most cases, the system is fully independent of the rest of the traffic sector.

The first example is the **tire train**, where the steel wheels are replaced by rubber tires. This system is applied, for instance, in the subway in Paris. The wheels do not roll on steel rails but on wooden and concrete beams. In principle, such trains are only used in local transport and their operational processes are extremely similar to those of the railways Zeit Online (2006).

Another special case is the **monorail**. Its main characteristic is the individual driveway, which is mostly elevated. This design allows easy separation from the rest of the traffic. They are particularly suitable for point-to-point connections. As a rule, operation is largely automated. Various solutions are used with regard to the specific design of the drive, lane guidance or guideway constructions. The most common application for such a train is at airports, as a connecting train between the different terminals (Fiedler and Scherz 2012).

The last type of railway is the **suspended funicular**. The most famous example is the Wuppertal suspension railway, which is the only funicular in Europe that fulfils an essential function in local public transport. Although if this system could, in principle, also work over longer distances, the construction of new monorails is extremely unlikely, above all due to the lack of competitiveness compared to conventional railways (Fiedler and Scherz 2012).

With the exception of special applications in local transport, such as for shuttle connections at the airport, **these railways hardly seem to be competitive with rail transport.** Conventional railways have several advantages over these rail systems. Examples include the existing rail network, the greater flexibility of the system, significantly higher driving speeds, cheaper maintenance and, above all, the lower overall costs.

5 Magnetic Levitation Trains

The magnetic levitation train is a special type of train with no contact between the track and the vehicle. It is planned for high-speed long-distance traffic, as an alternative to air traffic. According to proponents of the technology, it would make the overall transport system more effective and reduce the negative environmental impacts. The Shanghai example argues that they can also be effectively used in local and regional transport services. The Maglev is carried and guided by magnetic fields.

Two different kinds of magnets are used (Filipović 2015):

- **Support magnets**: they are the equivalent of a wheel tread in a standard rail-wheel system;
- **Guide magnets**: they guide the train on the track by exerting horizontal forces, just like the wheel flange on a traditional railroad. They also take on the role of a brake.

The trains can be established (suspended) by three different systems:

- **Permanent magnets**: there are mutually repelling permanent magnets on the vehicle and the road. Theoretically, these magnets have the great advantage that the attractive or repulsive effect can be used without additional energy expenditure. However, due to physical instabilities, such a solution generally requires additional stabilization, which can consist of lateral guide rollers or electromagnets. Compared to the other two types of magnetic levitation (discussed later on), systems that operate by means of permanent magnetic levitation are unsuitable for high speeds.

- **Electromagnetic levitation**: the "attractive force between a regulated DC magnet and an armature rail" is used to create the levitation state. A system that works by means of electrodynamic levitation, on the other hand, uses the repulsive magnetic forces between a "moving magnetic field and a conductor".
- **Electrodynamic hovering**: the hovering state is possible in any operating state, with the vehicle hovering approximately 15 cm above the road. It makes no difference whether the vehicle stops at the platform or is moving at whichever speed range. As a rule, the vehicles in the station are lowered onto the driveway. Vehicles that hover using the electrodynamic system, on the other hand, have a wheel drive and only start to hover at speeds of around 100 km/h. Another difference between the two systems is the distance between the support magnets on the vehicle and the road. While this is only 10 mm for the EMS, it is 10 cm in case of the EDS (Fiedler and Scherz 2012).

The consequence of contactless driving technology is that only contact-free electric linear motors are suitable for the drive in all three system types. Since it is difficult to supply power to the vehicle with larger outputs, the long-stator motor is particularly advantageous because it does not require any power supply to the vehicle. This type of motor also has the great advantage that it is independent of static friction, which makes it easy for vehicles to overcome large inclinations. This drive can be compared to an electric rack due to its mode of operation (Matthews 2007).

The Infrastructure for Maglev

The basic infrastructure for all Maglevs is a track made of concrete, reinforced concrete or steel, erected on supports with an average distance of 30 m. The elevation is favoured by the relatively low static and dynamic loads. The junctions are usually designed as electromechanical bending points and can be passed over with a speed of 200 km/h. In contrast to conventional rail traffic, the switches can also be designed relatively easily as three-way switches. However, due to the significantly higher total weight of the switch elements, the energy required for an adjustment process is much higher than that of a classic rail switch (Matthews 2007).

The shape of the vehicles and the roadway prevents the vehicles from derailing. A major risk to the operation is power failures, as this would lead to an immediate breakdown of the magnetic fields of the carrying and guiding systems. This would be particularly critical with electromagnetic levitation systems, since they do not have any additional wheel drives. In order to minimize this risk, the systems generally have additional batteries that are charged while driving and would take over the energy supply of the magnetic fields for a short period in the event of a power failure.

Transrapid in Germany

The Transrapid magnetic levitation train was developed in Germany in 1979 and is designed for speeds of up to 500 km/h. A test track was built in Emsland in 1984 with the purpose of developing various system components for carrying, guiding and accelerating, or braking, until they were ready for series production.

In addition, the route has become a tourist attraction and it was also possible to take part in the regular test drives. In September 2006, however, a serious accident occurred during such a test drive, resulting in a head-on collision with a workshop trolley forgotten on the route. 23 people lost their lives. Following an extensive investigation, the test operation on the route was resumed in July 2008, although no more visitor rides were allowed.

In the 1990s, some Maglev development projects were already in advanced stages. The feasibility studies were completed, and significant sums of money were invested. These were lines such as Köln-Düsseldorf or Berlin-Hamburg. However, incorrect cost estimation, a lack of investment sources, and the German reunification put an end to these projects (Zeit Online 2006, 2013; Süddeutsche Zeitung 2011; Spiegel Online 2000).

Shanghai Maglev Train

The only commercial application of the Transrapid system at the moment is found in the Chinese metropolis of Shanghai. Since the beginning of 2004, the Transrapid has been operating on a 30 km route between Shanghai Pudong Airport and an underground station in the city of Shanghai. The maximum speed in operation is 300 km/h. Only a couple of trains during rush hour still run at the originally planned maximum speed of 430 km/h. One of the main reasons for the reduced speed may be the higher energy consumption at higher speeds. Originally, today's route was only planned as a first section, with extensions to Beijing or other major Chinese cities envisaged for later. However, these plans never came to fruition due to a lack of political will and strong opposition from the population (Zeit Online) (NDR.de 2012, 2019; Focus Online 2008; Kolonko 2013).

JR Maglev

The Japanese JR Maglev is an electrodynamic levitation system. A great distinguishing feature of the system is the fact that the trough-shaped route embraces the train, while in case of the Transrapid, for example, the train encloses the route. According to the engineers, the advantage of this system is the improvement in earthquake safety and greater driving comfort. During the development stage of levitation technology, the arrangement of the coils, like other system components, was changed several times. The essential feature of the JR Maglev is the equipment of the vehicles with superconducting magnetic coils. Due to the electrodynamic system, the vehicle's hovering condition only begins at speeds of around 150 km/h. The vehicle therefore runs on rubber wheels at lower speeds.

Under the name Shinkansen, a new route is currently being built between Tokyo and Osaka, whose construction work started in 2014. The opening of this 285.6 km section is planned for 2027, and 2045 for the entire route. The total construction costs for both sections are estimated at 70 billion euros and will be borne by JR Central. The JR Maglev is currently the fastest train in the world and was able to improve the world speed record to 603 km/h in 2013 as part of a test drive. The route is designed for a maximum speed of 500 km/h (China.org.cn 2016; International Railway Journal—IRJ 2014, 2016; The International Maglev Board 2019).

Advantages…

Compared to the conventional railroad, the advantages of the magnetic levitation train are, above all, the smoother operation, higher driving speed (up to 500 km/h), faster acceleration and largely wear-free technology. In addition, the driving noise at low speeds is significantly lower, since a magnetic levitation train does not generate any rolling noise. At higher speeds, however, this advantage no longer applies because the airstream generates more noise than the rolling noise.

Considering the infrastructure, the biggest advantage of the Maglev is the low space consumption, mostly due to the narrower curves. Because of the elevation of the track, the area under the track can be used for other purposes. In addition, the technology is independent of static friction, which is why significantly speeds can be reached than in standard rail transport.

… and Disadvantages

However, due to the expansion of the railway high-speed lines and the lower ticket prices for air traffic, the Maglev does not seem particularly economic or cost-effective. The costs increase significantly with the speed of the train. For the 37 km route planned from Munich Airport to Munich city, costs of well over three billion euros were estimated once the significantly too low initial cost estimates had been corrected. As a result, these also led to the failure of the project.

All expenditure and calculations with regard to magnetic levitation are not based on empirical values, but only assumptions. In principle, the environmental balance of magnetic levitation trains is controversial in research. On the one hand, an analysis of the energy consumption for the line between Berlin and Hamburg has shown that the energy requirements of the Transrapid are higher than the ICE 3 at speeds below 300 km/h. Only at speeds significantly above 300 km/h is the energy consumption at MGS lower. On the other hand, other studies show that the energy demand increases exponentially with growing speed because of the air resistance.

Since the Hyperloop technology is very similar to magnetic levitation, it is possible to draw conclusions regarding probable challenges for this new means of transport before the Hyperloop is dealt with in more detail. Similar to the Maglev, one of the main questions will also be whether it is more economical than conventional high-speed rail transport.

6 The Hyperloop

Its Predecessors...

Long before the Hyperloop concept was first introduced, there were ideas and research into similar systems. One of the first mentions of a train system in a vacuum tunnel originated in the early twentieth century from American student Robert Goddard and was further developed by Robert M. Salter in the 1970s (Miller 2014). The system was designed with steel wheels, and the train was not intended to hover. A tube system was planned in which vacuum-like conditions were to be created, and individual trains or cabins were to be transported. The acceleration of the trains was supposed work via electromagnetic coils, similar to today's magnetic levitation systems. The "Planetran" finally failed due to the enormously high cost estimates. At that time, the network proposed by Salter was assumed to cost 750 billion dollars. (Salter 1972, 1978; Businesstech 2017)

The physicist Gerard K. O'Neill designed a Vactrain system, where the vehicles would float electromagnetically and reach a speed of 4000 km/h with an acceleration of 0.5 g. Vacuum-like conditions were also supposed be created in the tube. The drive was to be provided by means of electromagnetic forces in this concept. (O'Neill 1991)

Some other concepts, such as Swissmetro (Gast 2016; Krummenacher 2016)—a magnetic levitation train in a tube with reduced air pressure—were also introduced at the beginning of the twenty-first century but failed due to a lack of funding and doubts about profitability. In his book (2014), Janic (2014) concludes that one of the main advantages of the Evacuated Tube Transport (ETT) systems is, above all, the better environmental balance. However, the question of the actual economic feasibility and usefulness of the system is described as a major uncertainty. There may also be technical barriers and obstacles that are not yet known, especially because of the very long development time of such systems (estimated for at least 20 years). This also applies to the possibility that other, more suitable and efficient, means of transport could possibly be developed during the period up to commissioning.

The Concept

The entrepreneur Elon Musk, CEO of Tesla and SpaceX, presented the Hyperloop concept in August 2013 in a White Paper titled "Hyperloop Alpha". He developed the concept as an alternative to the planned California High-Speed Rail project between Los Angeles and San Francisco, which was expensive, inefficient and slow. He also decided to develop a prototype in order to prove the basic feasibility. It is an open source concept, which means that everyone is invited to contribute to the design process. Since then, numerous universities and research institutions have shown an interest, and a number of companies have been founded that want to put the Hyperloop into practice.

To make sense, the new means of transport should be an environmentally friendly alternative to airplanes and cars. The ideal application would be strongly frequented connections between two cities that are a maximum of 1500 km apart. Musk considers the aircraft, or long-term supersonic aircraft, to be more sensible and cost-effective for longer distances. In the Whitebook 2013, the following requirements compared to the existing transport systems are formulated (Musk et al. 2013):

- Safer
- Faster
- Lower costs
- Greater convenience and traveling comfort
- Complete independence from weather
- Sustainably self-powering
- Resistant to earthquakes
- No burden for the population living along the route

Technical Implementation

The principle of the Hyperloop is based on the creation of a vacuum or a vacuum-like state in a tube. As a result, the air resistance could be kept low, which would allow for very high speeds. However, small cracks or leaks along the tubes would stop the entire system. This is why low air pressure instead of practically no air pressure or a vacuum-like state is introduced. The air pressure in the tube is supposed to be 100 Pa. This would reduce drag by about 1000 times compared to sea level conditions. At the same time, the air pressure would still be sufficiently high for air pumps to keep the air pressure at this level in the event of a leak.

However, this solution means that a certain amount of air always remains in the tube. With a moving capsule in a narrow tube, the air can only flow between the capsule and the inner walls of the tube. When the capsule reaches a sufficiently high speed, the air currents around the capsule reach the speed of sound. As a result, there is a certain "clogging" effect, which leads to a much higher air resistance. In theory, this so-called Kantrowitz Limit could be solved by designing the tube diameter many times larger than the capsule. However, this solution would be extremely expensive. Instead, the Hyperloop provides a solution using electrical compressors to transport the air behind the capsule. The compressors are also intended to generate air cushions under the capsule on which the capsule slides. This would also keep the friction extremely low. The original Hyperloop concept would therefore not be a magnetic levitation train, since the levitation state is not achieved by magnets.

In keeping with the various magnetic levitation trains, linear motors are provided for acceleration. These should enable high acceleration along the route (especially at the stations), while the compressors should generate the necessary propulsion once the cruising speed has been reached. In principle, a maximum speed of 1223 km/h is provided for the Hyperloop. The implementation of supersonic speeds on longer distances is not completely excluded, but aviation would be cheaper in this case.

One of the great advantages of the Hyperloop concept is the elevated construction. The sections of the tube are pre-built in a factory and are only erected and connected directly along the route. Due to the elevated construction, the space consumption of the Hyperloop infrastructure is greatly reduced, and significantly less land has to be bought. For the planned Los Angeles—San Francisco track, this need will also be reduced by the fact that the route will be built over large parts along the California Motorway 5.

California is a region with a relatively high risk of earthquakes, which is why special security measures are necessary for the transport systems. With the Hyperloop, higher earthquake security is achieved primarily by elevating the track. Since the tube is not fixed directly on the surface of the earth but rests on the supports, the load is better distributed in case of an earthquake and the impact is minimized as much as possible. Damping elements are also provided in the supports. In addition, the capsules are to be stopped by an emergency braking system in the event of an earthquake.

The tubes themselves are made of steel. In order to keep the material expenditure as low as possible, the inner diameter of the tube, depending on the capsule size, will be only 2.23 or 3.30 m. The reinforced concrete pillars should support the pipes at intervals of approximately 30 m. The energy requirements of the complete Hyperloop system will be covered by solar panels, which are placed directly on the tube. Storage of the generated solar energy in batteries is planned for operation in cloudy weather and at night.

Operational Implementation

The Hyperloop is a transport system in which the capsules are transported in a closed steel tube. The floating state is to be achieved using electromagnets. However, there is currently no information regarding the exact implementation of a hyperloop operation. This is probably due to the fact that the companies are still researching the actual technology and that an extensive test phase will probably also be necessary before actual operation. For this reason, no operating concept for a specific project is presented, but instead the essential points are dealt with in general.

However, the White Paper describes a possible implementation of the Hyperloop concept for the Los Angeles—San Francisco route. A journey time of approximately half an hour is planned here. The capsules are supposed to depart at intervals of up to 30 s and could transport 28 passengers each. According to these figures, the Hyperloop would theoretically have a maximum capacity of 3,360 passengers per hour. However, the study actually assumes an average interval of two minutes. The concept envisages the use of up to 40 capsules, with five minutes of waiting in the stations for boarding and alighting. The regular capacity should therefore be 840 passengers per hour, respectively 20,000 per day.

The passenger capsule would have a maximum height of only 1.10 m with a width of 1.35 m. Due to the lack of space, it is intended that the passengers remain sitting throughout the entire journey. Since the capsule only moves in a closed tube, no windows are provided. Instead, exterior views of the passing landscape are to be projected onto the interior walls of the cabin. In addition, each seat is to be equipped with a screen. Similar to many long-haul aircrafts, this should enable the use of an entertainment system.

Basically, two stations are planned on the Californian route: Los Angeles and San Francisco. Further stations along the route would also be possible, with branches from the tube, but would still have to be planned. To avoid possible pressure losses or leaks, the stations will be separated from the main tubes by means of a kind of airlock. The design of the stations is to be kept as minimalistic and practical as possible. The boarding process should be similar to that of an aircraft and also provides for mandatory security checks for all passengers.

Operations

Due to comfort reasons, the acceleration is limited to a maximum of 0.5 g. All capsules are equipped with emergency call devices for monitoring personnel as well as first aid kits. Similar to aviation procedures, the capsule should continue to the end station in case a passenger becomes ill during the journey. In the event of a power failure, it is provided that all capsules also continue to the next station. This is possible because the capsules are to be equipped with batteries and are therefore independent of the energy supply after reaching cruising speed. As in the aircraft, oxygen masks are provided to deal with a sudden drop of pressure in the capsule. The probability that a capsule will remain stationary in the tube is said to be extremely low. Nevertheless, there are considerations for an emergency system that would take the vehicle to the next station using an electric motor. Emergency exits along the route are not provided.

The Hyperloop should be operated as autonomously as possible. The capsules themselves will operate fully autonomously without a driver and will only be monitored by a control center. Obviously, there is no need for signalling along the tube. In normal operation, each capsule runs completely the same route, and intervention by the monitoring personnel in the control center is not necessary. In an emergency and in the event of other deviations, this staff should be able to control the capsule. The tasks of this control center would be to communicate with the passengers in the capsules, to monitor ongoing operations, to monitor the boarding and alighting process in the stations and to support the personnel there, as well as to coordinate the emergency services in the event of a possible accident. It is also important that fixed operating procedures are planned for various emergency scenarios, which start automatically (without intervention by the personnel). For example, in the case of a defective capsule, all capsules traveling behind it must be braked immediately and transported back to the starting station (Fig. 2).

Fig. 2 Capacity comparison of several railway systems, *Source* own work

System	Interval [min]	Train capacity	Capacity in 1h in one direction
Hyperloop Variant 1	2 5	28	**840** 336
Hyperloop Variant 2	2 5	40	**1,200** 480
Standard Railway	3 5	900	**18,000** 10,800
Maglev	2	500	**15,000**

To determine the intervals between the different capsules, it is important to take into account the time required for passengers to get on and off, as well as the safety distance for emergency braking. According to Elon Musk's Hyperloop White Paper, departure intervals of up to 30 s should theoretically be possible. In normal operation, however, it assumes an average interval of two minutes. A traditional high-speed railroad can reach a maximum frequency of 5 min, mainly due to breaking distances. By using a capsule with 28 or 40 seats (those two models are currently proposed), the difference with regard to the possible (theoretical) total capacity compared to the conventional train and the Maglev is enormous. This significantly lower performance of the Hyperloop is a major disadvantage compared to other means of transport, which will automatically have an impact on the profitability and the accessibility.

Route Planning and Stations

Although there is currently hardly any specific information available on the actual routes or planned routes, certain infrastructural requirements and conditions are known due to the system-inherent properties of the Hyperloop.

At the start of planning, it is important to decide whether the planned project is a point-to-point connection between two cities or whether several stations should be connected in a network. In the first variant, it makes sense to differentiate between routes without intermediate stations and routes with such stations along the route. In principle, it should be noted that with increasing complexity of the system, the demands on the route, the safety and, for example, the operational monitoring also increase.

A point-to-point Hyperloop route is theoretically very easy to implement. At both ends of the route are the stations that are connected by the Hyperloop tube. Practically all Hyperloop systems provide for the erection of two tubes lying side by side or one above the other. This means that each tube can be used in one direction. The stations are usually separated from the tube by a kind of airlock. After reaching the station, the passengers get off on a platform. Subsequently, the capsule is turned around and the passengers who want to travel in the opposite direction get in. After completing the entry process, the capsule departs from the station in the other tube. In the concept studies for Hyperloop stations, there are always several platforms. This makes it possible to switch passengers for several capsules at the same time.

So far, there is hardly any information in the concepts on how the capsules should get to the platforms. To address this, there should probably be a solution with switches. However, since the stations themselves are implemented "openly" (not in the Hyperloop tube), this implementation should be relatively easy from a technical perspective. It would also be conceivable that the capsules in the stations use normal wheels instead of levitation, which would probably make the switches cheaper. Wheels were already proposed in the original Hyperloop concept. At that time, however, these were intended primarily as a fallback option in the event of a suspension technology failure. In such a system, no switches are required in the Hyperloop tube itself. This significantly simplifies operation, reduces the construction and operating costs of the system, and probably also increases the system's reliability.

Limits...

However, as soon as branches are planned along the route, for example because a city is to be connected to the system between the start and end point of the connection, switches are also required directly in the Hyperloop tunnel. This will particularly affect planned Hyperloop networks. Junctions are much more complex with a magnetic levitation train than with conventional rail traffic. The basic problem is that, unlike the railroad, the Hyperloop has no rails on which the capsules roll. Although research on junction systems without any moving elements is already being conducted, it will be quite difficult to steer the capsules on branches (Bilton 2013).

Each intermediate station along a Hyperloop route would also lead to a considerable increase in the total travel time between the two end stations due to the time-consuming braking or acceleration process. In addition, the energy requirements increase with each additional stop, since the acceleration and braking processes are particularly energy-intensive.

Another problem with branched Hyperloop networks is ensuring the consistently low air pressure in the entire system. If, for example, a Hyperloop network were to be built in Europe, it would probably have a total length of at least several 1000 kms. With multiple branches and junctions, the complexity of the system would increase further. It would be worth considering whether there should be possible (emergency) airlocks between individual tube sections. In this way, only a section of the overall system would be affected in the event of a serious leak along the tube.

The size and diameter of the tubes must be logically adapted to the capsules. Whether the tubes are built one above the other or next to each other seems secondary and probably has an impact on the design of the stations. Putting the tubes side by side would possibly make more sense here, as this would mean that the capsules would not have to change levels at the end stations.

Theoretically, a narrower radius would be possible with a Hyperloop route than with the railway. In addition, significantly higher gradients can be negotiated. Despite the theoretical feasibility, however, it is doubtful whether excessively narrow radii and strong inclinations make sense. In tight curves, the driving speed would have to be reduced due to the otherwise excessive g load. Therefore, building many curves with tight radii would lead to a significant increase in travel time and thus compromise one of the main advantages of the Hyperloop system.

According to Doppelbauer (2018), the maximum speed of the Hyperloop would require curve radii with a diameter of at least 100 km. In the case of smaller curves, the maximum lateral acceleration permitted by the ICE or Transrapid of 0.1 g would be exceeded. It is, therefore, extremely important that the route is as straight as possible for the maximum speed sections of Hyperloop routes. Taking these restrictions into account, the advantage of the supposedly possible narrower radii and steep gradients appears to be significantly smaller.

Compared to conventional high-speed traffic, the space requirement for a Hyperloop route is fundamentally lower, since large parts of the space under the elevated route can be used for other purposes. For most of the concepts, a tunnel is provided for connecting the city centers. A new Hyperloop route has the disadvantage compared to the railway that the track has to be completely rebuilt. In conventional rail traffic, at least in Europe, there is a train station in practically all major cities that is close to the center. The establishment of a Hyperloop station near the center would be a long-term and, above all, expensive project in many cities. Alternatively, it would also be possible to set up the Hyperloop stations on the outskirts of the city. In this case, however, care should be taken to ensure a functioning connection to the rest of the public transport system. In general, when implementing the Hyperloop, it should be noted that this will only ever be part of a multimodal transport chain, since passengers need other means of transport to travel to and from the stations.

Capsules and Driving Comfort

Although the Hyperloop technology is mainly developed for passenger transport, freight transport for courier services or other small quantities can also be considered. Overall, the Hyperloop does not appear to be competitive with trucks (for smaller quantities) and the conventional railroad (for larger quantities). In passenger transport, all companies are planning passenger capsules that can transport around 28–40 people, depending on the concept. The travel time for all proposed connections is less than one hour. Compared to other means of transport, however, passengers have to accept certain losses in comfort. The small diameters of the capsules, the lack of toilet facilities or the obligation to remain seated with fastened seatbelts throughout the entire journey are examples of restrictions of this means of transport.

Since the capsule moves in a closed tube, no windows are provided. Instead, however, the companies are planning to provide special lighting for the inner walls of the capsules, which could project landscape views, for example. In any case, the possibility of a built-in screen would be conceivable for every seat where the passenger would have access to the on-board entertainment system. Basically, the Hyperloop operation is planned to be as autonomous as possible. Therefore, no personnel are intended for the capsules. Nevertheless, communication with a control center should be made possible for every passenger (e.g. via radio equipment).

Safety

There are still important open questions regarding the necessary safety measures and the procedures in emergencies. As there is currently no existing Hyperloop legislation, there are no specific requirements that explicitly relate to the Hyperloop. However, this is likely to change in the medium term. Many Hyperloop companies are already in contact with public institutions in this regard and are trying to influence the legislative process. At least in the EU, it can be assumed that the safety requirements for the Hyperloop tube will probably have some similarities to the regulations for railway tunnels and road tunnels. In emergency cases, the vehicle should return to the station independently.

The protection against danger to humans and terrorist attacks will be similar to the measures at the airport, with controls and identity checks before boarding. Therefore, the entire entry process needs to be reconsidered. In contrast to most rail connections, passengers have to arrive at the Hyperloop station well before departure in order to have enough time for the security check. However, this once again increases the total travel time, which is why the security check is a disadvantage compared to the conventional train. As part of this inspection, the passenger's ticket is also checked.

Maintenance

If the tube is not in operation, the capsules should wait in the stations or in workshops, where inspection and maintenance work could be carried out. There must be the option of taking capsules out of operation and transferring them to the workshop or coach station/depot. The simplest implementation would be to branch off from the actual Hyperloop route. For this, however, a separate lock for the workshop would be required in addition to the airlocks. Therefore, it is alternatively conceivable to branch off only after leaving the Hyperloop tube, directly linked to the stations. In analogy to some subway end stations, for example, a branch would be conceivable in the area of the turning systems, where the capsule is rotated.

Cost Estimate and Financing

The costs are listed in detail in the White Paper and divided according to the various system components. The expenditure for the route infrastructure, in particular the tube and any tunnels, seems to be the largest part. The easy mounting of the tubes, the low space consumption, especially compared to rail traffic, can significantly reduce the costs, especially by buying land and building space.

Overall, the White Paper assumes total costs of 6 billion dollars to implement the Hyperloop with pure passenger capsules. This includes only the costs for the construction of the track itself. The development costs are not considered. The variant with the larger capsules, which could also transport vehicles, would increase the costs to 7.5 billion dollars. Elon Musk announced ticket prices of only about 20 dollars, which would cover a payback after 20 years.

Shortly after the first publication of the Hyperloop concept, it was already discussed among economists that Musk's cost estimates were too low. When planning the high-speed rail line in California, costs were significantly higher because a large number of cities and areas wanted bypasses and tunnels, or a connection to their city. These requirements would probably also apply to a Hyperloop route, which is why the cheapest route in terms of costs would hardly be feasible. In addition, the costs of connecting the city centers, for which a tunnelling or bridging of San Francisco Bay would be necessary in San Francisco, must also be taken into account. Instead of the advertised costs of only 6 billion dollars, there are estimates by experts that assume total costs of up to 100 billion dollars.

A large amount of costs have been discovered that were not originally included in the plans of the Hyperloop. Above all, the very low assumed cost of buying space is criticized. On the one hand, it is still unclear how the Hyperloop should be built from the respective cities to the planned route along the Motorway 5. On the other hand, implementation along this highway is not that easy. The area along this street is owned by the regional authority responsible for road traffic which would have to agree to the sale of the area. A construction directly next to the interstate would also lead to considerable road traffic disruptions.

In addition, the cost calculation for the operation of the Hyperloop is also questionable. Musk's White Paper states that the Hyperloop transports at least 840 passengers an hour, each paying a ticket price of 20 dollars. However, Dan Sperling, Director of the Institute of Transportation Studies at the University of California, believes it is absolutely out of the question that such figures can be used to cover costs.

The financing could be even more problematic than the viability of the project. The question remains as to which institutions or companies would finance a Hyperloop project such as the proposed Los Angeles—San Francisco connection. The technology of the Hyperloop, which has not yet been put into practice, represents a significant risk. Financing through public funds is also extremely unlikely, since investments in the "classic" infrastructure, such as road or rail, are already undertaken. This accentuates the criticism in the reception of the Hyperloop concept.

Reception and Professional Criticism

The basic technological feasibility of the concept is generally not in doubt. Instead, the criticism focuses primarily on questions pertaining to the concrete implementation of the project and the assumed costs (Bilton 2013). Since a concept such as the Hyperloop has not yet been implemented, there are a large number of uncertainties regarding the exact technological implementation. Experts describe the capsules and their acceleration as technology that has never been used in practice before and therefore represents a challenge. According to John Hansmann, director of the MIT International Center for Air Transportation, the question of the system's actual energy requirements has not yet been resolved, either. The decision to only lower the air pressure and forego an (almost) complete vacuum in the tube is seen as a sensible idea and an effective measure to keep the energy requirements as low as possible. Despite this measure, it is assumed that the energy requirements of the system cannot, as Musk claims, be met by the solar panels on the tube (Bullis 2013; Wolverton 2013).

As previously explained, there are already considerations regarding the earthquake safety of the system in the original White Paper. However, the effectiveness of the planned emergency braking system, which immediately triggers an emergency stop when the first earthquake waves are detected, is doubtful. Even with immediate emergency braking with very high g-forces, the Hyperloop capsule would need at least 15 s for a full stop and would cover a distance of more than two kilometres. This period of time could be too long in the case of a strong earthquake.

Markus Hecht, professor at the Technical University of Berlin, thought that implementation of the Hyperloop concept was generally unrealistic in 2013. Apart from financing, he sees the biggest problems in rescuing passengers in the event of a malfunction, and waste heat, which is difficult to solve in a tube with low air pressure. The latter question has also been raised by other experts. Due to the compressor, there would be a lot of heat in the capsule and in the tube. There is further criticism regarding other details, such as the assumptions regarding the weight of the passengers and the luggage, which are too low. Apart from that, regulatory questions regarding the concept and the consideration of the various legal requirements for security have not been considered (Becker 2013; Melendez 2013; Lavanchy 2013).

The statements concerning the actual travel time are also doubtful, as factors such as the necessary security checks are not taken into account. It is also unclear where the Hyperloop connection between Los Angeles and San Francisco would end and therefore how long the actual travel time would be between the two city centers. With a maximum of 3,360 passengers, the capacity of the Hyperloop concept is also significantly less than the capacity of conventional high-speed rail traffic (up to 12,000 passengers per hour), or even less than the capacity of Motorway 5. The acceleration would take more than 5 min and would therefore probably be rather uncomfortable or physically demanding for many passengers (Blodget 2013).

Dutch expert Carlo van de Weijer from the Technical University of Eindhoven also believes that flexible transport systems and models will become increasingly important in the future. The success of companies such as the low-cost airline EasyJet, the long-distance bus company Flixbus or the travel provider Uber can always be explained by the fact that they were able to adapt flexibly to the existing infrastructure and system properties. In contrast, the Hyperloop technology requires completely new infrastructure. According to Van de Weijer, a Hyperloop connection may be suitable for point-to-point connections but the lack of integration into the existing traffic infrastructure is a major disadvantage of the system.

7 Other Possible Implementations of the Hyperloop

Vienna—Bratislava

In 2016, the Slovak government proposed building a Hyperloop system. In addition to domestic Slovakian connections, it was proposed to establish a connection from Vienna to Bratislava and further on to Budapest. The travel time between the capitals would be only 8 min for a distance of 55 km. To explore the potential of the new transportation system, the government signed an agreement with Hyperloop TT. The stated goal at the time was to make Slovakia an innovation hub for new technology. A feasibility study was announced by the end of 2016, which was to analyse the concrete implementation options in more detail. However, the partnership between Hyperloop TT and the Slovak government already expired eight months after the signing without any results. Since then, there has been no information on the resumption of work. Apparently there have been no considerations regarding the construction costs, the exact route, the connection to existing traffic systems, the financing or the exact technological implementation Futurezone (2016), (2018); The Slovak Spectator (2018).

There are currently different transport options between Vienna and Bratislava. In addition to individual transport, there are several long-distance bus connections and two train connections per hour. The train connection between Vienna Central Station and different train stations in Bratislava takes approximately 1 h. After the expansion of one of the tracks, the driving speed should increase to 160 km/h and shorten the travel time to around 40 min.

With regard to the possible travel time of only eight minutes, it must be borne in mind that this is only the net travel time (not the total travel time) and the duration of the boarding and alighting process is not taken into account. Experts emphasize that the duration of this process in the Hyperloop should not be underestimated due to the tightness of the capsule and will likely take several minutes. Assuming that the Hyperloop reaches speeds of almost 1200 km/h, an actual travel time of at least 14 min can be assumed. Nevertheless, the short total travel time of the Hyperloop is the great advantage of the system compared to the existing traffic connections between the two cities (Wien 2019; Doppelbauer 2018).

Compared to the Hyperloop, conventional rail transport has several advantages for this connection. An expansion of the routes, as planned for the Marchegger Ostbahn, can be carried out relatively easily and comparably inexpensively. However, the main argument in favour of rail transport is the integration into the existing rail network. In contrast to the Hyperloop, an expansion of conventional rail traffic between cities would not only benefit the point-to-point connection but would generally improve the accessibility of cities from other cities in the rail network. It must also be considered that a longer tunnel section would be inevitable if the two city centers were connected. The expected resistance of the local population to the project, the very long planning phase, and any organizational or financial difficulties in the cooperation between the two countries should bring the speculation concerning the project to an end (Visit Dubai 2019) ; Hyperloop 2018a).

Abu Dhabi—Dubai

Another possible application of the Hyperloop appears in the United Arab Emirates. In 2018, Hyperloop TT announced the signing of an agreement to build a commercially operated Hyperloop route. In the first step, the construction of a 10 km route on areas of the real estate company Al Dar is planned. Subsequently, the Hyperloop route is to be expanded in sections and will connect Abu Dhabi with Dubai for a total length of 150 km. A connection to the city of Al Ain is also being considered. The company anticipates construction costs of 20 to 40 million dollars per kilometre. According to this estimate, the total costs of the entire route between the two cities could amount to 6 billion dollars. However, details of the exact route planning and financing have not yet been released (Maceda 2019).

The project is intended to massively reduce travel times between cities, although the exact future travel times and speed are not yet known. There are currently only road connections between Abu Dhabi and Dubai. The travel time from city center to city center is generally 1:20 h but is heavily dependent on the current traffic situation. In addition to various public bus routes and motorized individual transport, the connection is also used by various tourist shuttle services and for excursions. Since there is no flight connection and no conventional rail link between the cities, practically all traffic is handled via the road connections. However, the United Arab Emirates are currently building an extensive rail network which should also connect Abu Dhabi and Dubai by 2024. In addition to use by freight transport, services in rail passenger transport are also provided. However, there will be no dedicated high-speed track (Railway Gazette International (Hrsg.) 2018, 2019).

The Hyperloop project would therefore lead to significant time savings, especially when compared to road traffic. The route is also well-suited for a point-to-point connection, as there are hardly any larger cities between Abu Dhabi and Dubai, which means that no additional stations are necessary. The opening of the first section was originally planned in time for the Expo 2020 in Dubai. Construction work was specifically scheduled to start in the third quarter of 2019. It is currently unclear whether this construction has already begun.

The project seems realistic and has several advantages over other projects, such as Vienna—Bratislava. Compared to projects in the EU or the USA, the regulatory and legal hurdles appear to be much lower. The route could be built in a significantly shorter period of time than would be possible in the western world. Furthermore, there is no elevation, which makes tunnels unnecessary. A further advantage for route planning is the fact that large sections of the area between the two cities are relatively undeveloped and route planning is therefore hardly restricted. Unlike traditional trains and Maglevs, the Hyperloop is completely independent of the weather. This means that sandstorms would have no impact on traffic (Expo 2020).

The planned rail link between the cities is not necessarily an argument against the project, as this will not be a high-speed track and the travel time for the Hyperloop would still be significantly shorter. Since the route would be approximately three times longer than the Vienna—Bratislava connection, the time gained compared to the train is also significantly higher. The question of whether there is enough traffic demand for a Hyperloop connection is difficult to answer at this stage. Due to the lack of data or studies, it is not possible to make a well-founded statement on this topic. According to the author, the Hyperloop connection would probably be of particular interest to tourists and business travelers (Kolonko 2013).

8 Comparison with Other Systems

To allow for a better assessment of the Hyperloop system, it will be compared to the other systems mentioned in this article.

This is done by comparing the systems using the route of Vienna to Bratislava, which is one of the proposed Hyperloop routes.

The systems are compared based on their travel time, average speed, estimated construction costs, ticket price, and passengers per vehicle. The detailed numbers can be seen in Fig. 3.

In terms of travel time, the Hyperloop is the fastest because of its top speed of 1200 km/h, but it is only four minutes faster than a Maglev train and only ten minutes faster than the high-speed railway. The construction costs of a Hyperloop system are the second-highest in this comparison and are surpassed only by the Maglev. However, it has to be said that the Maglev construction costs were calculated using the estimated costs of the Chūō-Shinkansen line which will run in underground tunnels on 86% of its length (Asahi and Shimbun 2013).

An above-ground Maglev system will therefore most likely be closer to the construction costs of the Hyperloop. Both the high-speed rail und the normal railway are much cheaper. Elon Musk claims that the ticket price of the Hyperloop will be 17 euros (Investopedia 2019), which puts it right in line with the other systems. Experts claim that the actual ticket price will be much higher because of the large investment costs and the very small vessel size compared to the other systems (Huffington and Post 2013).

Route	Vienna - Bratislava					
System	Time [h]	Average speed [kph]	Track length [km]	Estimated construction costs [€]	Ticket price [€]	Passengers per vehicle
Railway	01:06	59.36	65.30	195.8 million €	€ 16.00	570 seats (ÖBB-REX with 5 BiLevel cars type Viaggio Twin)
HS Railway	00:13	300.00	65.30	867.2 million €	€ 16.00	460 seats (Siemens Velaro D)
Maglev	00:07	500.00	65.30	10.2 billion € (Chūō-Shinkansen)	€ 17.66	252 seats (Shinkansen L0)
Hyperloop	00:03	1200.00	65.30	2.9 billion € to 4.2 billion €	€ 17.00*	30-40 seats (Hyperloop Transportation Technologies)

Fig. 3 Comparison of transportation systems

This comparison shows that the hyperloop does not make sense on shorter routes such as the one used in this example. It only makes sense on very long routes where it can utilize its high speed to gain a better advantage over the other systems. In this example, the best system is the high-speed rail because it takes only ten minutes longer but has a lower ticket price, a higher passenger capacity, and much lower construction costs.

9 Conclusions

The transport system is currently facing major challenges. The need for transportation services is constantly increasing in the globalized world of transport. Compared to aviation and trucks, the railways had the advantage of a significantly better environmental balance. The railways are also currently experiencing a kind of "renaissance" in many countries. Following this impulse, new train technologies such as the magnetic levitation train were developed in the last few decades. Compared to conventional rail transport, the advantages of the magnetic levitation train are, above all, the smoother operation, the higher driving speed, the faster acceleration, and the largely wear-free technology. Despite intensive testing and numerous project proposals, only one commercial Transrapid line was built in Shanghai. Other countries preferred to invest in the continuous expansion of conventional high-speed rail transport.

First proposals for a train system in a vacuum tunnel were made in the 1970s. These concepts were concretized by the entrepreneur Elon Musk in 2013. The Hyperloop is a traffic system which features capsules transported in a closed tube with reduced air pressure. The resulting reduced drag should make speeds of up to 1200 km/h possible. The compressors are also intended to generate air cushions under the capsules on which the capsules slide. This would also keep the friction extremely low. The concept was made open and numerous companies have therefore begun to develop the technology. The two largest companies are Hyperloop TT and Virgin Hyperloop One. They have already set up test tracks and plan to put the first Hyperloop tracks into operation in the coming decade.

However, there are still some open questions regarding the implementation of the Hyperloop technology. The basic technological feasibility of the concept is undoubted. Instead, the criticism focuses primarily on the concrete implementation and on the economical or financial aspects of the system. Amongst other things, the cost estimates published by the Hyperloop companies are highly doubtful. Expert calculations usually arrive at significantly higher total costs. It is generally doubted whether the Hyperloop would actually be cheaper than conventional rail transport in terms of total costs. A major issue is the planning of emergency exits along the tube, which makes the system very complex to implement, and thus greatly increases the overall costs. Another open question is the fact that there is currently no legislation or regulation regarding Hyperloop technology. The legislative process in the USA and the EU could take several years and significantly delay the development of Hyperloop routes.

Compared to conventional rail transport or the magnetic levitation train, the maximum capacity of the Hyperloop is also many times lower. At best, the capsules with a maximum of 40 seats will run at intervals of two minutes. Even in this case, a capacity of only 1200 passengers per hour and direction would be achieved. The question, therefore, arises whether the advantage of the higher driving speed with the Hyperloop is sufficient, or whether the concept will fail to be implemented in the end, as was the case with the Transrapid.

The rough concept for a possible Hyperloop operation has shown that at least in the first implementation phase, the focus should be on possible 'point-to-point' connections. This does not do justice to the European traffic demand, where numerous cities have to be connected to each other, with a lot of intermediate stations and junctions. Furthermore, there is already a very dense rail network and, in some cases, also high-speed rail links. For routes such as Vienna—Bratislava, the implementation of a Hyperloop connection makes little sense. According to the current status, however, implementation of the Hyperloop could still make sense for certain connections under certain circumstances. The Dubai—Abu Dhabi link, for example, seems to be ideal to meet the requirements. Support from the government also helps in the development of the project. For this reason, the construction of the first Hyperloop route in the United Arab Emirates currently seems most likely.

Sources

Allgemeines Eisenbahngesetz 1993 (GER) (idgF); Auch Online unter www.gesetze-im-internet.de/aeg_1994/index.html (30.07.2019)

Becker K (2013) Luftkissenzug "Hyperloop"– Risiko in der Röhre In: Spiegel Online (13.08.2013); Online unter https://www.spiegel.de/wissenschaft/technik/projekt-hyperloop-wie-elon-musk-das-zugwunder-wahrmachen-will-a-916316.html (20.07.2019)

BGBl. Nr. 113/1963 (idgF): Eisenbahngesetz 1957; Auch Online unter https://www.ris.bka.gv.at/GeltendeFassung.wxe?Abfrage=Bundesnormen&Gesetzesnummer=10011302 (30.06.2019)

Bilton M (2013) Could the hyperloop really cost $6 billion? critics say no. In: The New York Times (15.08.2013); Auch Online unter https://bits.blogs.nytimes.com/2013/08/15/could-the-hyperloop-really-cost-6-billion-critics-say-no/ (18.07.2019)

Blodget H (2013) Transport blogger ridicules the hyperloop--says it will cost a fortune and be a terrifying 'Barf Ride'. In: Business Insider (20.08.2013); Oline unter https://www.businessinsider.com/hyperloop-barf-ride-2013-8?IR=T (20.07.2019)

Bullis K (2013) Experts raise doubts over elon musk's hyperloop dream. In: MIT Technology Review (12.08.2013); Online unter https://www.technologyreviewcom/s/518076/experts-raise-doubts-over-elon-musks-hyperloop-dream/ (20.07.2019)

Businesstech (2017) Meet the 'flying train' concept–30 years before Elon Musk made it a reality (17.07.2017); Online unter https://businesstech.co.za/news/technology/185313/meet-the-flying-train-concept-30-years-before-elon-musk-made-it-a-reality/# (16.07.2019)

China.org.cn (2016) China's maglev transit initiative picks up steam (01.11.2016); Online unter http://www.china.org.cn/business/2016-11/01/content_39610565.htm (14.07.2019)

Doppelbauer J (2018) Hyperloop—eine Innovation für den globalen Verkehr? —In: ZEVrail 6–7(142), S.218–224; Auch Online unter https://www.era.europa.eu/content/hyperloop-innovation-global-transportation_en (→ related documents) (abgerufen am 28.07.2019)

EU (2001) RL 2001/12/EG: Entwicklung der Eisenbahnunternehmen der Gemeinschaft; Auch Online unter https://eur-lex.europa.eu/LexUriServ/LexUriServ.do?uri=OJ:L:2001:075:0001:0025:DE:PDF (03.07.2019)

EU (2012) RL 2012/34/EU zur Schaffung eines einheitlichen europäischen Eisenbahnraums; Auch Online unter https://eur-lex.europa.eu/LexUriServ/LexUriServ.do?uri=OJ:L:2012:343:0032:0077:DE:PDF (30.07.2019)

Expo 2020 (2019) Discover—understanding our expo—themes and subthemes. Online unter https://www.expo2020dubai.com/de/discover/themes (29.07.2019)

Fiedler J, Scherz W (2012) Bahnwesen: Planung, Bau und Betrieb von Eisenbahnen, S-, U-, *Stadt- und Straßenbahnen.* Werner Verlag, Düsseldorf

Filipović Z (2015) Elektrische Bahnen: Grundlagen, Triebfahrzeuge, Stromversorgung. Springer, Berlin Heidelberg

Focus Online (2008) Der Transrapid fährt nur in China (27.03.2008); Online unter https://www.focus.de/finanzen/news/transrapid/transrapid_aid_230366.html (13.07.2019) Fockenbrock D (2019) Durchbruch für Wasserstoff-Züge—Alstom setzt auf Wachstum in Deutschland. In: Handelsblatt (21.05.2019); Online unter https://www.handelsblatt.com/unternehmen/handel-konsumgueter/bahnverkehr-durchbruch-fuer-wasserstoff-zuege-alstom-setzt-auf-wachstum-in-deutschland/24364778.html (29.07.2019)

Futurezone (2018) Bau der ersten hyperloop-Teststrecke in Europa beginnt (12.04.2018); Online unter https://futurezone.at/science/bau-der-ersten-hyperloop-teststrecke-in-europa-beginnt/400020055 (27.07.2019)

Futurezone(2016) Hyperloop plant Wien-Bratislava mit 8 Minuten Fahrzeit (10.03.2016); Online unter https://futurezone.at/science/hyperloop-plant-wien-bratislava-mit-8-minuten-fahrzeit/186.072.377 (27.07.2019)

Gast R (2016) Hype um den Hyperloop. In: Neue Zürcher Zeitung (26.08.2016); Online unter https://www.nzz.ch/mobilitaet/aktuell/zukunft-des-personentransports-hype-um-den-hyperloop-ld.112946 (27.07.2019)

Hyperloop TT (2018) Hyperloop transportation technologies signs agreement for commercial system in Ukraine. In: PR Newswire (14.06.2018); Online unter https://www.prnewswire.com/news-releases/hyperloop-transportation-technologies-signs-agreement-for-commercial-system-in-ukraine-300666420.html?=hyp (22.07.2019) Ilgmann G, Polatschek K (2006) Wunder ohne

Wirklichkeit: Der wahre Fluch des Transrapid. In: Frankfurter Allgemeine Sonntagszeitung 38:S. 69 (24.10.2016); Auch Online unter https://www.faz.net/aktuell/wissen/physik-mehr/wunder-ohne-wirklichkeit-der-wahre-fluch-des-transrapid-1360048.html?printPagedArticle=true#pageIndex_0 (14.07.2019)

International Railway Journal—IRJ (2014) JR Central starts construction on Chuo maglev (18.12.2014); Online unter https://www.railjournal.com/passenger/high-speed/jr-central-starts-construction-on-chuo-maglev/ (14.07.2019)

International Railway Journal–IRJ(2016) Changsha airport maglev line opens maglev (18.12.2014); Online unter https://www.railjournal.com/regions/asia/changsha-airport-maglev-line-opens/ (04.04.2016)

Investopedia (2019) Hyperloop versuss high-speed rail travel: what's the difference?: Online unter https://www.investopedia.com/articles/investing/050815/elon-musks-hyperloop-economically-feasible.asp (1.10.2020)

Janić M (2014) Advanced transport systems: analysis, modeling, and evaluation of performances. Springer, London

Kolonko P (2013) Transrapid in Schanghai—Der Zug endet hier. In: Frankfurter Allgemeine Zeitung (26.09.2013); Auch Online unter: https://www.faz.net/aktuell/gesellschaft/transrapid-in-schanghai-der-zug-endet-hier-12591451.html (13.07.2019)

Krummenacher J (2016) Die Swissmetro taucht wieder auf. In: Neue Zürcher Zeitung (01.12.2016); Online unter https://www.nzz.ch/schweiz/unterirdische-magnetschwebebahn-die-swissmetro-taucht-wieder-auf-ld.132067 (17.07.2019)

Lavanchy R (2013) Sorry, Elon Musk—your hyperloop is going nowhere. In: The Guardian (16.08.2013); Online unter https://www.theguardian.com/commentisfree/2013/aug/16/elon-musk-hyperloop-going-nowhere (20.07.2019)

Maceda C (2019) Hyperloop in Abu Dhabi to cost up to Dh1.4 billion, to be ready by 2020. (17.01.2019). In: Gulf News (17.01.2019); Online unter https://gulfnews.com/business/hyperloop-in-abu-dhabi-to-cost-up-to-dh14-billion-to-be-ready-by-2020-1.1547722844596 (29.07.2019)

Matthews V. (20077) Bahnbau. Wiesbaden, B. G. Teubner Verlag

Melendez ED (2013) Hyperloop would have 'astronomical' pricing, unrealistic construction costs, experts say. In: Huffpost (14.08.2013); Online unter https://www.huffpost.com/entry/hyperloop-experts_n_3749756? (20.07.2019)

Miller R (2014) The SciFi story Robert H. Goddard published 100 years ago. In: Gizmodo (07.06.2014); Online unter https://io9.gizmodo.com/the-scifi-story-robert-h-goddard-published-100-years-a-1494959842 (16.07.2019)

Musk et al. (2013) Hyperloop Alpha (Case Study). Los Angeles, SpaceX; Auch Online unter https://www.spacex.com/sites/spacex/files/hyperloop_alpha-20130812.pdf (17.07.2019); Anm.: Die Autoren sind im Dokument nicht genannt, lt. Medienberichten, Vance A. und Musk selbst wurde das Whitepaper jedoch von ihm und Mitarbeitern von Tesla und SpaceX verfasst

NDR.de (2012) Transrapid-Teststrecke vor dem Abriss (10.01.2012); Online unter https://web.archive.org/web/20161202150026/https://www.ndr.de/nachrichten/niedersachsen/osnabrueck_emsland/Transrapid-Teststrecke-vor-dem-Abriss,transrapid295.html (Archiv-Link/13.07.2019)

NDR.de (2019) Lathen: Neues Zuhause für Hyperloop-Forschung? (23.05.2019); Online unter https://www.ndr.de/nachrichten/niedersachsen/osnabrueck_emsland/Lathen-Neues-Zuhause-fuer-Hyperloop-Forschung,hyperloop150.html (13.07.2019)

O'Neill GK (1991) High speed transport system. US-Patent US 5433155A; Auch online unter https://patentimages.storage.googleapis.com/4f/07/17/e549842c885f84/US5433155.pdf (16.07.2019)

Pachl J (2018) Systemtechnik des Schienenverkehrs—Bahnbetrieb planen, steuern und sichern. Wiesbaden, Springer Vieweg

Railway Gazette International (Hrsg.) (2018) Etihad rail awards stage 2/3 consultancy contract (24.05.2018); Auch Online unter https://www.railwaygazette.com/news/freight/single-view/view/etihad-rail-awards-stage-23-consultancy-contract.html (29.07.2019)

Railway Gazette International (Hrsg.) (2019) High speed rail plan scaled back. In: Railway Gazette International 3(175):S. 8; Auch Online unter https://www.railwaygazette.com/news/single-view/view/california-high-speed-rail-plan-scaled-back.html (10.06.2019)

Salter RM (1972) The very high speed transit system. Santa Monica, USA, RAND Corporation; Auch Online unter https://www.rand.org/pubs/papers/P4874.html (16.07.2019)

Salter RM (1978) Trans-planetary subway systems: a burgeoning capability. Santa Monica, USA: RAND Corporation; Auch Online unter https://www.rand.org/pubs/papers/P6092.html (16.07.2019)

Schienen-Control GmbH (2018) Jahresbericht 2017. Wien; Auch Online unter http://www.schienencontrol.gv.at/files/1-Homepage-Schienen-Control/1g-Presse/Pressemappen/Pressemappen2018/SC-Jahresbericht-2017_Web.pdf (03.07.2019)

Siemens (o. J.) (2019) Desiro ML—ÖBB Cityjet eco für die ÖBB-Personenverkehr AG; Online unter https://assets.new.siemens.com/siemens/assets/api/uuid:f48fa9d1-2abd-41ab-b2f9-f37cfd6d59e6/db-desiro-ml-oebb-cityjet-eco-d.pdf (30.07.2019)

Spiegel Online (2000) Aus für die Strecke Berlin-Hamburg (05.02.2000); Online unter https://www.spiegel.de/wirtschaft/transrapid-aus-fuer-die-strecke-berlin-hamburg-a-63042.html (13.07.2019)

Spiegel Online (2008) Transrapid in München wird nicht gebaut (27.03.2008); Online unter https://www.spiegel.de/wirtschaft/aus-fuer-prestigeprojekt-transrapid-in-muenchen-wird-nicht-gebaut-a-543688.html (13.07.2019)

Stadt Wien (2019) Verkehr & Stadtentwicklung—Öffentlicher Verkehr—Streckenausbau Wien—Bratislava; Online unter https://www.wien.gv.at/verkehr/oeffentlich/bratislava.html (28.07.2019)

Süddeutsche Zeitung (2011) Bewährungsstrafen im Transrapid-Prozess (03.03.2011); Online unter https://www.sueddeutsche.de/panorama/landgericht-osnabrueck-bewaehrungsstrafen-im-transrapid-prozess-1.1067506 (10.07.2019)

The Asahi Shimbun (2013) New maglev Shinkansen to run underground for 86% of initial route: Online unter https://web.archive.org/web/20141226124352/https://web.archive.org/web/20141226124352/http://ajw.asahi.com/article/behind_news/social_affairs/AJ201309190078 (01.10.2020)

The Huffington Post (2013) Hyperloop would have 'astronomical' pricing, unrealistic construction costs, experts say: Online unter https://www.huffpost.com/entry/hyperloop-experts_n_3749756 (1.10.2020)

The International Maglev Board (2019) Maglevboard–Facts–Linimo Urban Maglev; Online unter https://www.maglevboard.net/en/facts/systems-overview/linimo-urban-maglev (14.07.2019)

The Slovak Spectator (2018) Is Slovakia's hyperloop dream over? (14.08.2018); Online unter https://spectator.sme.sk/c/20891772/is-slovakias-hyperloop-dream-over.html (28.07.2019)

UIC (2018) High speed rail: fast track to sustainable mobility. Paris; Auch Online unter https://uic.org/IMG/pdf/uic_high_speed_2018_ph08_web.pdf (07.07.2019)

Visit Dubai (o.J.) Das weltweit erste hyperloop-system soll Dubai mit Abu Dhabi verbinden. Online unter https://www.visitdubai.com/de/business-in-dubai/why-dubai/news-and-insights/dubai-abudhabi-hyperloop-link (27.07.2019)

Wolverton T (2013) Wolverton: Elon Musk's hyperloop hype ignores practical problems. In The Mercury News (13.08.2013); Online unter https://www.mercurynews.com/2013/08/13/wolverton-elon-musks-hyperloop-hype-ignores-practical-problems/

Zeit Online (2006) Menschliches Versagen (22.09.2006); Online unter https://www.zeit.de/online/2006/39/Transrapid-menschliches-Versagen (12.07.2019)

Zeit Online (2013) Warum Deutschlands Transrapid in Shanghai floppte (21.03.2013); Online unter https://blog.zeit.de/china/2013/02/21/der-gefloppte-transrapid/ (14.07.2019)